Acta Physica Austriaca
Supplementum XV

Proceedings of the
XV. Internationale Universitätswochen für Kernphysik 1976
der Karl-Franzens-Universität Graz
at Schladming (Steiermark, Austria)
16th February—27th February 1976

Sponsored by
Bundesministerium für Wissenschaft und Forschung
Steiermärkische Landesregierung
Sektion Industrie der Kammer der
gewerblichen Wirtschaft für Steiermark
International Centre for Theoretical Physics, Triest
CERN, Genève

1976

Springer-Verlag
Wien New York

Current Problems in Elementary Particle and Mathematical Physics

Edited by Paul Urban, Graz

With 59 Figures

1976

Springer-Verlag

Wien New York

Organizing Committee

Chairman

Prof. Dr. Paul Urban
Institut für Theoretische Physik
der Universität Graz

Committee Members

Prof. Dr. H. Latal
Dr. A. Mas-Parareda
Dr. L. Pittner
Doz. Dr. F. Widder
Dr. H. Zankel

Secretary

M. Krautilik

Library of Congress Cataloging in Publication Data

Internationale Universitätswochen für Kernphysik der
    Karl-Franzens-Universität Graz, 15th, Schladming,
    Austria, 1976.
    Current problems in elementary particle and
mathematical physics.

    (Acta physica Austriaca : Supplementum ; 15)
    1. Particles (Nuclear physics)--Congresses.
2. Mathematical physics--Congresses. I. Urban,
Paul Oskar, 1905-   II. Austria. Bundesministerium
für Wissenschaft und Forschung. III. Title.
IV. Series.
QC793.I57 1976   539.7'21   76-29076

ISBN 978-3-7091-8464-6   ISBN 978-3-7091-8462-2   (eBook)
DOI 10.1007/978-3-7091-8462-2

CONTENTS

VI

Acta Physica Austriaca, Suppl. XV, 1–5 (1976)
© by Springer-Verlag 1976

OPENING ADDRESS

by

P. URBAN
Institut für Theoretische Physik
Universität Graz

It is a great pleasure for me to welcome you most
cordially at our fifteenth International Winter School
in Schladming. Again about 17o scientists from 2o countr-
ies will be participating this year, which number certainly shows that our efforts to provide an interesting
program of lectures were successful.

The program of this year's meeting is very versatile,
but only seemingly divergent. By my following brief sketch
of the topics of the various lectures I may show the underlying quite general view point. In general, during the
last decades different methods developed in scientific
research: there are, roughly speaking, abstract theoreticians which are already more akin to abstract
mathematicians; experimentalists, which reside often
more in the area of applied physics or science manage-
ment; and the so-called phenomenologists which try to

exploit the knowledge of the two other groups and to put it together into a coherent picture. None of these approaches by itself would make much sense; theory without experiment would soon become sterile, without theory the experimentalist would quickly loose conspectus, and the phenomenologist needs close contact with both groups. I can not, and therefore do not want to, make a value judgement because all of them are equally important for the progress of science, and out of this results our varied program.

In the first week mainly lectures on theoretical topics are scheduled and here I want to thank my former student Prof. Streit for the excellent cooperation with the "Zentrum für interdisziplinäre Forschung" of the University of Bielefeld, which enabled us to invite to Schladming prominent experts in the field of quantum statistics and field theory. It is impossible to go into the details of these lectures with just a few words, I must therefore restrict myself to a short outline. Prof. Streit himself introduced some of the necessary mathematical concepts which rightly were named "the wondrous things". These tools are essential for the attempts to formulate realistic theories of interacting fields, about which Prof. Challifour will give us a review. Mathematically relatively simple is fieldtheoretic life in spaces with just one or two dimensions, but also here much remains to be investigated and the final goal of all this research, as demonstrated by Prof. Fröhlich and Prof. Park, is that also abstract field theory can rise into three-dimensional space. In the course of investigating the structure of field theories one always falls back to classical theories, for instance to the field of statistics; Prof. Emch will introduce us to its extension to abstract quantum systems. Here one is concerned with the generalization of the important notions of ergodic systems; also the investigations of Prof. Thirring on the stability of matter fall into this area of problems.

The lectures of the second week deal mainly with new experimental discoveries and their theoretical interpretation in the area of high-energy particle physics. The basic realization of Salam and others, that by means of generalized gauge theories one may combine the so far seemingly disparate electromagnetic and weak interactions into a unified theory, had many consequences. For instance it almost necessarily predicts that also in weak interactions somewhat like a heavy photon exists. Its manifestation was called "neutral currents" and this phenomenon was then also discovered experimentally. This happened mainly in the technically difficult neutrino experiments, about which Prof. Faissner will talk. Prof. Sakurai will critically look upon the theoretical concepts of neutral currents, and here certainly an unprejudiced criticism is necessary to find the correct theory out of the many hypotheses.

Naturally soon the conjecture was put forward, that the traditional distinction between the interactions may be out of date altogether. The inclusion of the strong interaction again could be made via the application of gauge theories, and Prof. Kummer will discuss the current status of these investigations. In various models, which should demonstrate the fundamental possibility of such superunified theories, new quantum numbers had to be postulated in order to fit

theoretical results to the empirical findings. In the
meantime a whole family of particles which possibly are
carriers of this new quantum number have been found mainly in storage rings. These experimental results allow a
deeper insight into the structure of all interactions;
Prof. Gaillard will report the latest news of these
highly important developments. A complementary approach
to the unification of the interactions represents the
investigation of a very far-reaching symmetry, called
"supersymmetry". It was theoretically discovered by Salam,
Wess and Zumino, who also investigated its consequences;
Prof. Wess will inform us on the current status of these
investigations. This symmetry is so comprehensive that
superficially viewed upon nature does not show it. With
the phenomenon of this hiding of fundamental symmetries
we will be faced this year several times- as for instance
in the lecture of Prof. Budini.

Certain symmetries of the strong interaction led
Gell-Mann to the introduction of the so-called quarks
as basic elements of the nuclear particles, that is
protons and neutrons. Since free quarks so far have not
been found experimentally in spite of vigorous searches,
they must be strongly bound together and this raises a
still not fully understood and therefore even the more
interesting problem, the question for "quark-confinement".
I do not want to anticipate the exposition of Professor
Vinciarelli, but only point out some of the various
connections and ideas: the understanding of the problem
in the framework of generalized gauge theories, which
contain the phenomenon of strong binding and strong
symmetry breaking at common energies, and the transition to weak binding, free quarks and manifest symmetry

at highest energies; or a model treatment by means of
newly to be formulated field theories, which we will be
presented with during the first week. Here various connections to areas of physics follow which so far were
completely differently viewed upon: to statistical and
solid state physics.

With this outline I hope to have made clear to
you that one can not set the bounds of the program of
such a meeting too broad: the questions and problems
of the various areas of research are closely connected,
or at least often of equal value, as it is always the
case in the exploration of new aspects. Therefore it is
of much importance for the sound education of our
scientific young generation, that it doesn't specialize
itself too early and can retain also later on its view
of the common aspects of scientific research. I hope
that this meeting again contributes to the advancement
of our scientific knowledge and wish to all of you a
successful and pleasant stay in Schladming.

Acta Physica Austriaca, Suppl. XV, 7–43 (1976)

THE WONDROUS THINGS:[*]

A REVIEW OF PROBABILISTIC CONCEPTS IN

QUANTUM DYNAMICS[+]

by

L. STREIT

Fakultät für Physik der
Universität Bielefeld

§0. INTRODUCTORY REMARKS

Eleven years ago when I gave a set of lectures on path integrals - most prominently the Wiener and Feynman integrals - here in Schladming [1], this probabilistic approach played a rather marginal role in quantum physics, mainly as a compact but mostly formal reformulation of quantum (field) theory.

In the meantime the situation has changed drastically, largely as a result of two papers by Edward Nelson. The first one, in 1965 [2], laid the foundations for a

---

[*] Ref.[8] , p.4.

[+] Lecture given at XV. Internationale Universitätswochen für Kernphysik,Schladming,Austria,February 16-27,1976.

whole industry, namely for constructive quantum field theory, with the help of essentially probabilistic arguments. The second one in 1971 established a two way connection between the constructive work and that of Symanzik on Euclidean field theory, hence with equilibrium statistical mechanics the techniques of which have become immensely fruitful in the field theory context.

The intrinsic successes of constructive field theory are fascinating, and there are now indications that one is learning to control the existence of nontrivial models in the physical four-dimensional space time [4] but one may also expect methods of such effectiveness to play a role beyond the purpose for which they were first introduced - let me just mention the application of functional integration to dual resonance models [5].

For this reason it now seems worth while to present an account of some probabilistic concepts - in particular regarding stochastic processes - as applied in recent quantum field theory work. The specific goal of these lectures is to acquaint the non-experts with these methods and their applications. As a consequence emphasis will be not on detailed proofs and technical definitions but on simple illustrative examples. Characteristic functions and functionals of probability measures are particularly suited as a unifying concept because of their simple properties and also because of their affinity to expressions of quantum field theory.

We shall proceed by the following steps:

§1. Probability **Measures** - in Terms of Characteristic
   Functions

§2. Stochastic Processes -   - " -

§3. Quantum Theory   -   - " -

§4. The Renormalization Group -   - " -

There is a rich mathematical literature on the subjects
of §1 and 2, we list a couple of examples in refs. [6]
and [7]. In these lectures we cannot be encyclopedic,
all we aim to do is to prepare the ground for applications as in §3 and 4.

§1. PROBABILITY MEASURES - IN TERMS OF
CHARACTERISTIC FUNCTIONS

1.1 Characterization of probability measures

Visualize probability measures μ as all the ways
in which one can spread a unit mass over some set M.
For some subset $B \subset M$, $\mu(B)$ tells you the amount of
mass in the subset B.

In all of our applications M will be some linear
space, for the moment we shall focus on the simplest
case $M = R$.

Example 1: (probability density)

Let $\rho(x) \geq 0$ $\qquad \int_{-\infty}^{\infty} \rho(x)\,dx = 1$

$\qquad \mu(B) = \int_{B} \rho(x)\,dx$.

This is not sufficiently general:

Example 2: (point mass at x_o):

$$\mu(B) = \begin{cases} 1 & x_o \in B \\ 0 & \text{otherwise} \end{cases} .$$

Formally $\rho(x) = \delta(x-x_o)$, μ "Dirac measure". There is a very handy characterization of probability measures on R which encompasses both cases (and more). Let us ask for the amount of mass at and to the left of a given point x.

More precisely we want to introduce

$$\mu((-\infty,x]) \equiv F(x)$$

so that

1) $F(x) \nearrow$ (monotone increasing)

2) $F(x) = F(x+0)$ (continous from the right)

3) $F(-\infty) = 0, F(\infty) = 1$ (normalized).

Real functions obeying 1-3 are called <u>distribution funct-ions.</u>

Example 1: $F(a) = \int_{-\infty}^{a} \rho(x)\,dx$

Example 2: $F(a) = \begin{cases} 0 & a < x_o \\ 1 & a \geq x_o \end{cases} .$

In probability terminology we say that F is the distribution function of a <u>"random variable"</u> X:

F(a) is interpreted as the probability that the random variable takes a value less than or equal to a. Labelling F and μ by X

$$F_X = \mu_X((-\infty,a]) .$$

$\mu_X(B)$ is considered as the probability for X to take a value in B.

Example 3: Given a random variable X. An integrable function y defines a new random variable Y

$$F_Y(a) = \mu_X(\{x:y(x) \leq a\}) .$$

The __expectation__ of Y is defined as $<Y> = \int_{-\infty}^{\infty} y(x) dF(x)$. More generally n-tuples of random variables are characterized by functions

$$F_{X_1 \cdots X_n}(a_1,\ldots,a_n)$$

and if

$$F_{X_1 \cdots X_n} = \prod_{i=1}^{n} F_{X_i}$$

we call the random variables __independent__.

Example 4: $$F_X(\vec{a}) = \prod_{i=1}^{n} \int_{-\infty}^{a} dx_i \, \rho_X(x_1,\ldots,x_n)$$

with $\quad \rho_X = \prod\limits_{i=1}^{n} \rho_{X_i}(x_i)$.

In particular consider the sum of such random variables

$$Y = X_1 + X_2$$

$$\rho_Y(a) = \iint dx_1 dx_2 \delta(a-x_1-x_2)\rho_{X_1}(x_1)\rho_{X_2}(x_2) =$$

$$= \int dx_2 \, \rho_{X_1}(a-x_2)\rho_{X_2}(x_2)$$

$$\rho_{X_1+X_2} = \rho_{X_1} * \rho_{X_2} \quad .$$

Recall that convolutions such as this become products through Fourier transform. Hence it is useful to introduce

$$C_X(a) \equiv \int_{R^n} e^{i(x,a)} \, dF_X(x) = \langle e^{i(X,a)} \rangle \quad .$$

C is called the <u>characteristic function</u> of the random variable X, its properties will be discussed in the next section.

1.2 Characteristic Functions: Properties and Examples

1) $\quad C(0) = \int_{R^n} dF(\vec{x}) = 1$ \hfill (normalized)

2) $\quad \sum\limits_{\nu,\mu=1}^{m} z_\nu^* C(\vec{a}_\nu - \vec{a}_\mu) z_\mu = \int | \sum\limits_\mu e^{i(\vec{x},\vec{a}_\mu)} z_\mu |^2 dF(\vec{x}) \geq 0$

\hfill for all complex m-tuples

\hfill (z)

and all m-tuples of vectors $\vec{a}_\mu \in R^n$.

(positive definite).

3) $C(a)$ is continuous.

Conversely we have

"Bochner's Theorem": Every normalized positive definite continuous function is the characteristic function of a probability measure.

A list of some standard densities and their characteristic functions:

	$\rho(x)$	$C(a)$
Dirac	$\delta(x-x_o)$	e^{iax_o}
uniform	$\begin{cases} \ell^{-1}, & |x| < \ell/2 \\ 0 & \text{otherwise} \end{cases}$	$\dfrac{\sin a\,\ell}{a\,\ell}$
Gaussian	$\dfrac{1}{\sqrt{2\pi\lambda}}e^{-x^2/2\lambda}$	$e^{-\frac{\lambda}{2}a^2}$
Gamma	$\theta(x)\dfrac{c^r}{\Gamma(r)}x^{r-1}e^{-cx}$	$(1-\dfrac{ia}{c})^{-r}$ $(r,c>0)$
Cauchy	$\pi^{-1}\dfrac{1}{1+x^2}$	$e^{-|a|}$

Examples (Quantum mechanics):

$$C(\lambda) = (\psi, e^{i(\vec{\lambda},\vec{Q})}\psi) = \int d^n x \; e^{i(\vec{\lambda},\vec{x})}|\psi(\vec{x})|^2 \quad.$$

In particular for the ground state of

$$H = \frac{1}{2} (\vec{p} \cdot \vec{p}) + \frac{1}{2} (\vec{q}, \omega^2 \vec{q}) \qquad \omega > 0$$

$$C_\omega (\vec{a}) = e^{-\frac{1}{4}(\vec{a}, \omega^{-1} \vec{a})} \qquad .$$

Further properties:

$C(a)$ is a generating function for the moments $\langle x^n \rangle$
(if they exist)

$$C(a) = \sum_{n=0}^{\infty} \frac{i^n}{n!} \langle x^n \rangle a^n$$

$$\langle x^n \rangle = (-i)^n C^{(n)}(0) \quad .$$

Similarly the "second characteristic"

$$\log C(a) = \sum_{n=1}^{\infty} \frac{i^n}{n!} a^n \langle x^n \rangle_T$$

generates the truncated expectation values or "cumulants".
How do we get new characteristic functions from given ones?
Let $C_i(a)$ be a characteristic function. Then the follow-
ing are characteristic functions too:

$C(Aa)$ for linear maps A, $X' = A^{\dagger}X$,

in particular $C(-a) = C^*(a)$,

$$\sum_i \lambda_i \ C_i(a) \quad \text{if } \lambda_i > 0, \ \sum_i \lambda_i = 1 \qquad <f(X')> \ = \ \sum_i \lambda_i <f(X_i)>:$$

$$\prod_i \ C_i(a) \qquad\qquad\qquad\qquad X' = \sum_i X_i \ .$$

Concerning limits one has the

Continuity Theorem: Let C_n be characteristic functions
s.t. $C_n(a) \rightarrow C(a) \quad \forall a$

and $C(a)$ continuous at $a = 0$. Then $C(a)$ is again a
characteristic function.

1.3 Infinitely divisible characteristic function [6]

From the previous section we know that $C^n(a)$ is
a characteristic function $\forall n = 0,1,2,\ldots$. What about
non-integer powers?

$C_X^{1/n}(a) = C_Y(a)$ a characteristic function implies
that the random variable X can be written as the sum of
n equally distributed independent random variables $Y_i (=Y)$

$$X = \sum_{i=1}^{n} Y_i \ .$$

We call $C(a)$ infinitely divisible if $C^{1/n}(a)$ is a characteristic function $\forall \ n = 1,2,3,4,\ldots$.

As a direct consequence of the definition it
follows that all $C^{n_1/n_2}(a)$ are infinitely divisible,
and so are all $C^r (r > 0)$, by the Continuity Theorem.
Of our examples the Dirac, Gaussian, Gamma, Cauchy

distributions are infinitely divisible. The uniform
distribution is not since

Theorem:
An infinitely divisible characteristic function is
everywhere non-zero.

Proof: Let

$$C_\infty (a) = \lim_{n \to \infty} (|C(a)|^2)^{1/n} = \begin{cases} 0 & C(a) = 0 \\ 1 & \text{otherwise} \end{cases} .$$

By the Continuity Theorem C_∞ is a characteristic function but then it is continuous, i.e. $C(a)$ has no zeros.
Products and limits of infinitely divisible characteristic functions are again infinitely divisible.

By considering in particular limits of products
of Poisson type characteristic functions

$$\lim_n \prod_n e^{\lambda_n (e^{iat_n} - 1)} = \lim_n e^{\sum_n \lambda_n (e^{iat_n} - 1)}$$

one finds that

$$C(a) = e^{ix_o a + c^2 \int_{-\infty}^{\infty} (e^{iat} - 1 - \frac{i \, at}{1+t^2}) \frac{1+t^2}{t^2} dF(t)}$$

is an infinitely divisible characteristic function for
all real x_o, c and all distribution functions $F(t)$.
Conversely

Theorem (Levy-Khinchin): Any infinitely divisible characteristic function has a (unique) representation of the above form.

In particular

	x_o	$\dfrac{dF}{dt}$
Gaussian	0	$\sim \lambda^2 \delta(t)$
Poisson	$\lambda/2$	$\sim \lambda \delta(t-t_o)$, $t_o > 0$
Gamma	$\lambda \displaystyle\int_o^\infty \dfrac{e^{-ct}}{1+t^2} dt$	$\sim r \dfrac{t\, e^{-ct}}{1+t^2}$
Cauchy	0	$\sim \dfrac{1}{1+(x+1/2)^2}$

1.4 Stable Characteristic Functions

Infinitely divisible random variables X can be viewed, for any n, as sums of n independent equally distributed random variables

$$X = \sum_{\nu=1}^{n} X_{n,\nu} \quad .$$

An important special case is the one where the $X_{n,\nu}$ do not depend on n in an essential way. To be more precise we observe that intuitively

$$Y_n = \sum_{\nu=1}^{n} X_\nu$$

gets "bigger and bigger" as $n \to \infty$. The formalism of characteristic functions exhibits this fact in the form

$$\exists C_Y = \lim_{n\to\infty} C_X^n = 1 \quad \text{iff} \quad C_X = 1 \text{ i.e.iff } X = Y = 0 .$$

But there is a fighting chance for a scaled down sum

$$Y = \frac{1}{c_n} \sum_{\nu=1}^{n} X_\nu$$

to exist for all n if we allow for $c_n \xrightarrow[n\to\infty]{} \infty$. We have

$$C_Y(a) = C_X^n (a/c_n) \quad \text{for all } n$$

and call these "stable distributions". Varying n one sees

$$C_{n+m} Y = c_n Y_1 + c_m Y_2$$

which gives rise to the relation

$$C_Y(\alpha a) = C_Y(\beta a) \, C_Y(\gamma a) \qquad (*)$$

for arbitrary β, γ with α depending on β, γ. One shows

1)   (*) is sufficient for Y to be stable;

2)   from the Levy-Khinchin representation

$$C_Y(a) = e^{-\lambda |a|^\alpha} \qquad 0 < \alpha \leq 2$$

note

$$C_Y(a) = (e^{-\lambda \left|\frac{a}{n^{1/\alpha}}\right|^\alpha})^n$$

so that $c_n = n^{1/\alpha}$ and $Y = X$.

I.e. up to a scaling factor $n^{1/\alpha}$ the sum of n random variables behaves just like a single one of them. Note the affinity with the Kadanoff description of critical behaviour in statistical Mechanics: at the critical point macroscopic variables ("block spins") Y become equivalent to their microscopic constituents X. But there is one important difference in the statistical mechanical case

$$<e^{i(\vec{a},\vec{x})}>_\beta = \frac{<e^{i(\vec{a},\vec{x})} \, e^{-\beta H(\vec{x})}>_0}{<e^{-\beta H(\vec{x})}>_0} \qquad :$$

the X are correlated through the interaction. This would fit in the framework of general random vectors, but as we are interested in aggregates with infinitely many degrees of freedom ("thermodynamical limit") we must consider probability measures on infinite dimensional vector spaces.

§2. STOCHASTIC PROCESSES IN TERMS OF CHARACTERISTIC FUNCTIONS

Example 6:

$$C(a) = e^{-\frac{\lambda}{2}(a,a)}$$

characterizes a collection of Gaussian random variables. But now let a be an infinite sequence!

E. g. $\quad a = (a_1, a_2, \ldots) \in \ell^2 \quad$ (square summable).

What about a corresponding probability measure μ?

$$C(\vec{a}) \stackrel{?}{=} \int_{\ell^2} e^{i(\vec{a},\vec{x})} d\mu(\vec{x}) \quad .$$

Note for n-dimensional Gaussian distribution that

$$d\mu \sim r^{n-1} e^{-\frac{1}{2\lambda}r^2} dr \, d\Omega \quad .$$

The larger n the more this is concentrated around the sphere $r^2 = \lambda(n-1)$ so that

$$\lim_{n \to \infty} \frac{1}{n} \sum_{i=1}^{n} x_i^2 = \lambda$$

for almost all $\vec{x}$ (law of large numbers) whereas on ℓ^2: $\sum_i x_i^2 < \infty$ (contradiction), i.e. we must enlarge the range M of the integral

$$C(a) = \int_{M > \ell^2} e^{i(a,x)} d\mu(x) \, .$$

(It suffices to set $M = \{ (x_\nu) : \sum_\nu \frac{x_\nu^2}{\nu^2} < \infty \}$, technically to choose M as the Hilbert space completion of ℓ^2 with respect to the metric

$$<x,y>_M = (x,Ty)_{\ell^2} \qquad\qquad \text{with T some}$$

positive trace class operator). Before we leave this example we observe that a change of λ in the exponent, harmless as it looks in the characteristic functions, changes the relevant set of sample sequences (x_ν) drastically: for any two λ differing by an arbitrarily small amount the corresponding probability measure have disjoint supports, as a practical consequence we cannot express one measure in terms of the other as in the finite dimensional case :

$$d\mu^{\lambda_1}(\vec{x}) = \rho(\vec{x}) \, d\mu^{\lambda_2}(\vec{x}) \quad \text{if and only if dim } \{\vec{x}\} < \infty \quad .$$

Can we write

$$C(a) = \int_{E'} e^{i(a,x)} d\mu(x)$$

with E' the (topological dual) space of continuous linear functionals on E? The answer is in the affirmative

in many important cases where E' is big enough. A well
known class is covered by the

Bochner-Minlos Theorem: C(a) is the Fourier transform
of a probability measure on E' if E is a nuclear space.

Most important examples for E and E': Test functions
S, D and generalized functions S', D'. A proof in a
more general setting can be found in [7].

In this fashion random variables are replaced by random
sequences $\{x_n\}$ or (generalized) random functions $x(t)$.
It is this set-up which we mean by (generalized) "random
processes".

Example 6:

$$C(a) = e^{-\frac{1}{2} \int_{-\infty}^{\infty} a^2(t)\,dt}$$ "white noise" .

We shall reserve the notation X(t) for the corresponding
random functions.

Example 7:

$$Y(t) = (T\,X)(t)$$ T a linear operator,

$$C_Y(a) = <e^{i(a,T\,x)}>_X =$$

$$= <e^{i(T^*a,x)}>_X =$$

$$= e^{-\frac{1}{2}(a,TT^*a)}$$. ("Gaussian process")

Example 5':

$$C(f) = (\Omega, e^{i\phi(f)}\Omega) = \int e^{i(\phi, f)} d\mu(\phi)$$

where ϕ could be a quantum field at a fixed time or e.g.
a "Euclidean field" smeared in space and imaginary time.
$C(f)$ is the generating functional of the n point functions, they are the moments of the measure μ. This gets
us to the topic of applications. We close this section
with a short remark on topologies for the infinite dimensional vector spaces. In applications it will often
be possible to establish the continuity of the characteristic function on a space like e.g. that of test functions, S. But then depending on the given characteristic
function one may find it possible to extend the definition to some larger space. This means in vague terms that
characteristic functions somehow generate their own intrinsic topology, and indeed Hegerfeldt and Klauder [10]
have exhibited such a C-dependent natural metric

$$d^2(a) = \frac{1}{\pi^{1/2}} \int_{-\infty}^{\infty} (1 - \text{Re } C(\lambda a)) e^{-\lambda^2} d\lambda .$$

§3. QUANTUM THEORY IN TERMS OF CHARACTERISTIC FUNCTIONS

It is best to start this paragraph by a "Non-Ex-
ample": the generating function τ_o of the free field
τ-functions

$$\tau_o(f) = \sum_{n=2}^{\infty} \frac{i^n}{n!} \prod_{\nu=1}^{n} \int dx_\nu f(x_\nu) \tau(x_1, \ldots x_n) =$$

$$= e^{-\frac{1}{2}\int dxdy \ f(x)\Delta_F(x-y)f(y)} \quad .$$

This has almost the form of the characteristic functional of a Gaussian process, except to the fact that

$$\Delta_F(x-y) = \langle 0|T \ A(x)A(y)|0\rangle =$$

$$= \frac{i}{(2\pi)^{s+1}} \int dp \ \frac{e^{-ip(x-y)}}{(p,p)-m^2}$$

is <u>not</u> a positive definite kernel. Note however the
effect of a rotation to imaginary time

$$- (p,p) + m^2 = - p_0^2 + \vec{p}^2 + m^2 \rightarrow \sum_{i=0}^{s} p_i^2 + m^2 > 0$$

and τ_0 becomes the generating functional of the free
field Schwinger functions σ_0

$$\sigma_0(f) = e^{-\frac{1}{2} (\hat{f}, \frac{1}{p^2+m^2} \hat{f})} =$$

$$= \int e^{i(f,\Phi)} \ d\mu_0(\Phi) \quad .$$

What can we say about the measure? We want to exploit
the Gaussian form of σ_0. To the extent that we can

approximate it by a finite dimensional Fourier transform of a density function ρ_o, the latter is again a
Gaussian but with the inverse kernel:

$$\rho_o(\Phi) \sim N_o \, e^{-\frac{1}{2}(\overset{\vee}{\Phi}, (p^2 + m^2)\overset{\vee}{\Phi})} \quad .$$

The exponent can be rewritten as

$$\frac{1}{2} \int dx \, (\dot{\Phi}^2 + (\nabla \Phi)^2 + m^2 \, \Phi^2) =$$

$$= \int dx \, h_o(x)$$

and one recognizes

$$\int d\mu_o(\ldots) \; ' = ' \; N_o \int d^\infty \Phi \, e^{-g\int dx h_o(x)} \, (\ldots)$$

as the Euclidean variant of Feynman's sum over histories.
The modifications introduced by an interaction like e.g.

$$h_I(x) = :\Phi^n:(x)$$

are obvious, at least formally

$$\sigma_g(f) = N \int e^{i\int dx\Phi(x)f(x)} \, e^{-g\int dx: \Phi^n(x):} \, d\mu_o =$$

$$= \frac{<e^{i\phi(f)} \; e^{-g\int dx:\phi^{n}(x):}>_{0}}{<e^{-g\int dx:\phi^{n}(x):}>_{0}} \; .$$

(Note the similarity with Gibbs states in statistical mechanics). Since one has learned to control relativistic theories in terms of their Schwinger functions - see John Challifour's lecture for a review - this formula for the characteristic functions of interacting Euclidean fields has become the point of departure for constructive quantum field theory. We shall not trace out this modern saga here. For an excellent general reference there is B. Simon's monograph [8], of the vast original literature we single out J.Fröhlich's paper [9] for its emphasis on the properties of characteristic functions in constructive quantum field theory.

One measure theoretical question however is of central importance in the above derivation. Even the relation

$$d\mu_{g} = N_{g} \; e^{-g\int dxh_{I}(x)} \; d\mu_{0}$$

should be suspicious after what we have learned about its failure in the white noise case: essential changes of the measure for an infinite number of degrees of freedom changes the support and a relation as the above must fail. This is a probabilistic version of Haag's theorem; we can circumvent it through a space time cutoff Λ in the interaction, and then exploit the

good continuity properties of characteristic functions
to establish

$$\sigma_g^\Lambda(f) \xrightarrow[\Lambda \to \infty]{} \sigma_g(f) \ .$$

As the parameters of some dynamical ansatz are varied,
the corresponding probability measures and characteristic functions can change in qualitatively different
ways:

- measures are absolutely continuous i.e. there are
  intertwining density functions.

  (Example: finite changes in the cut off parameter Λ)
- measures are disjoint, but characteristic functions
  vary continuously.

  (Example: cutoff $\Lambda \to \infty$, or e.g. for a free field
  characteristic function $\sigma_o(f)$: finite changes in
  the mass).
- the natural test function topology changes dis-
  continuously.

  (Example: $m \to 0$ in the free field functional, more
  generally: at critical points of the system [11].
- the characteristic function changes discontinously.

  (Example: Characteristic functions which do not
  return to the free field form as the interaction
  is turned off, i.e. "Klauder's phenomenon". A
  simple model is provided by the coupling $h_I(x) = j(x):\phi^2(x):$ to a singular external source; for this
  and further references c.f. [12]).

It seems worth while to note that there is an alternate
cutoff-free construction of characteristic functions

which has the potential to deal with each of the four
types of singular behaviour listed above. For the path-
integral formulation of (imaginary time) nonrelativistic
potential scattering H. Ezawa, J.R. Klauder and L.A.Shepp
(EKS, [13]) have recently proposed an interesting alternative. To deal with Schrödinger models such as H =
- Δ + gV one considers (c.f. e.g. ref. [13]) functional
integrals of the form

$$\langle F \rangle^{(gV)} = N_g \int F(x) e^{-g \int_0^T V(x(t))dt} d\mu_B(x)$$

where μ_B is the probability measure of the Wiener process
B formally related to white noise X through

$$\dot{B}(t) = X(t) \quad .$$

Instead of this weighted average over Brownian motion
paths B with weight function $\rho_g(x) = N_g \exp \{- \int_0^T Vdt\}$
EKS propose to employ a direct, unweighted average over
a distorted set of paths Y

$$\langle F(g) \rangle = \frac{\langle Fe^{-\int Vdt} \rangle_B}{\langle e^{-\int Vdt} \rangle_B} \quad .$$

The dynamics is incorporated not in a weight function
but in a transformation of the paths which occur in
the averaging. It is remarkable that this EKS formulat-

ion of dynamics does not hinge on the equivalence
of measures. Consequently, a transcription   of
this approach to Euclidean quantum field theory, if at
all possible, should not require the space-time cutoff
that we had to introduce previously in the construction
of the interacting measure. It is interesting to check
out this idea for a class of exactly soluble models

$$H_O = \frac{1}{2} \int dx : \Pi^2(x) + \nabla\phi^2 + m^2 \phi^2 (x):$$

$$H_I = \frac{1}{2} \int dx \ g(x) :\phi^2(x): \qquad .$$

These can be solved in closed form [14], yet by a suitable choice of g (decreasing, constant, or singular)
they provide a laboratory for the various degrees of
singularity from absolute continuity to occurence of
the Klauder phenomenon. Non-withstanding such singular
behaviour, corresponding "distorted" stochastic processes Y can in all these cases be obtained by a well
defined transformation of the white noise process X:

$$Y(t,\vec{x}) = \int_{-\infty}^{t} ds \ e^{-\omega_g (t-s)} X(s,\vec{x})$$

where ω_g is the linear operator

$$\omega_g = (- \Delta_{\vec{x}} + m^2 + g(\vec{x}))^{1/2} .$$

It is straightforward to verify that

$$C_g(f) = e^{-\frac{1}{2}(f, (-\Delta_{\vec{x},t} + m^2 + g(\vec{x}))^{-1}f)}$$

$$= \int e^{i(Y,f)} \, d\mu(X) \ .$$

The above solution Y(t) has the particular property
that it depends only on X(s), s ≤ t, i.e. it is causal;
the same is true for X as a function of Y. The differential equation for Y

$$\dot{Y} = - \omega_g Y + X$$

is the (well defined, cutoff-free) path space equivalent
of the field equation. Explicit expressions can also be
given which relate the processes X for different per-
turbations g [15].

Before turning to the renormalization group we
should like to touch briefly on an application of probabilistic concepts to general field theory. In the
quest for models of the Wightman or Osterwalder-Schrader
axioms it is tempting to try and generalize the construction of new characteristic functions from given ones.
Convex sums of characteristic functions such as

$$C(f) = \int d\mu(\alpha) \, C_\alpha(f)$$

amount to convex linear combinations of the n-point
functions and as such are known to violate uniqueness
of the vacuum [16].

On the other hand it has been observed that this deficiency is absent when one superimposes second characteristics [17]. Characteristic functions of the form

$$C(f) = e^{\int d\mu(\kappa) \, \log \, C_\kappa(f)}$$

should generate new sets of n-point functions by convex superposition of given truncated n-point functions, they indeed are characteristic functions.

Now if we know that

$$C_\kappa^{\mu(\kappa)} = e^{\mu(\kappa) \, \log c_\kappa}$$

is a characteristic function for all $\mu(\kappa) > 0$, then by taking products and limits we can prove the same for $C(f)$.

This is **tantamount** to saying that infinitely divisible functionals are adequate inputs for this type of superposition. A simple example is furnished by the generating functionals of free fields with masses κ; positive powers of Gaussians are again Gaussians. However the resulting model is not really new - it is nothing but the generalized free field with Lehmann mass distribution ρ given by

$$\rho(\kappa) \, d\kappa = d\mu(\kappa) \quad .$$

I.e. the method might not be so bad but the free field

input is not good enough. Now as we all know the list of known field theories is rather limited. There are still the Wick powers, in particular $:\phi^2:$ Indeed the square of a single Gaussian random variable is infinitely divisible, it obeys the Gamma distribution

$$C_Y(a) = N\int dx \; e^{iax^2-\frac{1}{2}x^2} \; dx = (1 - 2ia)^{-\frac{1}{2}} \; .$$

This is also useful to calculate the Schwinger functional of $:\phi^2:$

$$\langle e^{i:\phi^2(f):} \rangle_> = \int e^{i((x,Ax)-\langle(x,Ax)\rangle)} d\mu(x)$$

where X is white noise and A has the momentum space kernel

$$\tilde{A}(\vec{p},\vec{q}) = (\vec{p}^2 + m^2)^{-1/2} \; \tilde{f}(\vec{p} + \vec{q}) \; (\vec{q}^2 + m^2)^{-1/2} \; .$$

For any (hermitian) A with discrete spectrum the white noise integral factorizes into one dimensional ones if one diagonalizes A, and one obtains as in the one dimensional case

$$\langle e^{i:(x,Ax):} \rangle_> = \prod_\nu \frac{e^{-ia_\nu}}{(1-2ia_\nu)^{1/2}} =$$

$$= e^{-\frac{1}{2} \; \mathrm{Tr} \; (\log(1-2iA) + 2iA)}$$

This functional C(A) is indeed infinitely divisible
as long as we consider it as a function of the eigenvalues (a_ν). But this is not good enough, since it
means that we take into account a set of A which are
simultaneously diagonalizable, i.e. commutative. And
indeed infinite divisibility fails to hold if we allow
for a larger class of test variables A like e.g. the
natural one of all real self-adjoint Hilbert-Schmidt
operators, since it turns out that infinite divisibility fails already for the three-dimensional set of real
symmetric 2x2 matrices A [18]. As a consequence of this
observation one must take into account the particular
form of operators A that arise from smearing out a
local (relativistic or Euclidean) Wick product.

This has recently been done by Hegerfeldt [19]
who finds infinite divisibility for the $:\phi^2:$ Schwinger
functional in 2 space-time dimension. (The trick is to
use

$$\log z = \int\limits_{0}^{\infty} \frac{d\zeta}{\zeta} \{e^{\zeta(1-z)} - 1\} e^{-\zeta}$$

in the exponent.)

The corresponding relativistic field however
turned out <u>not</u> to be infinitely divisible for any
dimensionality. Hegerfeldt has also recently extended
the factorization of characteristic functions into infinitely divisible and indecomposable or "prime" factors
to Euclidean and Wightman fields and has shown that

$$S = 1$$

for infinitely divisible ones [20]. Thus the results of

[21] on the nontriviality of the S-Matrix

$$S \neq 1 \quad \text{in} \quad \phi_2^4$$

show that the characteristic functional for this model
must contain indecomposable factors.

§4. THE RENORMALIZATION GROUP IN TERMS OF
CHARACTERISTIC FUNCTIONS

Critical phenomena are modelled not only by statistical mechanics but also by quantum field theories.
"Symptoms" are that for certain values of the (input)
parameters

- some of the (output) quantities become singular
- the system has long range correlations (massless
  excitations)
- it exhibits scale invariance
- macroscopic, collective observables dominate its
  behaviour.

The latter two observations form the starting point of
the Kadanoff-Wilson description of critical phenomena
[22]. It has recently been cast into probabilistic
terms in a series of papers by Jona-Lasinio [23]. As
a simple model to visualize his ideas

a) consider a chain of points ν

$$\times \quad \times \quad \times \quad \times \quad \times \quad \times \quad \times$$
$$\nu \qquad \nu+1$$

with a (dynamical) random variable ("spin") X_ν attached
to each one of them such that we can view the collection
(X_ν) as a stochastic process.

b) construct "block spins" by averaging over n (= block
length) spins X

$$Y_\nu = \frac{\sum\limits_{i=1}^{n} X_{n\nu + i}}{c_n}$$

c) single out the fixed points of the map $T : X \rightarrow Y$
critical distributions being characterized by the fact
that

$$T X^* = X^* .$$

The critical probability measure μ_{cr}^* is then a fixed
point of the mapping on probability measures induced
by T. In terms of characteristic functionals [11]

$$C(a) \rightarrow C(T^+ a)$$

and critical systems are distinguished by

$$c^* (T^+ a) = c^* (a) .$$

The standard procedure to find such fixed points is to
consider the iterated mapping T^n ("renormalization group").

$$\lim_{n\to\infty} C((T^+)^n a)$$

should be the characteristic function of a critical measure because

$$T\ T^\infty = T^\infty$$

if that exists.

Note

a) the appropriate normalization c_n is a dynamical quantity; given an initial probability distribution the iteration can only be expected to converge for a particular normalization.

b) various initial distributions will lie in the "domain of attraction" of one limit distribution. This is critical point universality.

We want to illustrate these concepts by two simple examples:

1) "Ultralocal Fields" [24] of the form

$$C(f) = <e^{i\phi(f)}> = e^{-\int dx \int (1-\cos\lambda f(x))g^2(\lambda)d\lambda} \ ,$$

g is a structure function related to the interaction potential. A dilatation

$$T_\ell^+ : f(x) \rightarrow f_\ell(x) = \frac{1}{c_\ell} f(\ell x)$$

induces on the structure functions a mapping

$$g^2(\lambda) \rightarrow \ell c_\ell g^2 (c_\ell \lambda) \quad .$$

This map has fixed points

$$g^* = \text{const. } \lambda^{-\gamma}, \quad -\infty < \gamma < \infty$$

of which $\frac{1}{2} < \gamma \le \frac{3}{2}$ are eligible as structure functions. c_ℓ is then obtained from

$$g^2(\lambda) = \ell c_\ell g^2 (c_\ell \lambda)$$

$$c_\ell = \ell^{\frac{1}{2\gamma-1}} \quad .$$

The explicit form of the limiting characteristic function is

$$c^+(f) = e^{-k \int |f(x)|^{2\gamma-1} dx}$$

i.e. the field tends toward a pseudofree (unless $\gamma = \frac{3}{2}$) massless limit [25] if g lies in the domain of attraction of g_γ^*. This in turn is guaranteed by

$$g(\lambda) = \lambda^{-\gamma} S(\lambda)$$

where $S(\lambda)$ is a "slowly varying function" [26] so that

$$\frac{S(a\lambda)}{S(\lambda)} \xrightarrow[\lambda \to \infty]{} 1 \qquad \forall a > 0 \quad .$$

The interaction Hamiltonian is $V(\phi)_{ren}$ given by the structure function g [24], [25]:

$$g''/g = \frac{\gamma(\gamma+1)}{2\lambda^2} + V(\lambda)$$

so that the above considerations determine the ultra-local Hamiltonians of a given universality class γ. It should be clear that the construction of long range scaling limits is only one of many capabilities of the general fixed point strategy. Different physical quest-ions even about the same dynamical system will give rise to different mappings T, and possibly to different fixed points $c^*(a)$. To illustrate this we choose as our further example

2) "The Infinite Momentum Limit"
Loosely speaking, velocity transformations tilt the ca-nonical $x_0 = 0$ plane until in the limit it becomes tangent to the light-cone: a "null plane". A closer look though reveals that under such a limit transfor-mation the time zero field $\phi(x)$ would not just move up to the null plane but also off to infinity:

$$(0, x_1, x_2, x_3) \rightarrow (x_1 \sinh \lambda, x_1 \cosh \lambda, x_2, x_3)$$

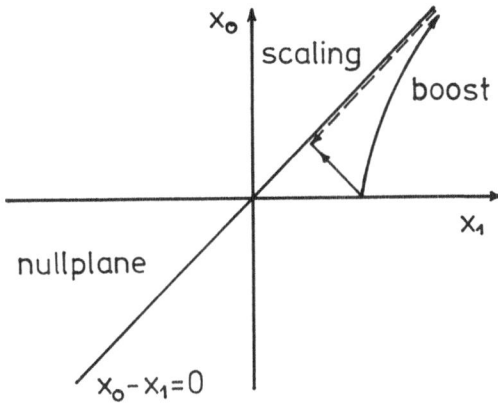

Fig. 1

The transformation T_λ on ϕ must consist of first a boost
and then a scaling of the coordinates x_0, x_1 by a factor ℓ,

$$\ell = (\cosh^2\lambda + \sinh^2\lambda)^{-1/2} \approx \frac{1}{\sqrt{2}} \, e^{-|\lambda|} \, .$$

On $C(f) = (\Omega, \, e^{i\phi(f)}\Omega)$

the boost acts as the identity so that T reduces to a
short distance scaling

$$\phi(x_1, \ldots) \to \frac{\phi(\ell x_1, \ldots)}{c_\ell} \quad , \, \ell \to 0$$

and of course a normalization c_ℓ to be chosen appro-
priately if a limit c^*

$$C(T_\ell^+ f) \to c^*(f)$$

with

$$(T_\ell^+ f)(x) = \frac{1}{\ell c_\ell} \; f(\frac{x_1}{\ell}, \; x_2, \dots)$$

is to exist.

For relativistic free fields in particular

$$C(T_\ell^+ f) = \exp \{-\frac{1}{4} c_\ell^{-2} \int d^s p \; \frac{|\tilde{f}(p)|^2}{\sqrt{p_1^2 + \ell^2 (p^2 + m^2)}}\}$$

so that we can choose c_ℓ = const. in this case. Focussing on symmetric functions f which give rise to a maximal abelian subalgebra on the null plane [27] we find a correlation function

$$K_\infty (\vec{x} - \vec{y}) \sim \log|x_1 - y_1| \; \delta^{(s-1)} (x-y)$$

in the limit.

Note:

a) the long range correlation in the x_1-direction.

b) the discontinuous change of the test function topology in keeping with the predictions of ref. [11].

$$\| f \|_m^2 = \int dp \; \frac{|\tilde{f}|^2}{\sqrt{p^2 + m^2}} \; \rightarrow \; \| f \|_0^2 = \int dp \; \frac{|\tilde{f}|^2}{|p|}$$

c) Under scale transformations K_∞ is not manifestly invariant,

$K_\infty (x_1 - y_1) \rightarrow K_\infty (x_1 - y_1) + a \log \ell,$

but the scale transformation changes $(f K_\infty f)$ by a term

$a \log \ell \cdot \int f(x) dx \cdot \int f(y) dy = 0$

if $\quad \|f\|_0 < \infty$,

so that dilatation and conformal invariance hold [28].

***

Discussions with V. Enss, T. Hida, F. Jegerlehner, J.R. Klauder are gratefully acknowledged.

REFERENCES

1. L. Streit, Acta Phys. Austr. Suppl. II, 2 (1966).

2. E. Nelson, A Quartic Interaction in Two Dimensions, in E.Goodman and I.Segal (eds.), Mathematical Theory of Elementary Particles (Cambridge 1966).

3. E. Nelson, Quantum Fields and Markoff Fields, Symp. in Pure Mathem. 23, 413 (1973).

4. J. Glimm and A. Jaffe, Phys. Rev. Lett. 33, 440 (1974).

5. e.g. D.I. Olive, Karpacz Lectures 1975.

6. E. Lukacs, Characteristic Functions, (London 1970).

7. T. Hida, Stationary Stochastic Processes, (Princeton 1970).

8. B. Simon, The $P(\phi)_2$ Euclidean Quantum field Theory, (Princeton).

9. J. Fröhlich, Helv. Phys. Acta 47, 265 (1974) and Adv. Math. (to appear).

10. G.C. Hegerfeldt, J.R. Klauder, Comm. Math. Phys. 16, 329 (1970).

11. J.R. Klauder, A Characteristic Glimpse of the Renormalization Group, (Bell Labs. preprint 1975, to appear in J. Math. Phys.).

12. L. Streit, Acta Phys. Austr. 42, 9 (1975).

13. H. Ezawa, J.R. Klauder, L.A. Shepp, Ann. Phys. (N.Y.), 88, 588 (1975).

14. J. Eachus, L. Streit, Rep. Math. Phys. 4, 161 (1973).

15. T. Hida, L. Streit (in preparation).

16. E.G. Sudarshan

17. J.R. Klauder, L. Streit, W. Wyss, unpublished.

18. Discussions with V. Enss were very helpful in clarifying this point.

19. G.C. Hegerfeldt, Contribution to the ZiF-Symposium Dec. 75, and K. Baumann and G.C. Hegerfeldt, Is the Wick Square Infinitely Divisible ?, ZiF preprint,1976.

20. G.C. Hegerfeldt, Comm. Math. Phys. 45, 133 (1975).

21. K. Osterwalder and R. Seneor, The S-Matrix is Nontrivial for Weakly Coupled $P(\phi)_2$ Models, preprint 1975.
also contributions of J.P. Eckmann and J. Dimock to Quantum Dynamics: Models and Mathematics (L.Streit, ed.), ZiF-Symposium Sept. 1975, to appear (Springer, Vienna + New York).

22. L.P. Kadanoff et al, Rev. Mod. Phys. 39, 395 (1967).
    F. Jegerlehner and B. Schroer, Acta Phys. Austr.
    Suppl. XI, (1973) has many further references.

23. G. Jona-Lasinio, Critical Behaviour in Terms of
    Probabilistic Concepts, at Int. Coll. on Math.
    Methods in Quantum Field Theory, Marseille, 1975.
    G. Jona-Lasinio, Nuov. Cim. 26B, 99 (1975).
    G. Jona-Lasinio, G.Gallavotti, Comm. Math. Phys.
    41, 3o1 (1975).

24. e.g. J.R. Klauder, Acta Phys. Austr. Suppl. VIII
    227 (1971).
    R. Klauder in Lectures in Theoretical Physics,
    Vol. XIV B, W. Brittin, ed., Boulder 1973.

25. J.R. Klauder, Acta Phys. Austr. Suppl. XI, 341
    (1973).

26. G. Feller, An Introduction to Probability Theory
    and its Applications, Vol. II, (New York 1971)
    Ch. VIII, 8.

27. H. Leutwyler, J.R. Klauder, L. Streit, Nuov. Cim.
    66A, 536 (1970).

28. F. Rohrlich, L. Streit, Nuovo Cim. 7B, 166 (1972).

Acta Physica Austriaca, Suppl. XV, 45–78 (1976)

SCHWINGER FUNCTIONALS AND
EUCLIDEAN MEASURES[+]

by

J.L. CHALLIFOUR
ZiF, University of Bielefeld
Germany

and

Depts. of Mathematics and Physics[++]
Indiana University
Bloomington, USA

1. INTRODUCTION

It is an understatement to say that the last ten years of constructive field theory have brought about a revolution in our understanding of relativistic quantum fields. Even though the models studied so far have been

---

[+] Lecture given at XV. Internationale Universitätswochen für Kernphysik,Schladming,Austria,February 16-27,1976.
[++]Permanent address.

super-renormalizable they have provided not only new
techniques but a new conceptual framework which unifies
quantum fields and statistical mechanics. Whether this
relationship will continue for the study of more complex
models is the immediate challenge of the next few years.

The first suggestion that a euclidean approach might
be useful in quantum electrodynamics was made by Schwinger
[35],[36] who wrote down the appropriate euclidean dynamical equations. It was Symanzik [43],[44] however who proposed and studied in perturbation theory a completely euclidean ϕ^4-model, isolating many of the underlying probabilistic ideas including the Markov property. Indeed throughout the study of the $(\phi^4)_2$-model by Glimm, Jaffe [11],
[12], [13] and Nelson [25] euclidean techniques and points
of view appeared in many places. The first abstract mathematical formulation for Markov euclidean fields and their
connection with the relativistic theory was provided by
Nelson [26], [27] in 1972. The converse question of for-
mulating the Wightman axioms for relativistic fields within a euclidean framework was solved shortly thereafter by
Osterwalder and Schrader [32]. In this direction, a direct
connection between relativistic fields and Nelson's work
was given by Simon [37] and studied in more generality by
Challifour and Slinker [4]. Recent work by Borchers and
Yngvason [2], [3] analyses necessary and sufficient conditions whereby the Wightman theory may be given by a
euclidean measure; though in general not positive. This
is the present stage of development in the axiomatic
study of euclidean field theory which we shall review in
the first part of these lectures.

In 1972 Glimm, Jaffe and Spencer [16] culminated
their study of $\lambda P(\phi)_2$-models with a euclidean field

theory which for weak coupling satisfies with its related
relativistic theory all the Wightman axioms and has much
of the mass spectrum that is expected on physical grounds
[14], [40], [42], [42]. Here P(x) is a polynomial bounded
from below for real x and weak coupling means λ/m_o^2 is sufficiently small. m_o is the mass of the free, neutral, scalar
field from which the construction starts. The technical
core of this work is a cluster expansion analogous to a
high temperature expansion in statistical mechanics whose
convergence requires the weak coupling. Subsequent results
by Dimock [7], Eckmann, Epstein and Fröhlich [8], Osterwalder and Seneor [33] show the existence and non-triviality of the scattering matrix for such models as well as
the fact that the usual renormalized Feynman perturbation
series is really an asymptotic expansion. In another direction, Guerra, Rosen, and Simon [20], [38] have studied the
$P(\phi)_2$-models from the point of view of classical statistical mechanics from 1973 on. They find not only existence
theorems and some mass spectrum results for the $\lambda(\phi^4)_2 +$
$a(\phi^2)_2 - \mu(\phi)_2$ model with arbitrary coupling $\lambda > 0$ for
many boundary conditions; but, also the role played by
approximating euclidean fields by lattice "Ising systems"
with continuous spin. It is here that the "ferromagnetic"
nature of the infinite volume euclidean theory appears
with the validity of many Ising magnet correlation inequalities such as these due to Griffiths [18], Kelley and
Sherman [22]. As shown by Simon and Griffiths [39] there
is a Lee-Yang theorem [23] for $(\phi^4)_2$ and correlation inequalities special to spin-1/2 systems; namely, of
Griffiths, Hurst, Sherman type [19]. The validity and
usefulness of these correlation inequalities is one more
strength of the euclidean point of view. Recently, Glimm,
Jaffe and Spencer [17] have shown that for λ/m_o^2 <u>sufficient-</u>

ly large, the $(\phi^4)_2$-model with Dirichlet boundary conditions can have two distinct phases. For the second
part of our discussion, we will illustrate the ferromagnetic nature of the euclidean measure by reporting
the results of Guerra, Rosen and Simon [20] and those
of Simon and Griffiths [39] in terms of the zeros of the
characteristic functional for the measure and a result
of Neuman [29], [30] modifying work by Glimm and Jaffe
[15] illustrating domination of the Schwinger functional
by Gaussian (generalized free field) estimates. It is
here that an axiomatic approach could make a useful contribution to quantum field theory by an abstract characterization of ferromagnetism for the infinite volume
euclidean measures and study of their general properties.
This is, as yet, an unsolved problem for conventional
Ising systems. The need to develop intuition at the infinite volume level is all the more apparent for the
more singular models where the eventual physical theory
lies at the end of numerous difficult and delicate renormalization arguments.

2. THE AXIOMATIC FRAMEWORK

In order to make clear the connection between the
Wightman axioms for a relativistic quantum field theory
and their representation by a euclidean field viewed as
a generalized stochastic process with respect to a
euclidean measure it is convenient to formulate the
axioms in terms of linear functionals on the Borchers
algebra of test functions [1]. The Schwartz space of
functions on R^{4n} will be denoted by S_n with S_0 stand-

ing for the complex numbers C. The Borchers algebra is then the direct sum $\underline{S} = \bigoplus_{n=0}^{\infty} S_n$ consisting of all finite sequences $\underline{f} = \{f_0, f_1, f_2, \ldots, f_N, 0, 0, \ldots\}$ where the nth component is $(\underline{f})_n = f_n \in S_n$. $\underline{S}$ becomes a $*$-algebra by the operations;

(1) addition: $\qquad \alpha \underline{f} + \underline{g} = \{\alpha f_0 + g_0, \ldots, \alpha f_n + g_n, \ldots\}, \alpha \in C$.

(2) $*$-operation: $\qquad (\underline{f}^*)_n(x_1, \ldots, x_n) = \overline{f_n(x_n, \ldots, x_1)}$.

(3) product: $\qquad (\underline{f} \times \underline{g})_n(x_1, \ldots, x_n) = \Sigma_{k=0}^{n} f_k(x_1, \ldots, x_k)$

$$\cdot g_{n-k}(x_{k+1}, \ldots, x_n)$$

with a unit element $\underline{1} = \{1, 0, 0, \ldots\}$.

It is easily checked that $(\underline{f} + \underline{g}) \times \underline{h} = (\underline{f} \times \underline{h}) + \underline{g} \times \underline{h}$, $(\underline{f} \times \underline{g}) \times \underline{h} = \underline{f} \times (\underline{g} \times \underline{h})$ and $(\underline{f} \times \underline{g})^* = \underline{g}^* \times \underline{f}^*$. Each S_n becomes a topological space with respect to a basis of continuous norms given by

$$p_n^{(N)}(f_n) = \sup_{\substack{x \in R^{4n} \\ |\alpha| \le N}} (1 + \|x\|^2)^{N/2} |D^\alpha f_n(x)|, \quad N \text{ a non-negative integer.}$$

We use here the conventional multi-index notation of distribution theory [34] with $x = (x_1, \ldots, x_n), \|x\|^2 = \Sigma_{i=1}^{n}(x_i^2 + \vec{x}_i^2)$, $\alpha = (\alpha_1, \ldots, \alpha_n)$ each α_i being a four component index,

$$|\alpha| = |\alpha_1| + \ldots + |\alpha_n| \quad \text{with} \quad |\alpha_i| = \Sigma^3_{k=0} \alpha_{i,k} \quad \text{and}$$

$$D^\alpha = D_{x_1}^{\alpha_1} \ldots D_{x_n}^{\alpha_n} .$$

$\underline{S}$ becomes a topological space upon taking as a continuous basis of norms

$$p(\underline{f}) = \Sigma^\infty_{n=0} p_n(f_n) \qquad \text{(finite sum)}$$

with $\{p_n\}$ varying over a basis for S_n. The positive elements of $\underline{S}$ are those given by

$$\underline{S}^+ = \{\underline{f} \in \underline{S} | \underline{f} = \Sigma^\infty_{k=1} \underline{f}_k^* \times \underline{f}_k, \ \underline{f}^k \in \underline{S}\};$$

just convergent sums of "squares" of elements in $\underline{S}$.

Continuous linear functionals on $\underline{S}$ are infinite sequences $\underline{T} = \{T_0, T_1, T_2, \ldots, T_n, \ldots\}$ where $T_n \in S_n'$ – the dual (topological) of S_n – and $\underline{T}(\underline{f}) = \Sigma^\infty_{n=0} T_N(f_n)$. $\underline{T}$ is real if $\underline{T}(\underline{f}) = \underline{T}(\underline{f}^*)$ and positive if $\underline{T}(\underline{f}) \geq 0$ for all $\underline{f} \in \underline{S}^+$.

Suppose now that $A(t,\vec{x})$ denotes the neutral, scalar relativistic field with vacuum (ground state) Ω_R. Then the euclidean theory of Osterwalder and Schrader [32] exploits the Wightman axioms for A by continuation to imaginary time in the Wightman distributions W_n thereby defining the non-coincident Schwinger functions $\mathcal{S}_n$; namely,

$$\mathscr{S}_n((t_1,\vec{x}_1),\ldots,(t_n,\vec{x}_n)) \equiv W_n((i\ t_1,\vec{x}_1),\ldots,(i\ t_n,\vec{x}_n))$$

$$-\infty < t_1 < t_2 < \ldots < t_n < \infty$$

where

$$W_n((t_1,\vec{x}_1),\ldots,(t_n,\vec{x}_n)) \equiv (\Omega_R, A(t_1,\vec{x}_1)\ldots A(t_n,\vec{x}_n)\Omega_R).$$

The $\mathscr{S}_n$ are real analytic functions of real arguments with time ordering as a consequence of the relativistic spectral property (positive energies). Using local commutativity and Lorentz covariance for A, the region of definition for $\mathscr{S}_n$ can be extended so that $\mathscr{S}_n$ is symmetric on the euclidean region

$$E_n = \{x \in R^{4n}|\ x_i \neq x_j\ 1 \leq i < j \leq n\}.$$

The subsequent properties of the $\{\mathscr{S}_n\}_{n=0}^{\infty}$ have been given by Osterwalder and Schrader to form an equivalence with the usual Wightman axioms. For the statement of those we need some more notation. Let Ω_n be an open set in R^{4n} and $S(\Omega_n)$ denote those functions in S_n which vanish with all derivatives unless their arguments lie in Ω_n. Define open sets

$$E_n^< = \{x \in R^{4n}|\ -\infty < t_1 < t_2 < \ldots < t_n < \infty\},$$

$$E_{n,+}^< = \{x \in R^{4n}|\ 0 < t_1 < \ldots < t_n < \infty\}$$

and subspaces of $\underline{S}$ by $\underline{S}_o = \bigoplus_n S(E_n)$, $\underline{S}_< = \bigoplus_n S(E_n^<)$ and $\underline{S}_+ = \bigoplus_n S(E_{n,+}^<)$; $\underline{S}_+ \subset \underline{S}_< \subset \underline{S}_o$. Also let θ denote time inversion with $\theta(t,\vec{x}) = (-t,\vec{x})$ and $(\theta\underline{f})_n (x_1,\ldots,x_n) = f_n(\theta x_1,\ldots,\theta x_n)$. The action of the euclidean group on test functions is $(\underline{f}_{(a,R)})_n (x_1,\ldots,x_n) = f_n(R^{-1}(x_1-a), \ldots, R^{-1}(x_n-a))$.

Osterwalder-Schrader Axioms [32]

In terms of the sequence $\underline{\mathscr{S}} = \{\mathscr{S}_o, \mathscr{S}_1, \ldots, \mathscr{S}_n, \ldots\}$;

EO: $\underline{\mathscr{S}} \in \underline{S}_o'$, $\mathscr{S}_o = 1$ and each $\mathscr{S}_n$ satisfies a regularity property.

E1: $\underline{\mathscr{S}}(\underline{f}_{(a,R)}) = \underline{\mathscr{S}}(\underline{f})$ for each $\underline{f} \in \underline{S}_o$, $(a,R) \in i$ SO(4).

E2: $\underline{\mathscr{S}}(\theta\underline{f}^* \times \underline{f}) \geq 0$ for each $\underline{f} \in \underline{S}_+$.

E3: $\mathscr{S}_n$ is symmetric in all variables.

E4: $\lim_{\lambda \to \infty} \underline{\mathscr{S}}(\theta\underline{f}^* \times \underline{g}_{(\lambda a,1)}) = \underline{\mathscr{S}}(\theta\underline{f}^*)\underline{\mathscr{S}}(\underline{g})$ for all $a \neq 0 \in R^3$ and $\underline{f}, \underline{g} \in \underline{S}_+$.

The conditions EO, E1, E2 are related in a one to one manner with the Poincaré covariance, positive definiteness and spectrum of the Wightman distributions. While together with these, E3 is equivalent to local commutativity and E4 with the cluster property or uniqueness of the vacuum Ω_R. The regularity property in EO, though important, is somewhat technical and not necessary to our discussion. The reader is referred to the original paper [32]. By a standard mathematical theorem (Hahn-Banach) the functional $\underline{\mathscr{S}}$ extends to a continous linear

functional $\underline{S}$ on $\underline{S}$; though in general not unique. To
summarize,

Theorem (Osterwalder-Schrader [32]). To every Wightman
quantum field theory there exists a state $\underline{S}$ on $\underline{S}$ satisfying E0 to E4, and conversely. ·

3. POSITIVE EXTENSIONS

Our next purpose is to relate the above framework
to a euclidean field theory as a stochastic process on
a measure space with a probability measure. For this we
require extensions $\underline{S}$ of $\underline{\mathcal{S}}$ which satisfy a positivity
requirement which permits construction of the measure.

Consider then a positive polynomial $P(x) =$
$P(x_1, \ldots, x_n) \geq 0$ on R^n and n real functions f_1, f_2,
$\ldots, f_n$ from S. Let S'_R denote elements of S' which are
real on real functions; namely, $\omega \in S'_R$ satisfies
$<\omega \ \bar{f}> = \overline{<\omega \ f>}$ for $f \in S$. Define a function on S'_R
with positive values by

$$P(\omega) \equiv P(<\omega, f_1> , \ldots, <\omega, f_n>) ;$$

and denote by P_+ the collection of all such polynomials
as $\{f_1, \ldots, f_n\}$ varies over all finite subsets of real
functions in S. P_+ is the cone of positive polynomials
on S'_R.

Definition Consider $\underline{S} = \{1, S_1, S_2, \ldots, S_n, \ldots\}$ an extension of $\underline{\mathcal{S}}$ to $\underline{S}$. $\underline{S}$ is called a Schwinger state if

$\underline{S0}$: $\underline{S} \in \underline{S}'$;

$\underline{S1}$: $\underline{S}(f_{(a,R)}) = \underline{S}(\underline{f})$ for $(a,R) \in i$ $SO(4)$, $\underline{S}(\Theta\underline{f}) = \underline{S}(\underline{f})$
for all $\underline{f} \in \underline{S}$;

$\underline{S2}$: $\underline{S}(\Theta\underline{f}^* \times \underline{f}) \geq 0$ for $\underline{f} \in \underline{S}_+$;

$\underline{S3}$: Each S_n is symmetric;

$\underline{S4}$: $\lim_{\lambda \to \infty} \underline{S}(\underline{f} \times \underline{g}_{(\lambda a,1)}) = \underline{S}(\underline{f}) \, \underline{S}(\underline{g})$ for all $a \neq 0 \in R^4$
and $\underline{f}, \underline{g} \in \underline{S}$.

$\underline{S5}$: $\underline{S}(P_+) \geq 0$.

This notation differs from that in [4] and conforms to usage in constructive field theory. States satisfying $S(P_+) \geq 0$ are called strongly positive or Nelson-Symanzik positive in contrast to positive states for which $S(\underline{S}^+) \geq 0$. Due to Θ invariance we have that $\underline{S}$ is real and the underlying relativistic theory Θ invariant. This restriction is a consequence of strong positivity [4].

As of yet interesting conditions on $\underline{S}$ which guarantee an extension to a Schwinger state are not known. There is a partial result for positive extensions. Define a Hilbert norm on $\underline{S}$ of the form

$$q^{(N_1,N_2)}(\underline{f}) = |\Sigma_{n=0}^{\infty} c(n)^2 \, h_n^{(N_1,N_2)}(f_n)^2|^{1/2}$$

with

$$h_n^{(N_1,N_2)}(f_n)^2 = \sum_{\substack{|\alpha_i| \leq N_1 \\ 1 \leq i \leq n}} \int dx_1 \ldots dx_n \prod_{j=1}^{n} (1+\|x_j\|^2)^{N_2/2} .$$

$$\cdot |D_{x_1}^{\alpha_1} \ldots D_{x_n}^{\alpha_n} f_n(x_1,\ldots,x_n)|^2$$

and notice that (N_1, N_2) is independent of n. This is the key to

Proposition (Challifour-Slinker [6]). Let $|\underline{\mathscr{S}}(\underline{f})| \leq q^{(N_1,N_2)}(\underline{f})$ for all $\underline{f} \in S_0$ and some pair of non-negative integers (N_1,N_2). Then there exists a positive symmetric extension of $\underline{\mathscr{S}}$ to $\underline{S}$. For $N_2 = 0$, this extension may be taken as translation invariant.

The choice of the norms $q^{(N_1,N_2)}$ is motivated by Osterwalder and Schrader [32] who show that if $c(n) \leq \alpha (n!)^\beta$ then the regularity property in EO for the $\mathscr{S}_n$ is a consequence of the other axioms and these bounds. Sufficient conditions on $\underline{\mathscr{S}}$ which lead to a strongly positive symmetric extension form an interesting mathematical problem.

4. THE EUCLIDEAN MEASURE

Before indicating the construction for the euclidean measure from a Schwinger state let us recall some terminology from probability theory. Consider a set Ω and family Σ of subsets of Ω.

Definition Σ is called an algebra of subsets if

(i) $\Omega \in \Sigma$; (ii) $A \in \Sigma$ implies the complement $\Omega - A \in \Sigma$;
(iii)   if $A_n \in \Sigma$ for $n = 1,2,\ldots,N$ then $\bigcup_{n=1}^{N} A_n \in \Sigma$ for
        any finite N.

Σ is called a σ-algebra if Σ is an algebra and the finite union property (iii) is improved to countable unions; namely,

(iv)   if $A_n \in \Sigma$ for $n = 1,2,..$ then $\bigcup_{n=1}^{\infty} A_n \in \Sigma$.

When Ω is also a topological space with respect to some collection of open sets there exists a unique smallest σ-algebra B containing all the open and hence closed sets. B is called the <u>Borel σ-algebra</u> for Ω.

<u>Definition.</u> A probability measure μ on (Ω,Σ) is a set function defined on Σ such that

(i) $0 \leq \mu(A) \leq 1$ for all $A \in \Sigma$;

(ii) $\mu(\Omega) = 1$; and

(iii)   if $\{A_n\}_{n=1}^{\infty}$ is a disjoint collection of sets in
        Σ then $\mu(\bigcup_{n=1}^{\infty} A_n) = \Sigma_{n=1}^{\infty}\mu(A_n)$.

This last requirement is called the countable additivity of μ on Σ and is basic to the general theory of integration for measure spaces. A probability space is then a probability measure μ on (Ω,Σ).

The strong positivity of $\underline{S}$ and its symmetry allows a commutative representation of $\underline{S}$ by random variables on a probability space. Throughout we shall choose a fixed representation for the measure space as (S_R',B). S_R' are the real tempered distributions introduced in section 3 and B is the Borel σ-algebra generated by any topology on S_R' which lies between the weak and strong topologies. This generosity arises because S is a example of a Fréchet-Montel space. More is in fact true, S is also a nuclear space in the sense of topological vector spaces. These properties have mathematical consequences which are basic to euclidean field theory viewed as a generalized stochastic process. For our specific construction

it should also be remarked that B is also the smallest
σ-algebra containing the algebra C of cylinder sets on
S_R'. Z ε C is a set of the form

$$Z = \{\omega \; \varepsilon \; S_R' \; | \; (<\omega,f_1>,\ldots,<\omega,f_N>) \; \varepsilon \; A \subset R^N\}$$

for some real functions $\{f_1,\ldots,f_N\}$ in S and Borel set
A in R^N. Our task is then to assemble this machinery
to find a μ such that

$$\underline{S} \; (\underline{f}) \; = \; \int \underline{f}(\omega) \; d\mu \, (\omega).$$

Here $\underline{f}(\omega) = f_0 + \Sigma_{n=1}^{\infty} <(\omega \otimes \omega \otimes \ldots \otimes \omega), f_n>$ with $\omega \; \varepsilon \; S_R'$
in which $\otimes^n \omega$ acts on functions $f_n \; \varepsilon \; S_n$ by continuous
extension from its definition on products $g_1 \otimes \ldots \otimes g_n$,
$g_i \; \varepsilon \; S$, whose span is dense in S_n; namely, $< \otimes^n \omega$,
$g_1 \otimes \ldots \otimes g_n> \equiv <\omega \, , \, g_1><\omega, \, g_2> \ldots <\omega, \, g_n>$.

The construction of μ follows [4]. $E(S_R')$ is the
class of real polynomially bounded functions on S_R'.
X ε E means there exists a finite collection $\{f_1,f_2,\ldots$
$\ldots,f_N\}$ of real functions in S such that

$$X(\omega) \equiv X(<\omega,f_1>,\ldots,<\omega,f_N>), \omega \; \varepsilon \; S_R'$$

where $X(x_1,\ldots,x_N)$ is a real polynomially bounded function on R^N. When X is a polynomial we shall also write
$P(f_1,\ldots,f_N)$ indicating the functions f_k, k = 1,...,N
explicitly and X ε $P(S_R)$. For polynomials we have by
[SO] that

$$I(P) \equiv \underline{S} \; (P(f_1,\ldots,f_N))$$

is well defined. Now I is extended to E by a method due to M. Riesz and M. Krein used in the classical moment problem on R^N. Choose $X \in E$ and let P be a polynomial and define

$$\text{ext } I(P + \lambda X) = I(P) + \lambda c \quad \text{for } \lambda \text{ real;}$$

with c to be chosen so that the extension is positive. Let Y, Z be polynomials so that $Z(\omega) \leq X(\omega) \leq Y(\omega)$. Then $Y(\omega) - Z(\omega) = (Y-X)(\omega) + (X-Z)(\omega) \geq 0$ so $I(Z) \leq I(Y)$. Take any c for which $\sup I(Z) \leq c \leq \inf I(Y)$.

$$I(Z) = \{Z \in P \cap E : Z \leq X\} \qquad I(Y) = \{Y \in P \cap E : X \leq Y\}$$

It is a short calculation to check that ext I is positive on the real linear span of $\{P, X\}$. Transfinite induction is used to extend I to a real positive linear functional on E. Suppose $l_Z(\omega) = 1$ for $\omega \in Z$ and 0 otherwise for a cylinder set Z. Clearly $l_Z \in E$ hence we may take;

Definition $\mu(Z) \equiv \text{ext } I(l_Z)$.

Exploiting this definition and properties of sets in C one readily shows that μ is a finitely additive normalized set function on C for which

$$\underline{S}(P(f_1,\ldots,f_N)) = \int P(<\omega,f_1>,\ldots,<\omega,f_N>) \, d\mu(\omega).$$

Unfortunately we are not finished since (S_R', C, μ) is not a probability space. μ must be extended to a probability measure on B. This is accomplished by yet another general theorem in measure theory and a theorem due to Minlos [24] exploiting the nuclearity for S remarked above. For details, one should see [4]. The final step to integration of general functions $\underline{f} \in \underline{S}$ uses the continuity of $\underline{S}$ and approximation to it by polynomials P_j. Then

$$\lim_{j \to \infty} \underline{S}(P_j) = \underline{S}(\underline{f}) \equiv \int \underline{f}(\omega) \, d\mu(\omega)$$

extends the definition of the integral in a consistent way. The original non-coincident Schwinger functions $\mathcal{S}_n^o$ from which we started our construction are given by this integral when $\underline{f} \in S_n(E_n)$. Questions relating to the uniqueness of μ are discussed in [4]. One classic result from the moment problem carries over here and is useful in applications.

__Carleman's Criterion.__ μ is unique if $\sum_{p=1}^{\infty} \underline{S}(f^{2p})^{-1/2p} = \infty$, for each real $f \in \underline{S}$.

At this stage it is natural to ask whether one can give conditions directly on $\mathcal{S}$ which lead to an integral representation by a measure μ. If we allow μ to be a complex-valued, normalized measure on (S_R', B) the answer is affirmative and has been given by Borchers and Yngvason as we shall now sketch [2]. Their result is again a manifestation of nuclearity for the space S. Since $|\underline{f}(\omega)|^2 = \overline{\underline{f}}(\omega) \, \underline{f}(\omega) = \overline{\underline{f}} \times \underline{f}(\omega)$ by the diagonal properties of point evaluation at $\chi_\omega = \{1, \omega, \omega \otimes \omega, \ldots \otimes^n \omega, \ldots\}$ we can map $\underline{S}$ into $L_2(S_R', B, \mu)$ by the identity mapping $(i\underline{f})(\omega) \equiv \underline{f}(\omega)$ for which by [SS]

$$\|i\underline{f}\|_2^2 = \int |\underline{f}(\omega)|^2 d\mu(\omega) = \underline{S}(\underline{f}^* \times \underline{f}) \geq 0 \, .$$

It is a standard argument in axiomatic field theory that $\sqrt{\underline{S}(\underline{f}^* \times \underline{f})}$ defines a continuous seminorm on $\underline{S}$. This is as follows:

$$\| i\underline{f} \| = \| i \sum_{n=0}^{N} f_n \| \leq \sum_{n=0}^{N} \| if_n \| = \sum_{n=0}^{N} S_{2n}(f_n^* \times f_n)^{1/2}$$

$$\leq \sum_{n=0}^{N} P_{2n}(f_n^* \times f_n)^{1/2}$$

where P_{2n} is a continuous norm on S_{2n}. By nuclearity of S_{2n} we can always arrange that $P_{2n} = P_n \otimes_\pi P_n$ for real symmetric continuous norms P_n on S_n. Hence as $P_n \otimes_\pi P_n$ $(f_n^* \times f_n) = P_n(f_n)^2$, $\| i\underline{f} \| \leq \sum_{n=0}^{N} P_n(f_n) = p(\underline{f})$. Now comes the dramatic part. Since $\underline{S}$ is a nuclear space <u>any</u> continuous linear mapping of $\underline{S}$ into L_2 is a nuclear mapping. Hence the scheme

$$\underline{S} \subset L_2 \ (S_R', \ B, \ \mu) \subset \underline{S}'$$

has L_2 as a rigged Hilbert space in the sense of Gelfand and Vilenkin [10, page 106]. From their analysis we can conclude there exists a positive function $u \in L_2$ and a continuous seminorm q on $\underline{S}$ such that $|i\underline{f}(\omega)| \leq u(\omega)q(\underline{f})$ for all $\underline{f} \in \underline{S}$. Define

$$q^o(x_\omega) = \sup \ \{|\underline{f}(\omega)|, |q(\underline{f})| \leq 1\} \quad \text{so that} \quad u(\omega) \geq q^o(x_\omega).$$

Then

$$|\underline{S}(\underline{f})| \leq \sup \frac{|\underline{f}(\omega)|}{q^o(x_\omega)} \int |u(\omega)| d\mu(\omega)$$

$$= (\text{Const.}) || \underline{f} ||_q .$$

$||\underline{f}||_q = \sup\limits_{\omega} \dfrac{|\underline{f}(\omega)|}{q^\circ(\chi_\omega)}$ is a seminorm on $\underline{S}$ and as q varies

over a basis of continuous seminorms for $\underline{S}$ generates a new topology say the $\hat{\tau}$-topology on $\underline{S}$, which is weaker than the original one. By decomposing a complex measure into its real and imaginary parts and each of these into positive and negative parts it is easy now to prove:

Proposition (Borchers-Yngvason [2]). $\mathcal{S}(\underline{f}) = \int \underline{f}(\omega)\, d\mu(\omega)$ for all $\underline{f} \in \underline{S}_o$ for a complex normalized measure μ on B if and only if $\mathcal{S}$ is continuous with respect to the $\hat{\tau}$-induced topology on $\underline{S}_o$ as a subspace of $\underline{S}$.

In summary, we see that it is possible to represent the Wightman theory by an integral representation upon going to the euclidean region and then requiring a topological property ($\hat{\tau}$-continuity) for the non-coincident Schwinger functions S_n. The nature of the singularities for these functions at coincident arguments is given in [2] and will not be reproduced here. For an integral representation with only positive measures it is necessary and sufficient to imposes useful conditions on the non-coincident Schwinger functions and what if any regularity properties are forced on the Wightman distributions.

5. THE EUCLIDEAN FIELD

Starting from a Schwinger state $\underline{S}$ let μ be the

probability measure constructed in §4. The euclidean
field is now given immediately by

Definition. For each $f \in S$ set $\phi(f)(\omega) = <\omega, f>$ for
$\omega \in S_R'$ and for $(a,R) \in i$ $SO(4)$ $[\eta_{(a,R)} \phi(f)](\omega) =$
$\phi(f_{(a,R)})(\omega)$.

   Then $\phi(f)$ is a stochastic process on (S_R', B) indexed
by S, called the euclidean field for the Schwinger state.
It satisfies;

(1) linearity: $\phi(\alpha f + g) = \alpha \phi(f) + \phi(g)$.

(2) continuity: $f_j \to 0$ in S implies $\phi(f_j) \to 0$ pointwise on S_R'.

(3) moments: $\underline{S}(P(f_1, \ldots, f_N)) = \int P(\phi(f_1), \ldots, \phi(f_N)) d\mu$.

(4) L_p, $0 < p < \infty$ continuity: $f_j \to 0$ in S implies $||\phi(f_j)||_p \to 0$.
   This is as consequence of $||\phi(f_j)||_p \le ||\phi(f_j)||_{2N} =$
   $\underline{S}(f_j^{2N})^{1/2N} \to 0$ for an integer N such that $p \le 2N$.

So far we have discussed only the implications of the
axioms [S0], [S3] and [S5]. The consequences of [S1]
are easy. Define two flows on (S_R', B) corresponding to
the euclidean group and θ-invariance by

$$<\eta_{(a,R)}^{-1} \omega, f> = <\omega, f_{(a,R)}>, \quad <\theta\omega, f> = <\omega, \theta f> \quad .$$

Then [S1] allows us to choose μ so that these flows are
measure preserving on B; namely, $\mu[\eta_{(a,R)}^{-1} B] = \mu[B]$ for
B Borel and similarly for θ [4]. As regards [S4] in the
situation that the polynomials $P(\phi(f_1), \ldots, \phi(f_N))$ are
dense in $L_2(S_R', B, \mu)$, the euclidean cluster property
implies mixing for the measure μ corresponding to trans-

lations;

viz, $\lim_{||a|| \to \infty} \mu(A \cap n_{(a,1)}^{-1} B) = \mu(A)\mu(B)$ for Borel sets A,B.

In turn this has the unique euclidean vacuum invariant under the action of translations as the function 1 on S_R'. Finally, the Osterwalder-Schrader positivity [S2] is more subtle to understand. In [5], it is shown with strong regularity requirements on the singularities of S_n at coincident arguments (local integrability) that [S2] allows representation of the physical Hilbert space as $L_2(S_R', B_o, \mu)$ where B_o is a sub σ-algebra of B corresponding to localization of the euclidean field at time zero. Such a representation is directly related to Nelson's Markov property for half-spaces and in the case of the $\lambda(\phi^4)_2$-model to the cyclicity of the time zero relativistic fields [9], [37], [38]. These questions remain open at this time both for $\lambda(\phi^4)_2$ with arbitrary coupling and for $\lambda P(\phi)_2$ with weak coupling. In the latter case a local Markov property has been demonstrated by Newman [31].

6. LATTICE APPROXIMATION FOR THE $\lambda(\phi^4)_2$-MODEL

I should now like to report on the lattice approximation to the $\lambda(\phi^4)_2$-model given in [20], [28] to see how ferromagnetism arises for these euclidean field theories. Our starting point is the free euclidean field ϕ_o of mass M_o. This is realized as a generalized Gaussian process over S with mean zero and covariance given by the Sobolev norm

$$E(\phi_o(f)\phi_o(g)) = \int \overline{\phi_o(f)}(\omega)\phi_o(g)(\omega)\, d\mu_o(\omega) = <f,g>$$

$$= (f,(-\Delta+m_o^2)^{-1}g)_{L_2} = \int_{R^2} dp\, \frac{\overline{\hat{f}}(p)\hat{g}(p)}{m_o^2+||p||^2}\,;$$

$d\mu_o$ denotes a Gaussian measure on S_R'. This means that if $\{f_1,\ldots,f_N\}$ are real linearly independent vectors from S the measure of a cylinder set $Z = \{\omega \in S_R'|$ $(<\omega,f_1>,\ldots,<\omega,f_N>) \in A\}$ is given by the N-dimensional Gaussian integral

$$\mu_o(Z) = \frac{1}{(2\pi)^{N/2}(\det C)^{1/2}} \int_A e^{-\frac{1}{2}(q,\, C^{-1}q)}\, d^Nq$$

in which $q = (q_1,\ldots,q_N)$; $(q,\, C^{-1}q) = \sum_{i,j=1}^N q_i(C^{-1})_{ij}q_j$ and $C_{i,j} = E(\phi_o(f_i)\phi_o(f_j))$ are the entries in the N x N covariance matrix. Let $\underline{S}^f$ denote the corresponding Schwinger state for this process. Then all the axioms are satisfied by the Schwinger distributions

$$S_{2n+1}^f = 0;\ \ S_{2n}^f(f_1 \otimes \ldots \otimes f_{2n}) = \sum_{pairs}\ \cdot$$

$$S_2^f(f_{i_1},f_{j_1})\ldots S_2^f(f_{i_n},f_{j_n})\ \cdot$$

The sum is over all $(2n)!/2^n n!$ ways of decomposing $(12\ldots2n)$ into n distinct unordered pairs $(i_1j_1)\ldots$ (i_nj_n). The fourier transform or characteristic functional for μ_o is given by

$$L(zf) = \int e^{iz\phi_o(f)} d\mu_o = e^{-\frac{z^2}{2} <f,f>} \quad \text{for real } f \in S.$$

$L(zf)$ is an entire analytic function of z of order 2.

A euclidean version of Fock space [20] can also be used to represent this process. Suppose $a(k)$, $a^*(k)$ are euclidean annihilation and creation operators with

$$[a(k),a(k')]=0=[a^*(k),a^*(k')] ; [a(k),a^*(k')]=\delta^{(2)}(k-k').$$

Then

$$\phi_o(f) = \int \tilde{f}(k) \frac{[a^*(k)+a(-k)]}{\mu(k)} d^2k, \mu(k) = \sqrt{||k||^2 + m_o^2}$$

according to the convention for fourier transform $\hat{f}(p) = (2\pi)^{-1} \int e^{-ipx} f(x) d^2x$. The Wick order of ϕ_o is given formally by

$$:\phi_o^4:(f) = (2\pi)^{-3} \int \hat{f}(k_1+k_2+k_3+k_4) \sum_{s=0}^{4} \binom{4}{s} a^*(k_1) \ldots a^*(k_s)$$

$$a(-k_{s+1}) \ldots a(-k_4) \times |\mu(k_1)\mu(k_2)\mu(k_3)\mu(k_4)|^{-1} d^2k_1 \ldots d^2k_4.$$

A series of calculations show that $:\phi_o^4:(f)$ with $f \in L_{1+\epsilon}(R^2)$, $0 < \epsilon \le 1$, is a random variable on (S_c', B, μ_o). The first deep result in the $(\phi^4)_2$-model concerns the existence of the euclidean action. We shall state only what is needed and refer the reader to [38] for the general result.

<u>Proposition</u> [11], [20], [25] $e^{-V_\Lambda} = \exp[-\lambda \int_\Lambda :\phi_o^4:(x) d^2x]$ is in $L_p(S_R', B, \mu_o)$ for $1 \le p < \infty$ wherever $\Lambda \subset R^2$ is a bounded region; $\lambda > 0$.

The $(\phi^4)_2$ euclidean field theory for the finite region Λ (or with cut-off Λ) is a generalized random process on S_R' corresponding to the cut-off measure

$$d\mu_\Lambda(\omega) = e^{-V_\Lambda} d\mu_0(\omega)/Z_\Lambda; \quad Z_\Lambda = E(e^{-V_\Lambda}),$$

with the cut-off Schwinger functions

$$S_\Lambda(f_1 \otimes \cdots \otimes f_n) = \int \phi_0(f_1) \cdots \phi_0(f_n) \, d\mu_\Lambda.$$

The analogy with statistical mechanics is clear. Z_Λ plays the role of the partition function for the "Gibbs" measure $e^{-V_\Lambda} d\mu_0$ which is a perturbed Gaussian. The first existence theorem due to Glimm, Jaffe and Spencer is:

Theorem (Glimm, Jaffe, Spencer [16]) $\lim_{\Lambda \to \infty} S_\Lambda(\underline{f})$ exists for all $\underline{f} \in \underline{S}$ and defines a Schwinger state $\underline{S}$ for λ/m_0^2 sufficiently small by the formula $\lim_{\Lambda \to \infty} S_\Lambda(\underline{f}) = \underline{S}(\underline{f})$.

Of course, as we have already stated in the introduction much more is proven in [16]. In particular, the construction in section 4 gives an infinite volume measure μ for which $\mu_\Lambda \to \mu$ in some weak sense.

Another type of cut-off can be used to initially define the $(\phi^4)_2$-model wherein Λ is replaced by a finite collection of points on a two-dimensional lattice with spacing $\epsilon > 0$. When one does this the remarkable phenomenon of ferromagnetism results. Following [20], let $L_\epsilon = \{\epsilon(n_1,n_2) = \epsilon n | n\epsilon \, Z^2\}$ denote the lattice and replace ϕ_0 by the corresponding Gaussian process on the lattice L_ϵ, $\phi_{0,\epsilon}(n)$, in which

$$\phi_{0,\epsilon}(n) = (2\pi)^{-1} \int_{-\pi\epsilon}^{\pi\epsilon} dk_1 \int_{-\pi\epsilon}^{\pi\epsilon} dk_2 \ e^{-ik\cdot n\epsilon} [a^*(k) + a(-k)] \mu_\epsilon(k)^{-1}$$

where $\mu_\epsilon(k) = |\epsilon^{-2}(4 - 2\cos(\epsilon k_1) - 2\cos(\epsilon k_2)) + m_0^2|^{1/2}$.
The covariance matrix C is given by

$$C_{n,n'}^{-1} = \epsilon^2 \cdot \begin{cases} m_0^2 + 4\ \epsilon^{-2} & n = n' \\ -\epsilon^{-2} & n = n' \pm(1,0) \text{ or } n' \pm(0,1) \\ 0 & \text{otherwise .} \end{cases}$$

The interaction term involves a finite sum over points
in $\Lambda_\epsilon = \Lambda \cap L_\epsilon$

$$e^{-V_{\Lambda,\epsilon}} = \exp\ [-\epsilon^2 \lambda \sum_{n\in\epsilon\Lambda_\epsilon} :\phi_{0,\epsilon}^4(n):]$$

with the corresponding change in cut-off measure $d\mu_{\Lambda,\epsilon} = e^{-V_{\Lambda,\epsilon}} d\mu_{0,\epsilon}/Z_{\Lambda,\epsilon}$ and cut-off Schwinger functions $S_{\Lambda,\epsilon}$ such that $\lim_{\epsilon\to 0} S_{\Lambda,\epsilon}(\underline{f}) = S_\Lambda(\underline{f})$, $\underline{f} \in \underline{S}$. The presence of a finite lattice spacing also serves to make the contractions in the Wick products finite so we can write the interaction polynomial as $\lambda > 0$ times

$$:\phi_{0,\epsilon}^4(n): = \phi_{0,\epsilon}^4(n) - 6\ \phi_{0,\epsilon}^2(n)\ C_{n,n} + 3\ C_{n,n}^2$$

$$= (\phi_{0,\epsilon}^2(n) - \phi_{\epsilon,+}^2(n))(\phi_{0,\epsilon}^2(n) - \phi_{\epsilon,-}^2(n))$$

with $\phi_{\epsilon,\pm}(n) = \sqrt{3\pm\sqrt{6}}\ \sqrt{C_{n,n}}$. The estimates given by Guerra, Rosen and Simon [20] show $(8 + m_0^2\ \epsilon^2)^{-1} \leq C \leq (m_0^2\ \epsilon^2)^{-1}$ for which $C_{n,\dot{n}}$ becomes large as $\epsilon\to 0$. The qualitative

68

features of this interaction polynomial are interesting
with two equal minima corresponding to the symmetry
$\phi \leftrightarrow -\phi$. at $\sqrt{3\ C_{nn}}$, each having positive curvature.
As $\varepsilon \downarrow 0$ these minima

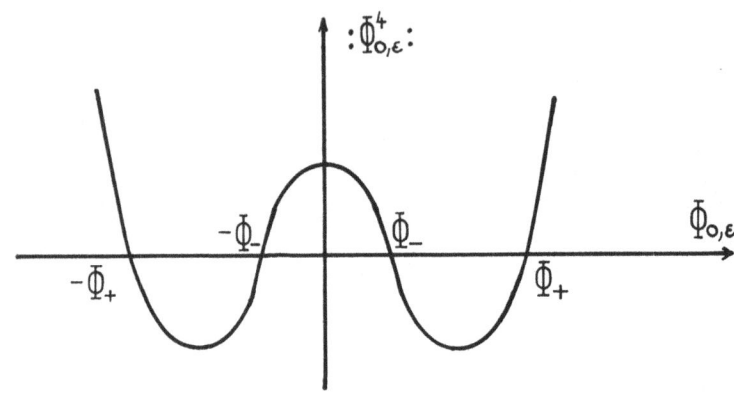

Fig. 1

became deeper allowing the field to become "localized"
in these troughs for sufficiently large coupling λ.
Glimm, Jaffe and Spencer [17] who proposed this picture
have analysed the corresponding infinite volume field
theory ($\Lambda \to \infty$) for Dirichlet boundary conditions confirm-
ing that for λ/m_o^2 large this theory has a doubly degenerate euclidean vacuum due to the existence of two phases.
These phases are separated by breaking the $\phi \leftrightarrow -\phi$
symmetry with an "external field"; namely an interaction
$\lambda : \phi_o^4 : -\mu\phi_o, \mu > 0$.

7. FERROMAGNETISM

Continuing our presentation of the Guerra, Rosen,
Simon [20] analysis for $\lambda(\phi^4)_2$ set $q_n = \phi_{o,\varepsilon}(n)$ for each
lattice site $n \varepsilon \Lambda_\varepsilon$. q_n lies in R and plays the role

of a continuous spin variable at the site $n\varepsilon$. In this notation the cut-off lattice measure becomes

$$d\mu_{\Lambda,\varepsilon}(q) = \frac{e^{-1/2(q,C_{\Lambda_\varepsilon}^{-1}q)}}{(2\pi)^{N/2}(\det C_{\Lambda_\varepsilon})^{1/2}} \prod_{n\varepsilon\Lambda_\varepsilon} \frac{e^{-[aq_n^4 + bq_n^2]}}{Z_{\Lambda,\varepsilon}} dq_n$$

where C_{Λ_ε} is the covariance matrix for the part of the lattice Gaussian measure in Λ_ε. It is shown that
$$(C_{\Lambda_\varepsilon})_{n,n'}^{-1} = -B_{\partial\Lambda_\varepsilon}n,n' + (C_{n,n'}^{-1})_{\Lambda\varepsilon} \text{ for } \varepsilon n, \varepsilon n' \varepsilon \Lambda_\varepsilon$$
where $B_{\partial\Lambda_\varepsilon}$ is a matrix supported on the boundary of Λ_ε consisting of these points in Λ_ε which have at least one nearest neighbour lying outside Λ_ε. Dirichlet boundary conditions for this model correspond to setting $B_{\partial\Lambda_\varepsilon} = 0$ in the measure for the case that Λ is a simple rectangle say. The full requirements on Λ may be found in the original paper cited. Finally, the Dirichlet lattice cut-off measure is

$$d\mu_{\Lambda,\varepsilon}^{D}(q) = \prod_{n\varepsilon\varepsilon\Lambda_\varepsilon} e^{\frac{1}{2}|n\stackrel{\Sigma}{=}n'| = 1 \, q_n \, q_{n'}} dv_n(q)$$

for which up to normalization

$$dv_n(q) \sim e^{-[aq_n^4 + b'q_n^2]} dq_n$$

is a Gaussian perturbed by e^{-aq^4} with $a > 0$. The off-diagonal terms in $C_{\Lambda_\varepsilon}^{-1}$ couple only nearest neighbours leading to the Ising type ferromagnetic coupling appearing above. The uncoupled measure at each site is also even.

Defining the lattice cut-off Dirichlet Schwinger distributions as moments of $\phi_{0,\varepsilon}$ with respect to $d\mu_{\Lambda,\varepsilon}^D$ the analogue of convergence theorems given by Nelson [28] for Half-Dirichlet boundary conditions results by exploiting monotonicity as $\Lambda \uparrow$ and boundeness from above by the free boundary condition field expressions.

Theorem (Guerra, Rosen, Simon [20]). Consider the $(\lambda\phi^4-\mu\phi)_2$; $\lambda>0$, $\mu>0$, model. Then for an open bounded rectangle $\Lambda \subset R^2$, $\lim_{\varepsilon\downarrow 0} \underline{S}_{\Lambda,\varepsilon}^D (\underline{f}) = \underline{S}_\Lambda^D (\underline{f})$ and as Λ increases through a series of such rectangles eventually enclosing every bounded region in R^2 $\lim_{\Lambda\to\infty} \underline{S}_\Lambda^D (\underline{f}) = \underline{S}^D (\underline{f})$ for $\underline{f} \in \underline{S}$ exists and defines a Dirichlet Schwinger state.

The need for $\mu \neq 0$ concerns the remarks in section 6 concerning the existence of phase transitions as $\mu\downarrow 0$ for this model. $\mu \neq 0$ ensures the uniqueness of the vacuum and hence the validity of axiom [S4].

In the same way that correlation inequalities hold for general Ising ferromagnets, they also hold for the lattice approximation above. The above convergence theorem is strong enough to carry these in inequalities over to the infinite volume theory. For example,

Theorem (Guerra, Rosen, Simon [20]). With the same hypotheses as above,

(a) $S_n(x_1,\ldots,x_n) \geq 0$

(b) $S_{n+m}(x_1,\ldots,x_{n+m}) \geq S_n(x_1,\ldots,x_n) S_m(x_{n+1},\ldots,x_{n+m})$

both inequalities holding as distributions; namely, when applied to positive test functions.

The implication of (a) by a classical result in distribution theory [33] is that for these models the Schwinger distributions are tempered measures at coincident arguments.

Thus ferromagnetism implies strong regularity properties on Schwinger states. We should also remark that the results above quoted from [20] hold in much greater generality than presented here.

In another direction, Simon and Griffiths [39] show that the measures v_n attached to each lattice site may be approximated by a sequence of ferromagnetically coupled spin-$\frac{1}{2}$ Ising systems in the sense of convergence for the distribution functions. This result seems to be particular to the $(\phi^4)_2$-model as opposed to the previous approximations. Apart from new correlation inequalities, application of the above limit theorems carries over a Lee-Yang theorem for the spin-$\frac{1}{2}$ approximants to the limiting euclidean field theory.

Theorem (Griffiths, Simon, [39]). Let μ be a euclidean measure corresponding to free or Dirichlet boundary conditions for the $\lambda(\phi^4)_2 + b(\phi^2)_2$-model and $f \geq 0$ be in S. Then $E(e^{\mu\phi(f)})$ has zeroes only when Re $\mu = 0$.

Applying this to general ferromagnetic Ising systems, Newman has sharpened a set of inequalities due to Glimm and Jaffe [15].

Proposition (Newman, [29]). Let $\underline{S}$ be one of the Schwinger states for suitable boundary conditions for the $\lambda\,(\phi^4)_2-$ model. Then the even order Schwinger distributions satisfy free field estimates; namely, for positive test functions

$$0 \le \underline{S}\,(f_1 \otimes \ldots \otimes f_{2n}) \le \frac{2n!}{2^n n!}\;\prod_{j=1}^{2n}\; S_2(f_j^* \otimes f_j)^{1/2}\;.$$

More recently, by a method which does not use the Lee-Yang theorem, Newman has extended this theorem further.

<u>Theorem</u> (Newman [30]). With the same assumptions as above,

$$0 \le \underline{S}(f_1 \otimes \ldots \otimes f_{2n}) \le \sum_{\text{pairs}}\; S_2(f_i \otimes f_j,) \ldots S_2(f_{i_n} \otimes f_{j_n})\;.$$

$\sum_{\text{pairs}}$ is the sum over all partitions into disjoint unordered pairs $(i_1,j_1)\ldots(i_n,j_n)$ of $(12\ldots2n)$.

These results give a precise meaning to the statement that Schwinger states for $(\phi^4)_2$-models are dominated by those of a generalized free field with the same two point (covariance) distribution.

In closing let us review the consequences of these estimates for the euclidean measure μ by examining its characteristic functional

$$L(zf) = \int e^{iz\,\phi(f)}\;d\mu,\quad f \text{ real in } S.$$

From Stirling's estimate and the positivity of the

Schwinger state $\underline{S}$, the Cauchy-Schwarz inequality

$$|S_{2n+1}(f_1 \otimes \cdots \otimes f_{2n+1})| \leq \underline{S}(f_{2n+1}^* \otimes f_{2n+1})^{1/2} \quad \times$$

$$\times S_{4n}(f_1^* \otimes f_1 \otimes \cdots \otimes f_{2n}^* \otimes f_{2n})^{1/2}$$

leads to bounds

$$|S_n(f_1 \otimes \cdots \otimes f_n)| \leq c_1 \, n^{n/2} \prod_{j=1}^{n} p(f_j)$$

for the continuous seminorm $p(f) \equiv c \, S_2(f^* \otimes f)^{1/2}$ on S.
This means in particular that the power series

$$L(zf) = \Sigma_{n=o}^{\infty} z^n \, i^n \, S_n(f^n)/n!$$

is absolutely convergent for all finite z to an entire
function which is of order at most 2. When $f \geq 0$, the
Lee-Yang theorem requires zeroes at most for real values
of z. These estimates may also be used in conjunction
with Minlos' theorem [24] that a normalized, positive
definite, continuous characteristic functional is in
one to one correspondence with the probability space
(S_R', B, μ) as shown by Fröhlich [9].

8. RECAPITULATION

It is, I believe, clear that the mathematical
physics of quantum field theory is enjoying a period
of great affluence with respect to new results,

techniques and ideas for the future. As the $(\phi^4)_3$-model moves forward to the same stage of technical development as for $(\phi^4)_2$ at present, interesting and difficult questions still remain for these models; for example, the general particle structure, scattering and asymptotic completeness, the number of pure phases that can exist in these models, general characterizations of ferromagnetism for euclidean measures. It is however true that one of the more immediately interesting questions is the basic existence and renormalization theory for $(\phi^4)_4$.

ACKNOWLEDGEMENT

It is a pleasure to thank Professors H. Satz and L. Streit for the hospitality of the Centre for Interdisciplinary Research During the Quantum Dynamics Project where this report was prepared as well as Professor P. Urban for the opportunity to share these matters with the Schladming Physicists.

REFERENCES

1. H.J. Borchers, On structure of the algebra of field operators, Nuovo Cimento <u>24</u>, 214 (1962).

2. H.J. Borchers, J. Yngvason, Necessary and Sufficient Conditions for Integral Representations of Wightman Functionals at Schwinger Points. Göttingen University Preprint, 1975.

3. H.J. Borchers, J. Yngvason, Integral Representations

for Schwinger Functionals and the Moment Problem over
Nuclear Spaces, Commun. Math. Phys. $\underline{43}$, 255 (1975).

4. J.L. Challifour, S.P. Slinker, Euclidean Field Theory.
   I. The Moment Problem. Commun. Math. Phys. $\underline{43}$, 41
   (1975).

5. J.L. Challifour, Euclidean Field Theory. II. Remarks
   on Embedding the Relativistic Hilbert Space. To appear
   in J. Math. Phys.

6. J.L. Challifour, S.P. Slinker, unpublished.

7. J. Dimock, The $P(\phi)_2$ Green's Functions: Asymptotic
   Perturbation Expansion. SUNY (Buffalo) Preprint,
   1975.

8. J.P. Eckmann, H. Epstein, J. Fröhlich, Asymptotic
   Perturbation Expansion for the S-Matrix and the Def-
   inition of Time Ordered Functions in Relativistic
   Quantum Field Models.

9. J. Fröhlich, Schwinger Functions and their Generating Functionals, I. Helv. Phys. Acta $\underline{47}$, 265 (1974).

lo. L.M. Gelfand, N.Ya. Vilenkin, Generalized Functions,
   Vol. 4. Academic Press, 1964 New York.

11. J. Glimm, Boson Fields with Non-linear Self-Interaction in Two Dimensions, Commun. Math. Phys. $\underline{8}$,
   12 (1968).

12. J. Glimm, A. Jaffe, Quantum Field Theory Models in
   Statistical Mechanics and Quantum Field Theory,
   Les Houches, 1970, ed. C. DeWitt and R. Stora, Gordon
   and Breach, 1971, New York.

13. J. Glimm, A. Jaffe, Boson Quantum Field Models, in
   Mathematics of Contemporary Physics, ed. R.Streater,
   Academic Press, 1972, New York.

14. J.Glimm, A.Jaffe, The n-Particle Cluster Expansion for the $P(\phi)_2$ Quantum Field Model, unpublished.

15. J.Glimm, A.Jaffe, Remark on the Existence of ϕ_4^4. Phys. Rev. Letters, 33, 440 (1974).

16. J.Glimm, A.Jaffe, T.Spencer, The Wightman Axioms and Particle Structure in the $P(\phi)_2$ Quantum Field Model. Ann. Math. 100, 585 (1974).

17. J. Glimm, A. Jaffe, T. Spencer, Phase Transitions for $(\phi^4)_2$ Quantum Fields, Commun. Math. Phys. 45, 203 (1975).

18. R.B. Griffiths, Correlation in Ising Ferromagnets, I., II. J. Math. Phys. 8, 478, 484 (1967).

19. R.B. Griffiths, C.A. Hurst, S. Sherman, Concavity of Magnetization of an Ising Ferromagnet in a Positive External Field. J. Math. Phys. 11, 790 (1970).

20. F. Guerra, L. Rosen, B. Simon, The $P(\phi)_2$ Euclidean Quantum Field Theory as Classical Statistical Mechenics. Ann. Math. 101, 111 (1975).

21. G.C. Hegerfeldt, Extremal Decomposition of Wightman Functions and of States on Nuclear *-Algebras by Choquet-Theory. Commun. Math. Phys. 45, 133 (1975).

22. D.G. Kelley, S. Sherman, General Griffiths' Inequalities on Correlations in Ising Ferromagnets. J. Math Phys. 9, 466 (1968).

23. T.D. Lee, C.N. Yang, Statistical Theory of Equations of State and Phase Transitions, II, Lattice Gas and Ising Model. Phys. Rev. 87, 410 (1952).

24. R.A. Minlos, Generalized Random Processes and their Extension to a Measure. Trudy. Mosk. Mat. Obs. 8, 497 (1959).

25. E. Nelson, A Quartic Interaction in Two Dimensions, in Mathematical Theory of Elementary Particles, ed. R. Goodman and I. Segal, M.I.T. Press, 1966 Cambridge, Mass.

26. E. Nelson, Construction of Quantum Fields from Markoff Fields, J. Funct. Analysis $\underline{12}$, 97 (1973).

27. E. Nelson, The Free Markov Field. J. Funct. Anal. $\underline{12}$, 211 (1973).

28. E. Nelson, Probability Theory and Euclidean Field Theory, in Constructive Quantum Field Theory, ed. G. Velo and A. Wightman. Lecture Notes in Physics, Vol. 25, Springer-Verlag, 1973 Heidelberg.

29. C.M. Newman, Inequalities for Ising Models and Field Theories which Obey the Lee-Yang Theorem. Commun. Math. Phys. $\underline{41}$, 1 (1975).

30. C.M. Newman, Gaussian Correlation Inequalities for Ferromagnets. Indiana University Preprint, 1975.

31. C.M. Newman, The Construction of Stationary Two-Dimensional Markoff Fields with an Application to Quantum Field Theory. J. Funct. Anal. $\underline{14}$, 44 (1973).

32. K. Osterwalder, R. Schrader, Axioms for Green's Functions. I. Commun. Math. Physics $\underline{31}$, 83 (1973) and II. ibid $\underline{42}$, 281 (1975).

33. K. Osterwalder, R. Seneor, The Scattering Matrix is non-trivial for weakly coupled $P(\phi)_2$ models. Harvard University Preprint, 1975.

34. L. Schwartz, Théorie des Distributions, Tome I. Herman, 1957, Paris.

35. J. Schwinger, On the Euclidean Structure of Relativ-

istic Field Theory. Proc. Nat. Acad. Sci. U.S.,
$\underline{44}$, 956 (1958).

36. J. Schwinger, Euclidean Quantum Electrodynamics.
Phys. Rev. $\underline{115}$, 721 (1959).

37. B. Simon, Positivity of the Hamiltonian Semigroup
and the Construction of Euclidean Region Fields.
Helv. Phys. Acta. $\underline{46}$, 686 (1973).

38. B. Simon, The $P(\phi)_2$ Euclidean (Quantum) Field Theory.
Princeton Series in Physics, Princeton University
Press, 1974 Princeton, N.J.

39. B. Simon, R.B. Griffiths, The $(\phi^4)_2$ Field Theory
as a Classical Ising Model. Commun. Math. Phys. $\underline{33}$,
145 (1973).

40. T. Spencer, The Absence of Even Bound States for
$\lambda(\phi^4)_2$. Commun. Math. Phys. $\underline{39}$, 77 (1974).

41. T. Spencer, The Decay of the Bethe-Salpeter Kernel
in $P(\phi)_2$ Quantum Field Models. Commun. Math. Phys.
$\underline{44}$, 143 (1975).

42. T. Spencer, F. Zirilli, Scattering States and Bound
States in $\lambda P(\phi)_2$. Rockefeller University Preprint,
1975.

43. K. Symanzik, Euclidean Quantum Field Theory, I.
Equations for a scalar model. J. Math. Phys. $\underline{7}$,
510 (1966).

44. K. Symanzik, Euclidean Quantum Field Theory, in
Varenna Course XLV, "Enrico Fermi" School of Physics, ed. R. Jost Academic Press, 1969 New York.

Acta Physica Austriaca, Suppl. XV, 79–131 (1976)
© by Springer-Verlag 1976

NON-EQUILIBRIUM QUANTUM STATISTICAL MECHANICS[+]

by

G. G. EMCH[++]

ZiF der Universität

Bielefeld, BRD

INTRODUCTION

Through the study of a particular model, the aim
of these lectures is to indicate how the so-called "algebraic approach" can help understanding the statistical
mechanics of quantum stochastic transport processes.

The first lecture is devoted to the thermodynamical
description of the diffusion of a quantum particle in a
harmonic well. We introduce here most of the algebraic notation to be used later on; for didactic purpose, we also
transcribed in the notes for this lecture several elementary computations.

[+] Lecture given at XV. Internationale Universitätswochen
für Kernphysik,Schladming,Austria,February 16-27, 1976.
[++] On leave of absence from the Depts. of Mathematics and
Physics, Univ. of Rochester, Rochester,NY 14627, USA.

The second lecture presents a statistical mechanical model for the stochastic process described in the
first lecture.

In the third lecture the structure of this model
is further analyzed. This provides a motivation for a
generalization to the quantum realm of the classical
concept of K-flow.

The fourth lecture discusses the structure of generalized K-flows; presents a few non-isomorphic examples;
and extends to generalized K-flows the concept of dynamical (or K-) entropy.

I. DIFFUSION OF A QUANTUM PARTICLE IN A
HARMONIC WELL

For simplicity we work in one dimension. The basic
observables of a spinless particle constrained to moving
on R are its momentum and its position. In the Schrödinger
representation these observables are represented by the
self-adjoint operators P and Q acting on $L^2(R)$ and determined by their restrictions to $S(R)$:

$$(P\psi)(x) = - i \frac{d\psi}{dx}(x); \quad (Q\psi)(x) = x\psi(x) . \qquad (1)$$

These operators are unbounded and their study is simplified by considering the continuous one-parameter groups
they generate:

$$U(a) = \exp(-iPa) \text{ and } V(b) = \exp(-iQb) \text{ with } a,b \in R. \quad (2)$$

When we place this particle in a harmonic well it is advantageous (see for instance formulas (12), (51) or (65) below) to introduce for every $z \in C$ the decomposition,

$$z = \omega^{1/2} a + i\omega^{-1/2} b \tag{3}$$

and the unitary operator

$$W(z) = \exp \{-i(aP + bQ)\} . \tag{4}$$

These operators satisfy the <u>canonical commutation relations in Weyl's form (CCR)</u>:

$$W(z_1)W(z_2) = W(z_1 + z_2) \exp\{\tfrac{i}{2} \text{ Im } (z_1^* z_2)\} \tag{5}$$

and the continuity condition:

$$W(\lambda z) \text{ weakly continuous in } \lambda \in R \text{ for all } z \in C . \tag{6}$$

We note in passing that from (5) and (6) follow:

$$W(0) = I; \quad W(z)^* = W(-z) \tag{7}$$

$$U(a) \, V(b) = W(z) \, \exp(\tfrac{i}{2} a b)$$

$$V(b) \, U(a) = W(z) \, \exp(-\tfrac{i}{2} a b) \tag{8}$$

and thus

$$U(a) \ V(b) \ U(-a) = V(b) \ \exp \ (i \ a \ b) \tag{9}$$

which is the original form of Weyl's CCR. Upon taking the derivative of this expression w.r.t. b at b = O we get

$$U(a) \ Q \ U(-a) = Q - a \ I \tag{10}$$

which says that U(a) is to be interpreted as the operator describing a translation by the amount a. Upon taking the derivative of this expression w.r.t. a at a = O we get

$$[P , Q] = - i \ I \tag{11}$$

which are the CCR in Heisenberg's form. We should notice that the expressions (10) and (11) contain unbounded operators and thus cannot be expected to hold on the whole of $L^2(R)$. It suffices for our purpose to know that they make sense at least on $S(R)$. No such mental restriction is necessary when we work with the unitary operators W(z); this is one of the reasons why in the sequel, we will use them systematically rather than the operators P and Q.

We now turn to the dynamics. Upon integrating the classical Hamilton equations for a harmonic oscillator of frequency $\nu = \omega/2\pi$ we find the following flow on the two-dimensional state-space C:

$$S(t) : z \ \epsilon \ C \to z_t = \exp \ (- i \ \omega \ t) \ z \ \epsilon \ C . \tag{12a}$$

The quantized form of this flow is then naturally guessed at as being:

$$\sigma(t) : W(z) \rightarrow W_t(z) = W(z_t) \qquad (12b)$$

with z_t given by eqn. (12a). From the defining expressions (compare with eqn. (4)):

$$W_t(z) = \exp\{-i(a P_t + b Q_t)\} \qquad (13)$$

we conclude (expectedly!) that

$$
\begin{bmatrix} P_t \\ Q_t \end{bmatrix}
=
\begin{bmatrix} \cos(\omega t) & -\omega\sin(\omega t) \\ \omega^{-1}\sin(\omega t) & \cos(\omega t) \end{bmatrix}
\begin{bmatrix} P \\ Q \end{bmatrix}
\qquad (14)
$$

i.e.

$$
\begin{bmatrix} \dot{P}_t \\ \dot{Q}_t \end{bmatrix}
=
\begin{bmatrix} 0 & -\omega^2 \\ 1 & 0 \end{bmatrix}
\begin{bmatrix} P_t \\ Q_t \end{bmatrix}
\qquad (15)
$$

or

$$\dot{P}_t = i \; [H, P_t]$$

$$\qquad (16)$$

$$\dot{Q}_t = i \; [H, Q_t]$$

with

$$H = \frac{1}{2} (P^2 + \omega^2 Q^2) \tag{17}$$

i.e.

$$W_t(z) = U_{-t} W(z) U_t \tag{18}$$

with

$$U_t = \exp(-iHt) . \tag{19}$$

We take thus the standpoint that the <u>algebraic</u> relations (5) and (12) (together with their interpretation given above) describe completely the physics of the quantum harmonic oscillator. We illustrate this point further with two computations. We first compute the ground-state functional

$$<\phi_o, W(z)> = (\phi_o, W(z)\phi_o) \text{ with } \phi_o \in L^2(R) \tag{20}$$

satisfying $H\phi_o = \frac{1}{2}\omega\phi_o$.

We then compute the canonical equilibrium functional

$$<\phi_\beta, W(z)> = Tr \rho W(z) \tag{21}$$

with $\rho = \exp(-\beta H)/Tr \exp(-\beta H)$; $\beta > 0$.

To compute the ground state functional ϕ_o we write (with eqn. (3) in mind):

$$<\phi_o;\ W(z)> \ =\ f(a,b)$$

$$=\ g(a,b)\ \exp\ (-\tfrac{i}{2}\ ab)\ . \tag{22}$$

From eqn. (7) we get:

$$g(a,b)\ =\ (\phi_o,\ U(a)\ V(b)\phi_o)\ . \tag{23}$$

We now notice that

$$\frac{\partial^2}{\partial a^2}\ g(a,b)\ =\ -\ (\phi_o,\ P^2\ U(a)\ V(b)\phi_o) \tag{24}$$

$$\frac{\partial^2}{\partial b^2}\ g(a,b)\ =\ -\ (\phi_o,\ U(a)\ V(b)\ Q^2\phi_o)\ . \tag{25}$$

We use (10) to rewrite (25) as:

$$\frac{\partial^2}{\partial b^2}\ g(a,b)\ =\ -\ (\phi_o;\ Q^2\ U(a)\ V(b)\phi_o)$$

$$+\ (2\ i\ a\ \frac{\partial}{\partial b}\ +\ a^2)\ g\ (a,b)\ . \tag{26}$$

Putting together eqns. (17), (20), (24) and (26) we get:

$$(\frac{\partial^2}{\partial a^2}\ +\ \omega^2\ \frac{\partial^2}{\partial b^2})g(a,b)\ =\ (\omega^2 a^2\ -\ \omega\ +\ 2i a\omega^2\frac{\partial}{\partial b})g(a,b)\ . \tag{27}$$

In the same way as we obtained (10) from (5) we get

$$V(b)\ P\ V(-b)\ =\ P\ +\ b\ I \tag{28}$$

which we use, in a manner similar to the above, to obtain

$$(\frac{\partial^2}{\partial a^2} + \omega^2 \frac{\partial^2}{\partial b^2}) g(a,b) = (b^2 - \omega + 2ib \frac{\partial}{\partial a}) g(a,b) . \qquad (29)$$

Comparing the RHS of (27) and (29) and taking into account (22) we conclude that:

$$(b \frac{\partial}{\partial a} - \omega^2 a \frac{\partial}{\partial b}) f(a,b) = 0 \qquad (30)$$

and thus

$$f(a,b) = \mu \exp \{\frac{1}{2} \lambda (\omega^2 a^2 + b^2)\} . \qquad (31)$$

From $W(0) = I$ we conclude that $\mu = 1$. Feeding then (31) into (24) we get $\lambda = -1/2\omega$. We obtain therefore for the ground-state functional of the harmonic oscillator:

$$<\phi_o, W(z)> = \exp \{- \frac{1}{4} |z|^2\} . \qquad (32)$$

To compute the canonical equilibrium functional ϕ_β (see eqn. 21) we proceed along a somewhat roundabout path which we nevertheless think to be instructive.

We begin by constructing a new representation of the CCR. For that purpose we define, for any $\beta > 0$, the positive numbers Θ , ζ_+, and ζ_- by:

$$\Theta = \coth (\frac{1}{2}\beta\omega) \qquad (33)$$

$$\zeta_+^2 + \zeta_-^2 = 0 \tag{34}$$

$$\zeta_+^2 - \zeta_-^2 = 1 . \tag{35}$$

We then define, for every $z \in C$, the unitary operator $W_\beta(z)$ acting on $L^2(R) \otimes L^2(R)$:

$$W_\beta(z) = W(\zeta_+ z) \otimes W(\zeta_- z^*) . \tag{36}$$

From (5) and (35) we see that these new operators satisfy again the canonical commutation relations in Weyl's form:

$$W_\beta(z_1) W_\beta(z_2) = W_\beta(z_1 + z_2) \exp \{\tfrac{i}{2} \text{Im}(z_1^* z_2)\} \tag{37}$$

as well as the continuity condition (6). With Φ_o defined as in eqn. (20), we construct the functional

$$<\psi_\beta, W(z)> = (\Phi_o \otimes \Phi_o, W_\beta(z)\Phi_o \otimes \Phi_o) . \tag{38}$$

From (32) and (34) we see that

$$<\psi_\beta, W(z)> = \exp \{-\tfrac{1}{4} \theta |z|^2\} . \tag{39}$$

We are now going to show that ψ_β is precisely equal to the canonical equilibrium functional which we are aiming to compute. To this effect we extend the mapping (12) to $W_\beta(z)$ as

$$\sigma_\beta(t) : W_\beta(z) \rightarrow W_{\beta,t}(z) = W_\beta(z_t)$$

$$= W(\zeta_+ e^{-i\omega t} z) \otimes W(\zeta_- e^{+i\omega t} z^*) . \tag{40}$$

We then compute

$$<\psi_\beta ; W(z_1) W_t(z_2)> =$$

$$\exp \{-\frac{1}{4} \Theta (|z_1|^2 + |z_2|^2)\} \times$$

$$\exp \{-\frac{1}{2} \zeta_-^2 z_1^* e^{-i\omega t} z_2 - \frac{1}{2} \zeta_+^2 z_2^* e^{+i\omega t} z_1\} . \tag{41}$$

The effect of replacing t by t+iβ in this expression amounts to the substitution

$$\zeta_-^2 \rightarrow \zeta_-^2 e^{\beta\omega} = \zeta_+^2$$

$$\zeta_+^2 \rightarrow \zeta_+^2 e^{-\beta\omega} = \zeta_-^2 \tag{42}$$

which in turn is equivalent to interchanging the roles of z_1 and z_2. We thus conclude that for all t ε R and all z_1, z_2 ε C:

$$<\psi_\beta ; W(z_1) W_{t+i\beta}(z_2)> = <\psi_\beta ; W_t(z_2) W(z_1)> . \tag{43}$$

This relation means exactly that ψ_β satisfies the Kubo-

<u>Martin-Schwinger condition</u> w.r.t. the time evolution $\{\sigma_\beta(t)|$ t ϵ R$\}$. Our last step is thus to verify that this condition determines uniquely a density matrix such that

$$<\psi_\beta; \ W(z)> \ = \ Tr \ \rho \ W(z) \tag{44}$$

and that this ρ is given by our expression (21). With A and B standing respectively for $W(z_1)$ and $W(z_2)$ we can rewrite the KMS condition (43) at t = O as:

$$Tr \ \rho \ A \ B_{i\beta} \ = \ Tr \ \rho \ B \ A \tag{45}$$

i.e.

$$Tr \ exp \ \{-iH(i\beta)\}\rho A \ exp \ \{iH(i\beta)\} \ B \ = \ Tr \ A\rho B \ . \tag{46}$$

Since the Schrödinger representation $\{W(z)|z \ \epsilon \ C\}$ constructed in the beginning of this section is irreducible, we have thus

$$exp(\beta H) \ \rho \ A \ exp \ (- \ \beta H) \ = \ A\rho \tag{47}$$

i.e.

$$[exp(\beta H)\rho, \ A] \ = \ O \ . \tag{48}$$

Upon using again the irreducibility of $\{W(z)| \ z \ \epsilon \ C\}$ we get

$$\exp(\beta H) \; \rho \; = \lambda I, \; \lambda \; \epsilon \; C \tag{49}$$

and thus with the normalization $\mathrm{Tr} \, \rho \; = \; 1$:

$$\rho \; = \; \exp(-\beta H)/\mathrm{Tr} \; \exp \; (-\beta H) \; . \tag{50}$$

We can now collect our results (39), (44) and (50) and conclude that the canonical equilibrium functional ϕ_β defined by (21) is given by:

$$<\phi_\beta; \; W(z)> \; = \; \exp \; \{ -\frac{1}{4} \; \theta | z |^2 \; \} \; . \tag{51}$$

To illustrate the fact that this formula contains implicitly all there is to know on the canonical equilibrium state of the harmonic oscillator, we now derive a few of its physical consequences. We first notice that with

$$X_{a,b} = aP + bQ; \; k \; \epsilon \; R \; \text{and} \; z \; = \; k(\omega^{1/2} \; a \; + \; i \; \omega^{-1/2} \; b) \tag{52}$$

we can rewrite (51) as:

$$<\phi_\beta; \; \exp(-i \; k \; X_{a,b})> \; = \; \exp \; \{ -\frac{k^2}{4} \; \theta \; (\omega \; a^2 \; + \; \omega^{-1} \; b^2) \} \; . \tag{53}$$

We recognize here a well-known result sometimes refered to as F. Bloch's theorem, according to which in canonical equilibrium, the probability law for an arbitrary linear combination (aP + bQ) is a gaussian. Explicitely we have:

$$<\phi_\beta; \; \exp(-i \; k \; X_{a,b})> \; = \; \int \exp \; (-i \; k \; \xi) \; d\mu_{a,b}(\xi) \tag{54}$$

with

$$d\mu_{a,b}(\xi) = \exp\{-\beta V_{a,b}(\xi)\}d\xi / \int \exp\{-\beta V_{a,b}(\xi)\}d\xi$$

$$V_{a,b}(\xi) = \frac{1}{2} \Omega_{a,b}^2 \xi^2$$

$$\Omega_{a,b}^2 = 2 \{\beta(\omega \ a^2 + \omega^{-1} \ b^2)\}^{-1} \tanh(\frac{1}{2}\beta\omega) \ . \tag{55}$$

In particular the _effective potentials_ $V_{1,0}$ and $V_{0,1}$ for P and Q are respectively given by:

$$\Lambda^2 = (\frac{1}{2}\beta\omega)^{-1} \tanh(\frac{1}{2}\beta\omega)$$

$$\Omega^2 = \omega^2 (\frac{1}{2}\beta\omega)^{-1} \tanh(\frac{1}{2}\beta\omega) \ . \tag{56}$$

As to be expected, the _high-temperature limit_ of these expressions are respectively 1 and ω^2 in agreement with the _classical_ result obtained from the hamiltonian function

$$H(p,q) = \frac{1}{2} (p^2 + \omega^2 \ q^2) \ . \tag{57}$$

We should also notice as a consequence of (51) that the gaussian measure $\mu_{a,b}$ is uniquely determined by its co-variance

$$\langle X_{a,b}^2 \rangle_\beta = \frac{1}{2}(\omega \ a^2 + \omega^{-1} \ b^2) \coth(\frac{1}{2}\beta\omega) \ . \tag{58}$$

In particular:

$$<P^2>_\beta = \tfrac{1}{2}\, \omega \;\; \text{coth} \;(\tfrac{1}{2}\beta\omega)$$

$$<Q^2>_\beta = (2\omega)^{-1}\; \text{coth} \;(\tfrac{1}{2}\beta\omega) \tag{59}$$

and thus:

$$<P^2>_\beta = <\omega^2\, Q^2>_\beta \tag{60}$$

which is Ehrenfest theorem for our harmonic oscillator; and

$$<H>_\beta = \tfrac{1}{2}\, \omega \;\; \text{coth} \;(\tfrac{1}{2}\beta\omega) \tag{61}$$

which is Plank's formula. We might also note that

$$<P^2>_\beta \;\; <Q^2>_\beta = \tfrac{1}{4}\theta^2 \geq \tfrac{1}{4} \tag{62}$$

in accordance with Heisenberg uncertainty relation. Actually we see from (62) that the minimum allowed by this uncertainty relation is reached in the low-temperature limit, as is also well-known. Incidentally we check that this limit gives

$$\lim_{T\to 0}<\phi_\beta;\; W(z)> \;=\; <\phi_o;\; W(z)> \tag{63}$$

which corresponds formally to

$$\lim_{T\to 0} W_\beta(z) = W(z) \otimes I \tag{64}$$

i.e. the Schrödinger representation $W(z)$ is "recovered" as the low temperature limit of the canonical equilibrium representation $W_\beta(z)$ of the CCR.

With a view to other applications we should draw here the attention on the role played by the KMS condition (43). This condition was originally expressed by Kubo [1] and by Martin and Schwinger [2] as a boundary condition on the analytic behaviour of thermal Green functions. Its importance for recognizing equilibrium states of infinitely extended systems, has been isolated by Haag, Hugenholtz and Winnink [3]; it has later been interpreted as a stability condition by Haag, Kastler and Trych-Pohlmeyer [4]. The fact that for a finite system such as our single harmonic oscillator this condition determines <u>uniquely</u> the state ϕ_β is linked to the absence of phase transition for such systems. The KMS condition has been the object of many publications in the mathematical physics literature, some of which are briefly reviewed in [5] (see in particular sections II.2. e & f). On the mathematical side, much activity has been devoted to the structure behind this condition; we only mention here the work of Takesaki [6] and Connes [7].

So much for the equilibrium situation. We now introduce a different "phenomenological" time evolution which we will interpret as describing the diffusion of a quantum particle in a harmonic well.

For a fixed positive constant λ, we consider the forward (i.e. $t \geq 0$) evolution described in the Heisenberg picture by:

$$\gamma(t) \; : \; W(z) \to W(e^{-\lambda t}z) \; \exp \; \{-\tfrac{1}{4}\theta|z|^{2}(1-e^{-2\lambda t})\} \tag{65}$$

or in the Schrödinger picture by:

$$\gamma_{*}(t) \; : \; \psi \to \psi_{t} \; \text{with}$$

$$<\psi_{t}; \; W(z)> \; = \; \text{Tr} \; \rho_{t} \; W(z) \; =$$

$$\text{Tr}\{\rho \; \; \gamma(t) \; [W(z)]\} \; = \; <\psi; \; \gamma(t) \; [W(z)]> \qquad \cdot \tag{66}$$

An easy computation shows that this new time evolution satisfies the Markovian property:

$$\gamma(t_{1}) \circ \gamma(t_{2}) \; = \; \gamma(t_{1} + t_{2}) \; \text{for all} \; t_{1}, \; t_{2} \; \epsilon \; R^{+} \quad \cdot \tag{67}$$

It is also trivial to verify from (65) and (51) that the canonical equilibrium state ϕ_{β} is invariant under this evolution:

$$<\phi_{\beta}; \; \gamma(t) \; [W(z)]> \; = \; <\phi_{\beta}; \; W(z)> \; \text{for all} \; z \; \epsilon \; C \; \text{and} \; t \; \epsilon \; R^{+}. \tag{68}$$

Actually this is the only (normal) state satisfying this property; and every other (normal) state ψ approaches ϕ_{β} as time proceeds; indeed one checks from (65) and (66) that:

$$\lim_{t \to \infty} \; <\psi_{t}; \; W(z)> \; = \; <\phi_{\beta}; \; W(z)> \; \cdot \tag{69}$$

The evolution $\{\gamma_*(t) \mid t \in R^+\}$ leads thus unmistakenly to an "approach to equilibrium". The question before us now is to interpret this process in a thermodynamical language. To this effect we notice (from eqn. 65) that, under $\{\gamma(t) \mid t \in R^+\}$ the random variables $X_{a,b}$ evolve independently of one another. We can therefore, without loss of information, restrict our attention to the time-dependent distribution functions $\rho_{a,b}$ defined by:

$$\langle \Psi_t ; \exp(-i k X_{a,b}) \rangle = \int \exp(-ik\xi) \rho_{a,b}(\xi,t) d\xi$$

$$= \tilde{\rho}_{a,b}(k,t) . \tag{70}$$

From (65) and (66) we get:

$$\tilde{\rho}_{a,b}(k,t) = \exp\{-\tfrac{1}{2}(\beta\Omega^2_{a,b})^{-1} k^2 (1-e^{-2\lambda t})\} \tilde{\rho}_{a,b}(e^{-\lambda t}k,0) \tag{71}$$

with $\Omega^2_{a,b}$ defined as in (55). A straightforward computation shows then that $\tilde{\rho}_{a,b}$ satisfies the differential equation:

$$\{\partial_t - \lambda[1 - \partial_k \cdot k - (\beta\Omega^2_{a,b})^{-1}k^2]\} \tilde{\rho}_{a,b}(k,t) = 0 . \tag{72}$$

Upon Fourier-transforming back we get finally:

$$\{\partial_t - D_{a,b}[\partial^2_\xi + \beta V'_{a,b}(\xi)\partial_\xi + V''_{a,b}(\xi)]\} \rho_{a,b}(\xi,t) = 0 \tag{73}$$

with

$$D_{a,b} = \lambda/\beta\Omega^2_{a,b} \qquad\qquad (74)$$

and $\Omega^2_{a,b}$, $V_{a,b}(\xi)$ given by (55) .

We therefore conclude that the time evolution $\{\gamma(t)\mid t \in R^+\}$ consists exactly in the collection of random processes $\{X_{a,b}(t)\}$ governed by the <u>diffusion equat-</u><u>ions</u> (73); we emphasize that the potentials $V_{a,b}(\xi)$ entering in the "phenomenological" eqns. (73) coincide for each a and b with the effective potentials (55) derived from the canonical equilibrium state ϕ_β.

We can thus summarize the results obtained in this lecture by the following assertion. <u>The diffusion of a</u> <u>quantum particle in a harmonic well</u> $V(x) = \frac{1}{2}\omega^2 x^2$ at <u>natural temperature β</u> is described by the triple

$$(N_\beta, \phi_\beta; \{\gamma(t)\mid t \in R^+\}) \qquad\qquad (75)$$

where N_β is the von Neumann algebra

$$\{W_\beta(z)\mid z \in C\}'' = \{B \in B(L^2(R) \otimes L^2(R)\mid B_{\cup\cup}W(z) \ \forall \ z \in C\} \qquad\qquad (76)$$

(with $B_{\cup\cup}A$ meaning $[B,C] = 0$ for all C such that $[C,A] = 0$); ϕ_β is the state on N_β defined by

$$<\phi_\beta; N> = (\Phi_0 \otimes \Phi_0, N\Phi_0 \otimes \Phi_0) \qquad \forall \ N \in N_\beta \qquad (77)$$

(with ϕ_o defined in eqn. 20); in particular this state is determined by:

$$<\phi_\beta; W_\beta(z)> = \exp(-\tfrac{1}{4}\theta|z|^2) \text{ with } \theta = \coth(\tfrac{1}{2}\beta\omega); \qquad (78)$$

and $\{\gamma(t)|t \in R^+\}$ is the semi-group of maps of N_β onto itself given by:

$$\gamma(t) : W_\beta(z) \to W_\beta(e^{-\lambda t}z) \exp\{-\tfrac{1}{4}\theta|z|^2(1-e^{-2\lambda t})\} . \qquad (79)$$

We recall that ϕ_β satisfies the KMS condition w.r.t. the time evolution

$$\sigma_\beta(t) : W_\beta(z) \to W_\beta(e^{-i\omega t}z) . \qquad (80)$$

We also mention, without proof (see however [5]) that ϕ_β is a faithful normal state on the factor N_β; that the vector $\phi_\beta = \phi_o \otimes \phi_o$ is cyclic and separating for N_β; and that the mappings $\gamma(t)$ are completely positive (the latter assertion will follow as a consequence of the main result of the next lecture).

II. STATISTICAL MECHANICS FOR A QUANTUM DIFFUSION PROCESS

The aim of this lecture is to construct:

- a von Neumann algebra N_R;
- a one-parameter group $\{\sigma_R(t) | t \in R\}$ of automorphisms of N_R;
- a faithful normal state ϕ_R on N_R satisfying the KMS condition w.r.t. $\{\sigma_R(t) | t \in R\}$;
- a one-parameter group $\{\alpha(t) | t \in R\}$ of automorphisms of $N = N_\beta \otimes N_R$ leaving $\phi = \phi_\beta \otimes \phi_R$ invariant;
- in such a manner that for every normal state ψ on N_β and every N in N_β we will have:

$$<\psi; \gamma(t)[N]> = <\psi \otimes \phi_R; \alpha(t)[N \otimes I]> \quad \text{for all } t \in R^+ \quad (81)$$

where $(N_\beta, \phi_\beta; \{\gamma(t) | t \in R^+\})$ is the triple (75).

This construction will thus show that it is <u>possible</u> to understand in the framework of Statistical Mechanics, the thermodynamical behaviour described in the first lecture. In particular, (N_R, ϕ_R) will have to be inter-preted as a thermal bath for the quantum stochastic process (75), whereas the total system $(N, \phi, \{\alpha(t) | t \in R\})$, constituted by the system of interest in interaction with the thermal bath, evolves according to the purely deter-ministic law of motion $\{\alpha(t) | t \in R\}$. When this mathematical construction is achieved we will indicate that a <u>physical interpretation</u> of the system $(N, \phi; \{\alpha(t) | t \in R\})$ can be given in terms of the van Hove limit of an infinite chain of harmonic oscillators. The interest of this model is further enhanced by the fact that it is "minimal" in a sense to be described at the end of this lecture.

We now proceed with our construction. Our first step is to exhibit the pair (N, ϕ). To this effect, let T denote the Hilbert space $L^2(R, dx)$. With θ defined as in (78) we form

$$\hat{\phi} \; : \; f \; \epsilon \; T \rightarrow \hat{\phi}\,(f) \; = \; \exp \; (-\tfrac{1}{4}\Theta\,||\,f\,||^{\,2}) \; . \tag{82}$$

We know (see for instance [5], Theorem III. 1.7) that there exists a triple (H, W, Φ) where:

- H is a separable Hilbert space;
- W is a mapping

$$W \; : \; f \; \epsilon \; T \; \rightarrow \; W(f) \; \epsilon \; U \; (H) \qquad \text{with} \tag{83}$$

$$W(f) \; W(g) \; = \; W(f+g) \; \exp \; \{\tfrac{i}{2} \; Im(f,g)\}$$

- Φ is vector in H (normalized to 1) such that

$$\hat{\phi}\,(f) \; = \; (\Phi, \; W(f)\Phi) \; \text{for all } f \; \epsilon \; T \tag{84}$$

and

$$\overline{Span} \; \{W(f)\,|\,f \; \epsilon \; T\} \; = \; H \quad . \tag{85}$$

This representation of the CCR for an infinite number of degrees of freedom is uniquely (up to unitary equivalence) determined by the above conditions. We then define N as the von Neumann algebra generated on H by the operators W(f):

$$N \; = \; \{W(f)\,|\,f \; \epsilon \; T\}'' \tag{86}$$

ϕ is further defined on N as the state

$$\phi \; : \; N \; \epsilon \; N \; \rightarrow \; <\phi; \; N> \; = \; (\Phi, \; N\Phi) \quad . \tag{87}$$

In particular we have thus:

$$\hat{\phi}(f) = <\phi; W(f)> = (\Phi, W(f)\Phi) . \tag{88}$$

The existence of this representation, and its uniqueness up to unitary equivalence, would suffice for our purpose. We will nevertheless give below an explicit realization of it.

For this purpose we start from the usual Fock space representation (H_F, W_F, Φ_F) of the CCR on T, associated to the vacuum functional

$$\hat{\phi}_F(f) = \exp \{-\tfrac{1}{4}||f||^2\} . \tag{89}$$

Incidentally we might note that the interpretation of $\hat{\phi}_F$ as the "vacuum functional" can be verified from (32); eqn. (20) is now replaced by:

$$\phi_F \varepsilon H_F; a_F(f)\Phi_F = 0 \text{ for all } f \varepsilon T;$$

$$\overline{\text{Span}} \{W_F(f)\Phi_F| f \varepsilon T\} = H_F . \tag{90}$$

We now form:

$$H = H_F \otimes H_F; \Phi = \Phi_F \otimes \Phi_F \tag{91}$$

and, with ζ_+ and ζ_- defined as in (33 - 35):

$$W(f) = W_F (\zeta_+ f) \otimes W_F (\zeta_- f^*) \tag{92}$$

with f^* defined by:

$$\tilde{f}^*(k) = \tilde{f}(k)^* \tag{93}$$

(where $\tilde{f}$ stands for the Fourier transform of f in T).

We now define on N two evolutions:

$$\sigma(t) : W(f) \rightarrow \sigma(t) [W(f)] = W(e^{-i\omega t} f) \tag{94}$$

$$\alpha(t) : W(f) \rightarrow \alpha(t) [W(f)] = W(u_t f) \tag{95}$$

with $(u_t f)(x) = e^{-i x t} f(x)$.

One checks that both leave ϕ invariant, i.e.

$$<\phi ; \sigma(t) [W(f)]> = <\phi ; W(f)> \tag{96}$$

$$<\phi ; \alpha(t) [W(f)]> = <\phi ; W(f)> \tag{97}$$

and in the same way as in the first lecture, that ϕ satisfies the KMS condition:

$$<\phi ; W(f) \sigma(t+i\beta) [W(g)]> = <\phi ; \sigma(t) [W(g)] W(f)> \tag{98}$$

$\{\alpha(t) \,|\, t \in R\}$ will be the one-parameter group of automorphisms of N required in the beginning of this lecture.

Our next step is to identify the decomposition

$$N = N_\beta \otimes N_R \quad . \tag{99}$$

Were it not for our request to recover $\gamma(t)$ from $\alpha(t)$, this decomposition could be obtained by picking any arbitrary vector $f_o \in T$. With a view to our ultimate aim in this lecture we however choose:

$$f_o(x) = [\frac{1}{\pi} (\frac{1}{\lambda^2 + x^2})]^{1/2} \tag{100}$$

and we define:

$$H_\beta = \overline{\text{Span}} \{W(z\ f_o)\phi \mid z \in C\} \tag{101}$$

$$W_\beta(z) = W(z\ f_o)\mid_{H_\beta} \quad \text{for every} \quad z \in C \tag{102}$$

$$N_\beta = \{W_\beta(z) \mid z \in C\}'' \tag{103}$$

$$<\phi_\beta;\ W_\beta(z)> = (\phi,\ W_\beta(z)\phi) = \exp (-\frac{1}{4}\Theta\mid z\mid^2) \tag{104}$$

$$\sigma_\beta(t) = \sigma(t)\mid_{N_\beta} \quad \text{for every} \quad t \in R \ . \tag{105}$$

Since the functional (104) determines uniquely (up to unitary equivalence) the cyclic representation of the CCR $\{W_\beta(z) \mid z \in C\}$ (see for instance [5] Thm. III.1.7) we have recovered the pair $(N_\beta,\ \phi_\beta)$ of the first lecture.

We now consider the subspace T_R of T

$$T_R = \{f \in T \mid (f,\ f_o) = 0\} \tag{106}$$

and form:

$$H_R = \overline{\text{Span}} \; \{W(f)\phi \,|\, f \; \varepsilon \; T_R\} \qquad (107)$$

$$W_R(f) = W(f)\big|_{H_R} \qquad \text{for every} \quad f \; \varepsilon \; T_R \qquad (108)$$

$$N_R = \{W_R(f) \,|\, f \; \varepsilon \; T_R\}'' \qquad (109)$$

$$<\phi_R; \; W_R(f)> \; = \; (\phi, \; W_R(f)\phi) \; = \; \exp(-\tfrac{1}{4}\theta||f||^2) \qquad (110)$$

$$\sigma_R(t) = \sigma(t)\big|_{N_R} \qquad \text{for every} \quad t \; \varepsilon \; R \; . \qquad (111)$$

This defines our decomposition (99) with $(N_R, \; \phi_R, \{\sigma_R(t)\,|$ $t \; \varepsilon \; R\})$ satisfying trivially the first four conditions requested in the beginning of this lecture. We thus only have to verify the last condition (81), which will establish the consistency of statistical mechanics with the thermodynamical behaviour considered in the first lecture. For this purpose let us denote by P the orthogonal projection from T to T_R. From the canonical commutation relations (83) we have:

$$W(f) = W((I-P)f) \; W(Pf) \qquad (112)$$

and thus for every f in T:

$$<\psi \otimes \phi_R; \; \alpha(t) \; [W(f)]> \; =$$
$$\qquad\qquad (113)$$
$$<\psi; \; W_\beta((I-P)u_t f)><\phi_R; \; W_R(Pu_t f)> \qquad .$$

Taking now into account the special form (100) of f_o

we have:

$$(f_o, u_t f_o) = \exp(-\lambda t) \text{ for all } t \in R^+ . \tag{114}$$

Upon feeding this in (113) with $f = z f_o$ ($z \in C$) we get:

$$<\psi \otimes \phi_R; \; \alpha(t) \; [\ddot{W}(zf_o)]> =$$

$$<\psi; \; W(e^{-\lambda t}z)> \exp\{-\tfrac{1}{4}\theta|z|^2 (1-e^{-2\lambda t})\} = \tag{115}$$

$$<\psi; \; \gamma(t) \; [W(z)]>$$

which extends trivially to our consistency consition (81). We have therefore achieved the construction we proposed in the beginning of this lecture.

Upon introducing the conditional expectation

$$E : N \otimes M \in N \rightarrow <\phi_R; M > N \in N_\beta \tag{116}$$

(conditionned w.r.t. ϕ) we can rewrite (81) [or (115)] in the form

$$E\alpha(t) \; E = \gamma(|t|) \tag{117}$$

and thus interpret $(N_\beta, \phi_\beta, \{\gamma(|t|)|t \in R\})$ as a "reduced description" of the complete system $(N, \phi, \{\alpha(t)|t \in R\})$. We emphasize again that this reduced description (being a diffusion process) is <u>dissipative</u>, whereas the total system evolves in a perfectly <u>deterministic</u> way, the reduction being effected by the projector E.

The existence of such a Markovian reduced description is a perenial problem in non-equilibrium
statistical mechanics. Having thus proven that it is
possible in a particular case, we must ask whether it
corresponds to any recognizable physical situation. The
answer is found by noticing that $(N, \phi, \{\alpha(t)|t \varepsilon R\})$
can be obtained as the van Hove "long-time,weak-coupling"
limit of a Hamiltonian system, namely an infinite chain
of harmonic oscillators interacting via a quadratic,
translation-invariant interaction of the Ford-Kac-Mazur
[8] variety. Specifically, one sees that, restricted to
any finite (but arbitrarily large, if one so wishes)
segment of the chain, the evolution $\alpha(t)$ coincides with
that found by Davies [9] in his treatment of the quantum
FKM model. In particular the fact that, for a single oscillator arbitrarily chosen in the chain, the van Hove
limit leads to the thermodynamical $(N_\beta, \phi_\beta, \{\gamma(t)|t \varepsilon R^+\})$
has been first rigourously established by Davies. In this
realization (N_R, ϕ_R) describes the rest of the chain,
which thus serves as a thermal bath for the single har-
monic oscillator considered. Note that $\{\sigma(t)|t \varepsilon R\}$ on
the other hand is the free evolution, obtained with the
interaction switched off; and that ϕ is the corresponding
canonical equilibrium state at the natural temperature β.

This result was the main point of this lecture.
From a mathematical point of view, it is also interesting
to note that (117) implies, via Stinespring's criterion
[10], that $\{\gamma(t)|t \varepsilon R^+\}$ is a continuous semi-group of
completely positive maps. Furthermore, we should note in
this respect that $\{\alpha(t) [N]|t \varepsilon R, N \varepsilon N_\beta\}$ generates N;

hence $(N, \phi, \{\alpha(t) | t \, \epsilon \, R\})$ is <u>minimal</u> with respect to the conditions which we imposed in the beginning of this lecture. The existence of such minimal extension of an arbitrary semi-group of completely positive maps from a von Neumann algebra onto itself follows for instance from Evans [11] who pointed out, in particular, the C^*-algebraic generalization of Nagy's theorem [12] on minimal unitary dilations of contraction semi-groups acting on a Hilbert space. It seemed thus worth noting here that, in the case of the diffusion of a quantum particle in a harmonic well, this minimal extension can be given physical interpretation.

III. K-STRUCTURE ASSOCIATED TO A QUANTUM
DIFFUSION PROCESS

We constructed in the previous lecture a <u>deterministic</u> dynamical system $(N, \phi, \{\alpha(t) | t \, \epsilon \, R\})$ which, from the point of view of non-equilibrium statistical mechanics, is canonically associated to a <u>dissipative</u> process, namely the system $(N_\beta, \phi_\beta, \{\gamma(t) | t \, \epsilon \, R^+\})$ studied in the first lecture. Our next task is to probe further into this structure with a view to elucidate its ergodic properties.

To do this, it is convenient to single out now two objects which will play a central role in the sequel.

First for each $t \, \epsilon \, R$ we introduce the von Neumann subalgebra A_t of N defined by:

$$A_t = \{W(f) | f \, \epsilon \, T_t\}'' \text{ with } T_t = \overline{\text{Span}} \{u_s f_o | s \le t\} . \qquad (118)$$

We simply write A for $A_{t=0}$ and notice that $A_t = \alpha(t)[A]$. A thus can be visualized as the "backward trajectory" of N_β. Second, we denote by N_ϕ the "centralizer" of N, defined by:

$$N_\phi = \{N \in N \mid \sigma(t)[N] = N \quad \forall t \in R\} \tag{119}$$

where $\{\sigma(t) \mid t \in R\}$ can be characterized abstractly as the only continuous one-paramter group of automorphisms of N w.r.t. which the faithful normal state ϕ satisfies the KMS condition (see [6]). In our model N_ϕ is interpreted as the algebra of the constants of the free motion.

We now introduce the main concept of this lecture. We define a <u>generalized K-flow</u> as the agregate $(N, \phi, \{\alpha(t) \mid t \in R\}, A)$ constituted by:

- a von Neumann algebra N acting on a Hilbert space H (we generally suppose H to be separable, and dim H $\geq$ 2);
- a faithful normal state ϕ on N such that

> every maximal abelian von Neumann sub-
> algebra of the centralizer N_ϕ of N w.r.t.
> ϕ be already maximal abelian as a von
> Neumann subalgebra of N; $\tag{120}$

- a continuous one-parameter group of automorphisms of N such that for all N $\in$ N and all t $\in$ R:

$$<\phi; \alpha(t)[N]> = <\phi; N> ; \tag{121}$$

- a von Neumann subalgebra A of N satisfying the following conditions:

$$A \subseteq A_t \equiv \alpha(t)[A] \qquad \text{for all} \quad t \in R^+ \tag{122}$$

$$\bigcap_t A_t = C\,I \tag{123}$$

$$\bigvee_t A_t = N \tag{124}$$

$$\sigma(t)[A] = A \qquad \text{for all} \quad t \in R \tag{125}$$

where $\{\sigma(t)\,|\,t \in R\}$ is the continuous one-parameter group of automorphisms of N w.r.t. which ϕ satisfies the KMS condition.

The aim of this lecture is to show that the system $(N,\phi,\{\alpha(t)\,|\,t \in R\})$ associated, in our second lecture, to a quantum diffusion process, with now A defined as in (118), satisfies all these conditions, and is thus a generalized K-flow. We postpone, to the last lecture of this series, the justification of this nomenclature.

The conditions (121), (122), (124) and (125) are quite easy to check in our model: Firstly, we already saw directly from the definition of ϕ and $\alpha(t)$ that (121) is satisfied (compare with 97). Secondly, the definition (94) of $\{\sigma(t)\,|\,t \in R\}$ for our model implies (see eqn. 98) that this is indeed the one-parameter group of automorphisms of N for which ϕ is KMS; from (94) and (118) we can therefore conclude that (125) is satisfied. Next, we notice that (122) follows immediately from (118). Finally, we remark that the vector f_o defined in (100) is cyclic in T w.r.t. $\{u_t\,|\,t \in R\}$ defined in (95); this allows to derive (124) from (118).

To prove (123) one can first remark that, because
of (125), a martingale type of theorem can be proven
(see for instance [13]) which establishes the equivalence of (123) with

$$\bigcap_t [A_t \phi] = C \phi \qquad (126)$$

where ϕ denotes the cyclic and separating vector (see
91) in H corresponding to ϕ; and $[A_t \phi]$ denotes the closed
subspace of H generated by $\{A\phi | A \quad \varepsilon \quad A_t\}$. One then gets,
upon using the explicite form (92) of our representation
of the CCR:

$$[A_t \phi] = H_t \otimes H_t \qquad (127)$$

where H_t is the subspace of H_F obtained as the Fock space
constructed over the test-function space T_t defined in
(118). We now note that the orthogonal complement $T_t^\perp$ of
T_t in T is:

$$T_t^\perp = \{g \varepsilon T \,|\, supp \,(g^* f_0)^\sim \subseteq [t,\infty]\} \qquad (128)$$

where $\sim$ denotes the Fourier transform. Clearly then:

$$\bigvee_t T_t^\perp = T \quad \text{and thus} \quad \bigcap_t T_t = \{0\} \quad . \qquad (129)$$

Joined to (127) this remark implies that (126) and, thus,
(123) are satisfied.

We thus only have still to check that condition
(120) is satisfied. This can be done in the following

roundabout, but hopefully instructive, way.

We define the two operators H^σ and H as the respective generators of the continuous one-parameter groups $\{U^\sigma(t) \mid t \in R\}$ and $\{U(t) \mid t \in R\}$ of unitary operators defined on H by:

$$U^\sigma(t) \ N\Phi \ = \sigma(t) \ [N]\Phi \tag{130}$$

$$U(t) \ N\Phi \ = \alpha(t) \ [N]\Phi \quad . \tag{131}$$

From eqns. (92-95) we get:

$$U^\sigma(t) \ = \ U^\sigma_F(t) \otimes U^\sigma_F(-t) \tag{132}$$

$$U(t) \ = \ U_F(t) \otimes U_F(t) \tag{133}$$

where the continuous one-parameter groups $\{U^\sigma_F(t) \mid t \in R\}$ and $\{U_F(t) \mid t \in R\}$ are defined on Fock space by:

$$U^\sigma_F(t) \ W(f)\Phi_F \ = \ W(e^{-i\omega t}f)\Phi_F \tag{134}$$

$$U_F(t) \ W(f)\Phi_F \ = \ W(u_t \ f)\Phi_F \tag{135}$$

with $\{u_t \mid t \in R\}$ defined in (95).

Upon denoting by H^σ_F and H_F the generators of these two groups, we obtain the relations

$$H^\sigma \ = \ H^\sigma_F \otimes I \ - \ I \otimes H^\sigma_F \tag{136}$$

$$H = H_F^\sigma \otimes I + I \otimes H_F \quad . \tag{137}$$

Clearly H_F^σ and H_F are the "second-quantized" form of the one-particle operators:

$$h^\sigma f = \omega f \tag{138}$$

$$(h \ f) \ (x) = xf(x) \ . \tag{139}$$

From this reasonning we obtain that the spectrum of the operators H^σ and H is given by:

$$Sp(H^\sigma) = \omega Z = \{n\omega \mid n = 0, \pm 1, \pm 2, \ldots\} \tag{140}$$

$$Sp(H) = Sp_d(H) \cup Sp_{ac}(H) \quad \text{with}$$

$$Sp_d(H) = \{0\} \quad \text{non degenerate} \tag{141}$$

$$Sp_{ac}(H) = R \quad \text{with countable infinite multiplicity.}$$

From (140) (or actually directly from 94) we conclude that

$$\{\sigma(t) \mid t \in R\} \text{ is periodic of period } 2\pi/\omega \quad . \tag{142}$$

Quite to the opposite, we see from (141) that, in the language of classical ergodic theory, H has "Lebesgue spectrum". We can thus use Lebesgue-Riemann's lemma to conclude from (141) that $\{\alpha(t) \mid t \in R\}$ is mixing, i.e. that for all M, N $\in$ N we have:

$$\lim_{t \to \infty} \ <\phi; \ M\alpha(t)[N]> = <\phi; \ M> <\phi; \ N>; \tag{143}$$

and the ergodic property:

ϕ is extremal $\{\alpha^*(t)|t \ \varepsilon \ R\}$ - invariant . (144)

Actually, since our system is (weakly) "asymptotically abelian" in time, we have (see for instance [5], Thm. II.2.8 and Cor. 2 to Thm. II. 2.7) the stronger ergodic properties:

ϕ is the only normal state on N

(145)

which is $\{\alpha^*(t)|t \ \varepsilon \ R\}$ - invariant;

$N \ \varepsilon \ N$ and $\alpha(t)$ [N] = N $\forall t \ \varepsilon \ R$

(146)

$\rightarrow N = <\phi; \ N > I$.

Together with (121) and (142), (146) implies that ϕ is periodic homogeneous in the sense of Takesaki [19] and thus that (120) is satisfied.

This concludes the proof of the result announced in the beginning of this lecture, namely that the deterministic system $(N,\phi,\{\alpha(t)|t \ \varepsilon \ R\},A)$ canonically associated to the quantum diffusion process $(N_\beta,\phi_\beta,\{\gamma(t)|t \varepsilon R^+\})$ studied in the first lecture, is a generalized K-flow.

As we shall see in the next lecture this structure is the reason behind the strong ergodic properties observed in the present model.

In closing this lecture we want to contrast the properties (120) and (146). Whereas (120) says that the

algebra N_ϕ of the constants of the free motion $\{\sigma(t)\,|\,$
$t \in R\}$ is very large in N, (146) says that there are
no non-trivial constants of the actual motion $\{\alpha(t)\,|$
$t \in R\}$. This indicates that the "perturbation"

$$V = H - H^\sigma \tag{147}$$

is a rather drastic one. This indication is further
supported by the remark that for every $\lambda > 0$, however
small one may choose it

$$H_\lambda = H^\sigma + \lambda V \tag{148}$$

would still generate a K-flow on (N,ϕ). Indeed we can
use (94-95) to see directly for our model (although this
actually follows in general from (121), see [6]) that:

$$\alpha(t)\sigma(s) = \sigma(s)\alpha(t) \quad \text{for all} \quad s,t \in R . \tag{149}$$

This implies that H_λ generates a one-parameter group
$\{\alpha_\lambda(t)\,|\,t \in R\}$ of automorphisms of N, with:

$$\alpha_\lambda(t) = \alpha(\lambda t)\sigma((1-\lambda)t) \quad \text{for all } t \in R . \tag{150}$$

It is then immediate, from the definition given in the
beginning of this lecture, that $(N,\phi,\{\alpha_\lambda(t)\,|\,t \in R\},A)$
is again a generalized K-flow.

This property reflects, in the model considered
here, the strong cumulative effects of the time evolution, registered in the van Hove "long-time, weak coupling" limit. As the above proof shows this stability

property is a genuine feature of generalized K-flows,
not limited to the particular model considered here.
We should take it here with some relief, since its
absence could have thrown serious doubts on the idea
that K-structures might belong in a genuine manner to
the realm of non-equilibrium statistical mechanics.

IV. GENERALIZED K-FLOWS AND K-ENTROPY

Our search of a statistical mechanical model for
the diffusion of a quantum particle in a harmonic well
has led us, in the first three lectures of this series,
to unearth a new mathematical structure which we called
"generalized K-flow".

The aims of this last lecture are: <u>firstly</u>, to show
how this structure is indeed a generalization of the
classical concept of "K-flow" in the sense of Kolmogorov,
Rohlin and Sinai (for textbook expositions of the theory
in the classical framework, see for instance [20], [21]
or [22]); and <u>secondly</u>, to point out that our generalization carries over to the quantum realm some of the essential properties of classical flows, such as ergodicity,
Lebesgue spectrum and strict positivity of the dynamical
entropy.

We recall that a classical K-flow is constituted
by: a probability space $(\Omega, \textstyle\sum, \mu)$; a measurable group
$\{T(t) | t \in R\}$ of measure preserving transformations of
$(\Omega, \textstyle\sum, \mu)$; and a partition of $\Omega, \xi \in \textstyle\sum$ such that: $\xi \subseteq$
$T(t)$ $[\xi]$ for all $t \in R^+$; $\bigcap_t T(t)$ $[\xi] = \{\emptyset, \Omega\} \pmod{\mu}$;

and $\bigvee_t T(t)[\xi] = \sum (\mathrm{mod}\ \mu)$.

To see that classical K-flows are special cases of the generalized K-flows defined in the preceding lecture we proceed with the following construction.

On the Hilbert space

$$H = L^2 (\Omega, \textstyle\sum, \mu) \tag{151}$$

we consider the von Neumann algebra

$$N = \pi[L^\infty(\Omega, \textstyle\sum)]\ \text{with}$$

$$\pi : f \in L^\infty(\Omega, \textstyle\sum) \to N_f \in N\ \text{where} \tag{152}$$

$(N_f\Psi)(x) = f(x)\Psi(x)$ for all $\Psi \in H$.

We further consider the element

$$\Phi \in H\ \text{defined by}\ \Phi(x) = 1\ \text{for all}\ x \in \Omega\ . \tag{153}$$

With this vector we form:

$$\phi : N_f \in N \to\ <\phi;\ N_f> = (\Phi, N_f\Phi) = \int f(x)\ d\mu(x)\ . \tag{154}$$

We then define for every $t \in R$ the mapping

$$\alpha(t) : N_f \in N \to\ \ \alpha(t)[N_f] = N_{f \circ T(t)}\ . \tag{155}$$

We finally define the von Neumann algebra A on H by

$$A = \{N_{X_\Delta} \mid \Delta \ \varepsilon \ \xi\}" \ . \tag{156}$$

We should remark that, since N is abelian, Φ satisfies the KMS condition w.r.t. $\{\sigma(t) = id \mid t \ \varepsilon \ R\}$; hence $N_\phi = N$.

It is then straightforward to verify that the aggregate $(N, \phi, \{\alpha(t) \mid t \ \varepsilon \ R\}, A)$ just constructed satisfies the conditions (120-125) of the preceding lecture, and is thus indeed a generalized K-flow.

We further recall that, conversely, starting from (N, ϕ) with N abelian, we can find back a triple $(\Omega, \sum, \mu)$ such that (151-152) are satisfied. Consequently, we can characterize classical K-flows as those generalized K-flows for which N is abelian.

Our generalization to the quantum realm of the classical concept of K-flow thus consists in dropping this additional condition that N be abelian. This generalization is genuine, in the sense that there exist generalized K-flows which are not classical: we in fact constructed explicitly in the preceding lecture a generalized K-flow where N is not abelian, but, quite to the opposite, is a factor (i.e. $N \cap N' = C \ I$). Actually, one can show [13] that this factor is of type III_λ, $\lambda \ \varepsilon \ ]0, 1[$ in the classification of Connes [7]. These factors moreover do not exhaust the class of generalized K-flows for which N is a factor, as the following examples will show.

First, we can form the tensor product of two generalized K-flows of the type constructed in the preceding lecture, with ω_1/ω_2 irrational. One can

verify [13] that we get again in this manner a generalized K-flow where N is a factor, but now of type III_1 in
the classification of Connes [7]. It is perhaps interesting to note that the generator $H_{1,2}^\sigma$ of the KMS evolution $\sigma(t) = \sigma_1(t) \otimes \sigma_2(t)$, corresponding to the state
$\phi = \phi_1 \otimes \phi_2$ on $N = N_1 \otimes N_2$, has for spectrum:

$$Sp(H_{1,2}^\sigma) = \{k_1\,\omega_1 + k_2\,\omega_2 \mid k_1, k_2 \in Z\} \tag{157}$$

which is a dense, discrete subgroup of R; in particular
the KMS time evolution for this system is not periodic,
but only "almost periodic", in the sense that for every
N and M in N

$$<\phi; N\sigma(t)\ [M]> \text{ is almost periodic in t.} \tag{158}$$

Physically this example can be interpreted as corresponding to an infinite chain of identical two-dimensional oscillators with irrational frequencies in the x and y directions.

Another example of a generalized K-flow, where N
is now a factor of type II_1, is obtained as follows.

For an arbitrary generalized K-flow we recall that
ϕ is a faithful normal state on N, invariant under
$\{\sigma^*(t)\mid t \in R\}$, so that N is $\{\sigma(t)\mid t \in R\}$-finite in
the sense of Kovacs and Szücs [23]. For each $N \in N$
there exists therefore an unique element $\underline{N}$ in

$$N_\phi \cap^{uw} \overline{co} \{\sigma(t)\ [N]\mid t \in R\} \tag{159}$$

and the mapping

$$E : N \varepsilon N \to \underline{N} \varepsilon N_\phi \qquad\qquad (160)$$

is a conditional expectation from N onto N_ϕ w.r.t. ϕ.
We now use (125) to conclude that

$$E (A) = A \cap N_\phi \qquad . \qquad\qquad (161)$$

We denote by A_ϕ this von Neumann subalgebra of N_ϕ. Finally, we recall that because (121) we have, via (149), that N_ϕ is stable under $\{\alpha(t) | t \varepsilon R\}$. For each $t \varepsilon R$, we denote then by $\alpha_\phi(t)$ the restriction of $\alpha(t)$ to N_ϕ.

   It is now easy to check that the aggregate

$$(N_\phi , \phi , \{\alpha_\phi (t) | t \varepsilon R\}, A_\phi) \qquad\qquad (162)$$

is again a generalized K-flow. Indeed from the definition of N_ϕ (see 119) we have that for every $t \varepsilon R$ the restriction of $\sigma(t)$ to N_ϕ is the identity map. Hence ϕ restricted to N_ϕ satisfies the KMS condition w.r.t. $\{\sigma_\phi(t) = \text{id} | t \varepsilon R\}$ and thus (120) and (125) are trivially satisfied for the aggregate (162). This system moreover inherits directly from our original flow the properties (121), (122) and (123). The proof of (124) can finally be done by using the continuity property of the conditional expectation (160).

   We now come back to the particular case of a generalized K-flow of type III constructed in the preceding lecture; one can check [13] that N_ϕ is a factor of type

II_1 in this case. The above construction leads us then
to a generalized K-flow of the form (162) where N_ϕ is
now a type II_1-factor.

We have thus obtained, besides the classical K-
flows, three nonisomorphic generalized K-flows which
already differ from one another by the type of the
factor N on which they are constructed.

In the third lecture of this series we have seen
that the nonabelian generalized K-flow, associated to
the diffusion of a quantum particle in a harmonic well,
enjoys rather striking ergodic properties. We now want
to show that these properties are actually a feature
common to all generalized K-flows.

In this series of lectures much emphasis has been
placed so far on the algebraic approach to non-equilibrium
statistical mechanics. It seems therefore appropriate
here to show explicitly that the Lebesgue spectrum property, from which the other ergodic properties follow,
can be proven as a somewhat unexpected (see however [24])
consequence of von Neumann's uniqueness theorem on the
representations of the CCR for one degree of freedom.
The proof given here will apply to all generalized K-
flows, including thus, in particular, classical K-flows.

Let us denote by Φ the cyclic and separating vector
of H corresponding to ϕ; and by $H_\perp$ its orthogonal complement in H. On $H_\perp$ we define for each $s,t \in R$ the operators
$E(s)$ and $V(t)$ as follows:

$$H(s) = [A_s \Phi] \cap H_\perp; \quad \text{and } E(s) : H \to H(s) \qquad (163)$$

$V(t) = U(t)$ restricted to $H_\perp$ (164)

where $U(t)$ is defined by:

$U(t) \ N\Phi = \alpha(t) \ [N]\Phi$ for all $N \ \epsilon \ N$ (165)

From the defining properties of a generalized K-flow we conclude via the martingale-type theorem (126), that $\{E(s) | s \ \epsilon \ R\}$ is a continuous, increasing family of projectors of $H_\perp$ with

$$\lim_{s \to -\infty} E(s) = 0 \quad \text{and} \quad \lim_{s \to +\infty} E(s) = I; \qquad (166)$$

and that $\{V(t) | t \ \epsilon \ R\}$ is a continuous, one-parameter group of unitary operators on $H_\perp$ satisfying moreover the relation:

$V(t) \ E(s) \ V(-t) = E(s + t)$ for all $s, t \ \epsilon \ R.$ (167)

This is just an equivalent form of the canonical commutation relations for one degree of freedom (compare with eqn. 9). Since H (and therefore $H_\perp$) is separable by assumption, we can use the von Neumann uniqueness theorem (see for instance [5] Thm. III.1.6) to conclude that $\{V(t), E(s) | s, t \ \epsilon \ R\}$ is a direct integral of Schrödinger representations of the CCR. In each component the generator P of $\{V(t) | t \ \epsilon \ R\}$ is absolutely continuous with respect to Lebesgue measure and covers R with multiplicity one. Since

$U(t) = I \oplus V(t)$ on $H = (C\Phi) \oplus H_\perp$ (168)

we have thus proven indeed that the generator H of
$\{U(t) \mid t \ \epsilon \ R\}$ satisfies

$$Sp \ (H) \ = \ Sp_d(H) \ \cup \ Sp_{ac}(H) \qquad with$$

(169)

$$Sp_d(H) \ = \ \{0\} \quad non\text{-}degenerate, \ and \ Sp_{ac}(H) = R$$

which is all one needs to conclude to the validity of
all the ergodic properties mentionned in the third
lecture, such as for instance (143-146).

For the multiplicity of $Sp_{ac}(H)$ see [13]. There
is one more property of classical K-flows which we have
not yet mentionned here with a view to extend it to
generalized K-flows. This property is the fact that
these flows, although deterministic in the sense that
$\{T(t) \mid t \ \epsilon \ R\}$ is a group of measure-preserving transformations, nevertheless exhibit an interesting dissipative behaviour. To express this notion precisely,
Kolmogorov introduced the concept of dynamical entropy.
In heuristic terms this object is linked to the quantity
of information retained, as time goes on, from a measurement of any partition $\xi \ \epsilon \ \sum$: classical K-flows have the
property that however often one might have measured
$T(t)[\xi]$ in the past, one still gains some information
by measuring ξ again.

In the classical context the measurement of an
arbitrary partition $\xi \ \epsilon \ \sum$ does not perturb the state ϕ.
As is well-known, this is no more the case in a typical
quantum situation. Indeed, in line with the orthodox
measurement theory due to von Neumann, we describe the
effect of the measurement of a partition F of the

identity into mutually orthogonal projectors $\{F_k\}$ in
N by:

$$\phi \to \phi_F = \sum_k \lambda_k \phi_k \qquad \text{where}$$

$\lambda_k = <\phi; F_k>$ and ϕ_k is the state

(170)

$\phi_k : N \in N \to <\phi_k; N>$ given by

$$<\phi_k; N> = <\phi; F_k N F_k> / <\phi; F_k> .$$

Notice that in the classical theory (where N is abelian)
we always have indeed $\phi = \phi_F$. In the theory of generalized
K-flows one can show (see for instance [16]) that:

$$\phi = \phi_F \quad \text{iff} \quad F \subset N_\phi \tag{171}$$

and

$$F \subset N_\phi \quad \text{iff} \quad \alpha(t)[F] \subset N_\phi \qquad \text{for all } t \in R . \tag{172}$$

Since we are only interested here in the dissipative
effects of the time evolution, and not in the additional
effects due to the measuring process (170), we will
restrict our attention to those partitions of the
identity which belong to N_ϕ, and call them "admissible"
for the purpose of this lecture. Notice again that in
the classical theory this is no restriction at all,
since $N = N_\phi$ in this case. Notice also that (172)
implies in particular that the von Neumann algebra

$$C_n[F] = \{\alpha(-k)[F] \mid 0 < k \leq n\}'' \, , \quad n \, \varepsilon \, Z^+ \tag{173}$$

generated from an admissible partition F, also belong to N_ϕ.

Let us denote by $\hat{F}$ (resp. $\hat{C}$) the class of all admissible partitions F (resp. all von Neumann sub-algebras C of N_ϕ).

Our next step will be to produce a "reasonable" mapping

$$H \, : \, (F,C) \, \varepsilon \, \hat{F} \times \hat{C} \rightarrow H(F \mid C) \, \varepsilon \, R^+ \tag{174}$$

such that: $H(F \mid C) = 0$ iff $F \subset C$ \hfill (175)

$$F_1 \subsetneq F_2 \rightarrow H(F_1 \mid C) \leq H(F_2 \mid C) \quad \forall C \, \varepsilon \, \hat{C} \tag{176}$$

$$C_2 \subsetneq C_1 \rightarrow H(F \mid C_1) \leq H(F \mid C_2) \quad \forall F \, \varepsilon \, \hat{F} \tag{177}$$

which we could interpret as the entropy of F conditioned by C w.r.t.ϕ.

With this in hand, we will then be able to define

$$H_n(F,\alpha) = H(F \mid C_n[F]) \tag{178}$$

which we will interpret as the information gained in measuring F when $\alpha(-1)[F], \alpha(-2)[F], \ldots, \alpha(-n)[F]$ have already been measured. We will then define:

$$H(F,\alpha) = \lim_{n\to\infty} H_n(F,\alpha) \tag{179}$$

(notice that this limit exists because of 174 and 177);
and:

$$H(\alpha) = \text{Sup}_{F \in \hat{F}} \ H(F,\alpha) \tag{180}$$

(which can be infinite). We will call this last quantity
the <u>dynamical (or K-) entropy</u> of the dynamical' system
$(N, \phi, \{\alpha(t) \mid t \in R\})$. Our task will then be to show that
this dynamical entropy is positive on generalized K-flows,
and that it reduces to the Kolmogorov dynamical entropy
when N is abelian.

Our first step is to define (174), which we shall
now proceed to do.

Although it is not really necessary for several of
the steps in the following argument, we shall suppose,
from here on and up to (189), that M is a von Neumann
algebra admitting a faithful normal trace ψ; this is
indeed all we need for (174) where we will make the
identification $M = N_\phi$ and $\psi = \phi$.

Let now $F = \{F_k\}$ be a (finite) partition of the
identity in M. As far as the entropy of F is concerned,
we can ignore M and restrict our attention to F". Since
F" is abelian (and atomic) we are exactly in the classical situation, and thus Khinchin's theorem gives uni-
voquely [25]:

$$H_\psi (F) = \sum_k h(<\phi; \ F_k>) \text{ with}$$

$$\tag{181}$$

$$h : x \in [0,1] \to -x\log x \in R^+ \ .$$

To proceed further in the quantum domain, we notice that
if M is finite-dimensional, and if F and G are two partitions of the identity into minimal projectors of M we
have:

$$H_\psi (F) = H_\psi (G) \; ; \qquad\qquad (182)$$

and we can therefore define the entropy of M finite-dimensional as

$$H_\psi (M) = H_\psi (F) \qquad \text{with} \quad F = \{F_k\}$$

$$\qquad\qquad (183)$$

and F_k minimal projectors in M .

A straightforward computation then shows that for an
arbitrary partition F of the identity in a finite-dimensional M we have:

$$\sum_k \lambda_k H_{\psi_k} (M_k) = H_\psi (M) - H_\psi (F) \qquad \text{where}$$

λ_k and ψ_k are defined as in (170); and $\qquad (184)$

M_k are the von Neumann algebras $F_k M F_k$ on $F_k H$.

We can thus interpret the LHS of (184) as the entropy
of M after the measurement of F; this is clearly in
line with the interpretation of (170) as a filtering
operation. We now return to our general M; let $F = \{F_k\}$ and $G = \{G_l\}$ be two partitons of the identity
in M. In analogy with (184) we define the entropy of
G conditioned by F w.r.t. ψ as:

$$H_\psi (G|F) = \sum_k \lambda_k H_{\psi_k} (G) \quad \text{where}$$

(185)

λ_k and ψ_k are defined as in (170) .

If we now denote by $E_\psi (\ldots |F'')$ the conditional expectation from M onto F'' w.r.t. ψ, we verify that (185) can be rewritten as:

$$H_\psi (G|F) = \sum_1 <\phi; h [E_\psi (G_1|F'')]> \quad .$$

(186)

If we denote, as in [26], by S the set of all finite families $x = \{x_k\}$ of positive elements in M with ($\sum_k x_k = I$) we can check that (185) can be also rewritten as:

$$H_\psi (G|F) =$$

$$\text{Sup}_{x \in S} \{\sum_k <\phi; h \quad [E_\psi (x_k|F'')]>$$

(187)

$$-\sum_k <\phi; h \quad [E_\psi (x_k|G'')]>\} \quad .$$

Up to this point there seems thus to be essentially no "reasonable" alternative to define $H_\psi (G|F)$ in a manner compatible with the orthodox interpretation (170) of the measurement process in quantum mechanics. We should also notice that the expression $H_\psi (G|F)$ defined above reduces to the classical conditional entropy for partitions (see for instance [20] or [21]) when F and G commute, i.e. when $[F_k,G_1] = 0$ for all (k,1). We still are quite bound by the preceding argument when we define for an arbitrary partition F in M, and an arbitrary von Neumann abelian subalgebra A of M, the entropy of F conditioned by A w.r.t. ψ as:

$$H_\psi (F|A) = \sum_k <\phi; \ h[E_\psi (F_k|A)]> \quad .\tag{188}$$

The next step however seems to offer some possibilities for alternate definitions. We choose to define for an arbitrary partition F in M and an arbitrary von Neumann algebra C in M the entropy of F conditioned by C w.r.t. ψ as:

$$H_\psi (F|C) = \mathop{\text{Inf}}_{\substack{A \text{ abelian} \\ (A \subseteq C)}} H_\psi (F|A)\tag{189}$$

with $H_\psi (F|A)$ defined by (188) .

One can then check (see [16]) that this H_ψ satisfies the conditions (175-177). We therefore take it as our definition of (174).

We now take back our argument where we left it, i.e. after the remark following (180).

We first notice as a consequence of (185-189) that $H(\alpha)$ reduces to the classical dynamical entropy of Kolmogorov when N is abelian.

One can further see [16] that $H(\alpha) = 0$ would imply for every von Neumann subalgebra C of N_ϕ that for all integers n and m

$$\{\alpha(k)[C]\,|\,k\epsilon Z,k\leq n\}" = \{\alpha(k)[C]\,|\,k\epsilon Z,k\leq m\}" \ . \qquad (190)$$

Upon applying this result to $C = A_\phi$ with A_ϕ defined by (161), we conclude from (162) and the K-properties (122-124) that $N_\phi = C \ I$. Because of (120) this would imply $N = C \ I$ which would contradict dim $H \geq 2$ since $H = [N\phi]$. We can therefore conclude that the dynamical entropy (180) is strictly positive on every generalized K-flow.

The last question, which we want to address ourselves to here, is whether one could compute $H(\alpha)$ for systems of physical interest. As with the classical K-flow of the Brownian motion [27], we can show [13] that the generalized K-flow, constructed and discussed in the second and third lectures of this series, contains, as subflows, Bernoulli flows of arbitrary large entropy. We therefore can conclude that this generalized K-flow has infinite dynamical entropy. The same reason would lead to the same result for the dynamical proposed by Connes and Størmer [26].

If we had to summarize these lectures with a heuristic catch-phrase, we would thus say that these results show how "very stochastic" indeed, any statistical mechanical description of the diffusion of a quantum particle in a harmonic well is bound to be.

REFERENCES

1. R. Kubo, Statistical Mechanical Theory of Irreversible Processes, I.J.Phys. Soc. Japan 12, 570-586 (1957).

2. P.C. Martin and J. Schwinger, Theory of Many-Particle
   Systems, I. Phys. Rev. <u>115</u>, 1342-1373 (1959).

3. R. Haag, N. Hugenholtz and M. Winnink, On the Equilibrium States in Quantum Statistical Mechanics, Commun.
   math. Phys. <u>16</u>, 81-104 (1967).

4. R. Haag, D.Kastler and E. Trych-Pohlmeyer, Stability
   and Equilibrium States. Commun. math. Phys. <u>38</u>, 173-
   193 (1974).

5. G.G. Emch, Algebraic Methods in Statistical Mechanics
   and Quantum Field Theory, Wiley-Interscience, New York,
   1972.

6. M. Takesaki, Tomita's Theory of Modular Hilbert Algebras and its Applications, Springer Lecture Notes
   in Mathematics No. 128, 1970.

7. A. Connes, Une classification des facteurs de type
   III, Ann. scient. Ec. Norm. Sup. (4e série) <u>6</u>, 133-
   252 (1973).

8. G.W. Ford, M. Kac and P. Mazur, Statistical Mechanics
   of Assemblies of Coupled Oscillators, Journ. Math.
   Phys. <u>6</u>, 504-515 (1965).

9. E.B. Davies, Diffusion for Weakly Coupled Quantum
   Oscillators, Comm. math. Phys. <u>27</u>, 309-325 (1972).

10. W.F. Stinespring, Positive Functions on C^*-algebras,
    Proc. Amer. Math. Soc. <u>6</u>, 211-216 (1955).

11. D.E. Evans, Positive Linear Maps on Operator Algebras.
    Preprint, DIAS-TP-75-39.

12. F. Riesz and B. Sz.-Nagy, Lecons d'analyse fonctionnelle
    Gauthier-Villars, Paris, 1955.

13. G.G. Emch, Generalized K-Flows, In preparation. This

paper extends the work done in [14], [15], and [16]
for summaries see [17] and [18].

14. G.G. Emch, Nonabelian Special K-flows, Journ. Funct.
    Analysis 19, 1-12 (1975); Zentralblatt f. Math.
    501130 (1975).

15. G.G. Emch, The Minimal K-Flow associated to a Quantum
    Diffusion Process. In Physical Reality and Mathematical
    Description, Ch. Enz & J. Mehra, Eds., Reidel Publ.,
    Dordrecht, 1974, 477-493.

16. G.G. Emch, Positivity of the K-entropy on Non-Abelian
    K-Flows, Z. Wahrscheinlichkeitstheorie verw. Gebiete
    29, 241-252 (1974).

17. G.G. Emch, An Algebraic Approach to the Theory of
    K-Flows and K-Entropy, In Proc. Intern. Symp. on Math.
    Problems in Theor. Phys. Springer Lecture Notes in
    Physics No. 39, 1975, 315-318.

18. G.G. Emch, Algebraic K-Flows, In Proc. Intern. Conf.
    on Dynamical Systems in Math. Physics, Rennes, 1975.

19. M. Takesaki, The Structure of a v.N. Algebra with
    Homogeneous Periodic State. Acta Math. 131, 79-121
    (1973).

20. V.I. Arnold and A. Avez, Ergodic Problems of Classical
    Mechanics, Benjamin, New York, 1968.

21. W.A. Parry, Entropy and Generators in Ergodic Theory.
    Benjamin, New York, 1969.

22. D.S. Ornstein, Ergodic Theory, Randomness, and Dynamical
    Systems, Yale University Press, New Haven, 1974.

23. I. Kovacs and J. Szücs, Ergodic Type Theorems in von
    Neumann algebras. Acta Sc.Math.(Szeged) 27, 233-246
    (1966).

24. P.D. Lax and R.S. Phillips, Scattering Theory, Academic Press, New York, 1967.

25. A. Khinchin, Mathematical Foundations of Information Theory, Dover, New York, 1957.

26. A. Connes and E. Størmer, Entropy for automorphisms of II_1 von Neumann algebras. Acta Math.

27. T. Hida, Stationary Stochastic Processes, Princeton University Press, Princeton, 1970.

Acta Physica Austriaca, Suppl. XV, 133–269 (1976)

PHASE TRANSITIONS, GOLDSTONE BOSONS AND

TOPOLOGICAL SUPERSELECTION RULES[*]

by

J. FRÖHLICH[+,x]

Department of Mathematics

Princeton University

Princeton, N.J. 08540

Instead of an abstract: Table of Contents

[+] present address: ZiF, Universität Bielefeld, D-4800 Bielefeld, Wellenberg 1, F.R.Germany

[x] supported in part by U.S. NSF under grant GP-39048 and by ZiF, Universität Bielefeld

[*] Lecture given at XV. internationale Universitätswochen für Kernphysik, Schladming, Austria, February 16-27, 1976.

1. INTRODUCTION AND PROGRAM

A warning and a reflection: The material I propose to cover in these four lectures is quite large, and ideas from different fields in mathematical physics must be combined. Therefore not all the details will be explained. I have tried to select proofs for presentation according to their technical simplicity and elegance. This should not mislead you to believe that mathematical physics is a simple thing. Some of the most outstanding and admirable recent results of, say, constructive quantum field theory (e.g. [GJ1] [GRS] [GJS1] [OS]; see also [CQFT]) require an enormous amount of sophisticated and hard analysis. These results concern the existence of relativistic quantum fields and their detailed properties, e.g. their non-triviality, (in the sense that the scattering matrix is different from the identity [EEF, OSé]). The fact that the proofs of many of these results are very hard and intricate may seem or be unpleasant. Yet it tells us something that I feel is important: The foundations of relativistic quantum field theory and statistical mechanics may neither be wrong nor do they necessarily require major modifications, but it could be that the

mathematical problems occuring in the construction of
quantum fields and models for systems with an infinite
number of degrees of freedom are just very difficult and
complicated, and that with many problems one has not yet
been successful on a mathematical level. If this should
indeed be true then the fact that one of two weeks of a
school on "Kernphysik" was devoted to mathematical physics
requires only minor defense, and it may then also seem
plausible that progress in physics does not only come
from the very important efforts of experimentalists and
theoreticians but even a little bit from the attempts of
mathematical physicists. Rather than presenting some
more reflections I should like to refer you to some nice
thoughts in the literature: [K] (some danger of producing
"dehydrated elephants" in mathematical physics), [St E]
(if what was intended to represent" a boa that has
swallowed an elephant" appears to you to represent "a hat"
(or worse an "old hat") it may be that I am a bad writer
and I wish to apologize myself for that).

Next I describe the program of these lectures: The
first part is centered around the phenomenon of phase
transitions which is sometimes accompanied by the spontaneous breaking of a (discrete or continuous, internal)
symmetry (of the "dynamics"). A simple (or very difficult)
example (depending on one's point of view) for a phase
transition is a ferromagnet: This is a macroscopic system
of matter which at high temperatures (i.e. above some
critical temperature T_c) does (or may) not have any
particularly exciting or extraordinary features; but, at
temperatures below T_c, it has the remarkable property
that it remains magnetized after an external magnetic
field has been turned off. Here "macroscopic" means that
the system consists of ~10^{23} elementary magnets - atoms

or molecules - (mathematically: infinitely many degrees
of freedom). It turns out that such a material may have
as many pure phases as there are directions in space.
What is a pure phase? In the case of H_2O one distinguishes
three pure phases: ice, water and vapor. These are especially pure states (or manifestations) of H_2O. Mathematically, the pure phases of a physical system correspond
to (time translation invariant) states of the system with
the property that the algebra of (time translation invariant) observables at infinity is trivial; see Sect. 3

The Hamilton function (the Hamilton operator, respectively) of a ferromagnet is generally assumed to be
invariant under an arbitrary, simultaneous rotation of
all the elementary magnets of the ferromagnet. Yet, below
the critical temperature, in a pure phase, there exists
a preferred direction; the direction of spontaneous
magnetization. We say that the state of the ferromagnet
in a pure phase breaks the symmetry of the Hamiltonian,
or: the symmetry is spontaneously broken in the pure
phases.

The main issue of the first part of my lectures is
the construction and analysis of simple models for phase
transitions and spontaneous symmetry breaking. Some of
these models are carricatures of (classical) ferromagnets,
one class of them are relativistic quantum field models
at temperature 0, the so called $(\vec{\phi} \cdot \vec{\phi})_3^2$ models. Mathematically,such models are related to models of classical
ferromagnets, e.g. the Ising model; see [GRS], [SG], [N],
[DN]. The role of the temperature is played by some
coupling constant (related to the field strength), and
phase transitions occur, as this coupling constant is
varied. - Some other models are merely of conceptual

interest: One serves to exemplify the concept that there
must exist phase transitions which are not accompanied
by the breaking of any symmetry, another one shows the
possibility of "triple points".

Field theorists are presently very much interested
in phase transitions and the spontaneous (or even more
the "dynamical") breakdown of continuous, internal (or
even nicer: "gauge") symmetries, because most current
theories of the fundamental interactions involve as a
central theoretical element the possibility that continuous symmetries may be broken by the physical vacuum.
(It could be that mathematical consistency of such theories
even requires symmetry breaking).

In the second part of my lectures I will illustrate,
in the context of Bose quantum field models in two space-
time dimensions, how at temperature 0 phase transitions
(as some coupling constant is varied) may be accompanied
by the occurrence of new superselection sectors, [St W],
[DHR], and non-trivial, dynamical (or "topological")
charges. In order to illustrate what I try to describe
I first sketch an example:

Consider an infinite quantum mechanical chain of
equally spaced, elementary dipoles (e.g. electric dipoles
that are anharmonically bound and ferroelectrically
coupled). Such a chain may have a degenerate groundstate,
namely two pure, spontaneously, polarized groundstates
with opposite polarizations ($\pm$ p): Pictorially

..... ↑↑↑↑↑↑↑ ....   or   .... ↓↓↓↓↓↓↓ ....
       +   -

Fig. 1

We may then ask the natural question whether there exist
states with a well defined, continuous time evolution
which very far to the left look like groundstate - and
very far to the right like groundstate +, or vice versa.
Pictorially

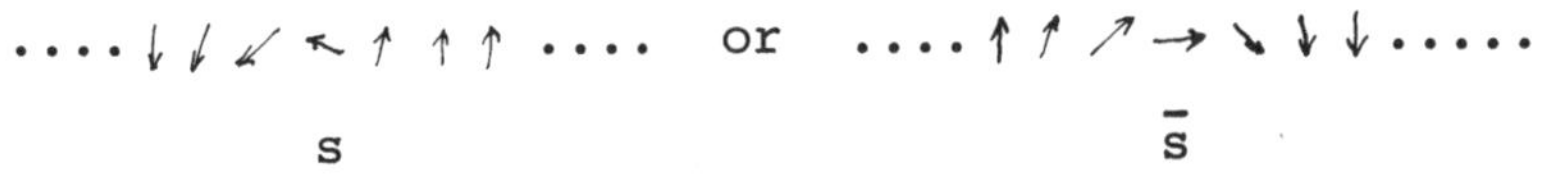

.... ↓ ↙ ↙ ↖ ↑ ↑↑ ....   or   .... ↑↑ ↗ → ↘ ↓↓ .....

　　　　　　s　　　　　　　　　　　　　　$\bar{s}$

Fig.2

(The pictures represent e.g. the expectation of the
"dipole field" p(i), i ∈ $\mathbb{Z}$, in the states +, -, s and $\bar{s}$)

What we will show in Part 2 is that the chain can
indeed be twisted over some bounded space region by an
angle of 180° to be in a state s (or $\bar{s}$) that <u>interpolates</u>
between the groundstates - and +. The state s ($\bar{s}$) so obtained is <u>not</u> a groundstate (not even a discrete eigenstate) of the Hamiltonian. The states +, -, s and $\bar{s}$ are
vectors in <u>mutually orthogonal</u> Hilbert spaces (super
selection sectors, denoted $\mathcal{H}_{+}$, $\mathcal{H}_{-}$, $\mathcal{H}_{s}$, $\mathcal{H}_{\bar{s}}$).

There exists a conserved charge Q such that Q $\mathcal{H}_{\pm}$
={$\vec{0}$}; Qψ = 2pψ, for all $\psi \in \mathcal{H}_{s}$; Qψ = - 2pψ, for all $\psi \in \mathcal{H}_{\bar{s}}$.

If the groundstate of the chain were unique there
would not exist any superselection sectors, and Q ≡ 0,
on all physical states.

On the other hand for a chain with n-fold degenerate
groundstate one can construct n(n-1) charged sectors.
The proper mathematical framework for the construction
and analysis of these sectors appears to be the <u>frame-
work of local observables and local morphisms</u> axiomati-
cally developped by Doplicher, Haag and Roberts, [DHR].

We will construct charged states by composing a ground-state with a charged local morphism (= a "generalized transformation") of the observables.

The analysis of the <u>spectrum</u> of the energy-momentum operator on the new sectors $\mathcal{H}_s$, $\mathcal{H}_{\bar{s}}$, however, must apparently be done in analogy to the analysis of the <u>surface tension</u> in classical ferromagnets; [GM].

Unfortunately the phenomenon described here is typically one (space) dimensional.

In more than one space dimension more complicated constructions of the kind described may be possible in much more complicated models (possibly only in models involving <u>gauge fields</u>, i.e. gauge theories).

In <u>one</u> space dimension, however, the phenomenon of spontaneous occurence of charged super selection sectors seems to be an observed fact (e.g. in linear (long, thin) systems of non-linear optics, where the twist regions in Fig. 2 are associated with pulses of the electromagnetic field [La]). Thus it is not merely a mathematical curiosity. Even in <u>two</u> space dimensions it can be observed experimentally: It describes the occurence of <u>vortices</u> in superconducters (described by a gauge theory!), and the charge Q is then related to <u>flux quantization</u>.

What I have said here is a rough picture of the subject material of my lectures. In the following sections we want to make that precise: I will describe the framework, formulate the results in a mathematically precise manner and present some of the proofs.
For brief overall information consult Section 2, 5-7 and 9

Remark:

    We will often add in between brackets some comments of mathematical or technical character directed towards the more mathematically inclined reader. If some reader finds such a comment confusing he should simply ignore it.

Acknowledgements:

    The reason why I can present some new results on phase transitions that I find rather exciting is that I had the luck of collaborating with two clever colleagues: Barry Simon and Tom Spencer. I wish to thank them for the joy of collaboration and for permission to present results that are not yet published; (Sections 3.1, 4-6).

    I am much indebted to Erhard Seiler and Sidney Coleman who have taught me many things about the material in Part 2. I am grateful to James Glimm, Elliott Lieb and Charles Pfister for useful discussions.

Part 1:

Section 2:

FERROMAGNETIC MODELS IN CLASSICAL STATISTICAL MECHANICS AND RELATIVISTIC BOSE QUANTUM FIELD THEORY: MAIN RESULTS

2.1 The framework and the class of models:

    All the models discussed in my lectures may be interpreted as models of classical statistical mechanics.

    A classical, physical system is specified by its phase space Γ, and a state of such a system is represented - mathematically - by a probability measure $d\mu$ on Γ.

(Technically, one can always choose a topology on Γ such that Γ is a compact Hausdorff space and then choose as a σ -algebra Σ_Γ the Borel sets; dμ is then assumed to be a regular Borel probability measure). The dynamics of the system is given in terms of a (Σ_Γ -measurable, often once continuously differentiable) Hamilton function H on Γ. Let {Q},{P} be canonical coordinates for some local (e.g. some bounded, open) region in Γ.

We will always deal with a Hamilton function of the form

$$H(\{Q\},\{P\}) = H_o(\{P\}) + V(\{Q\}) \qquad (2.1)$$

(with H_o e.g. a quadratic form in {P} the canonical momenta, V some potential only depending on the coordinates {Q}). The existence of solutions of the Hamilton equations of motion is <u>not</u> discussed at all. We limit our attention to the construction and analysis of the <u>Gibbs equilibrium states</u> which are formally given by

$$Z^{-1} e^{-\beta H(\{Q\},\{P\})} d\{Q\}d\{P\} =$$
$$\qquad \qquad (2.2)$$
$$= Z_o^{-1} e^{-\beta H_o(\{P\})} d\{P\} \cdot Z_V^{-1} e^{-\beta V(\{Q\})} d\{Q\},$$

where d{P} d{Q} is some factorizing a priori measure on Γ. We note that this state factorizes with respect to {P}, {Q}, (and that, for H_o a quadratic form in {P}, the first factor is simply a Gaussian measure, denoted $d\wp_o(\{P\})$. If F is some observable, i.e. F = F ({Q},{P}), a Σ_Γ- measurable function on Γ, we define

$$\hat{F}(\{Q\}) = \int F(\{Q\},\{P\}) \, d\wp_o(\{P\})$$

The expectation of F in the Gibbs state is then formally given by

$$z^{-1} \int \hat{F}(\{Q\}) \ e^{-\beta V(\{Q\})} \ d\{Q\} \qquad (2.3)$$

These observations permit us to eliminate the discussion of the momenta {P} completely, and hence forth we limit our attention to the construction and analysis of

$$Z_V^{-1} \ e^{-\beta V(\{Q\})} \ d\{Q\}$$

The observables of the system are functions $F(\{Q\})$ of the coordinates {Q} alone. We also change our notations:

$$\{\sigma\} \equiv \{Q\}$$

$$(2.4)$$

$$H \equiv H(\{\sigma\}) \equiv V(\{Q\})$$

For details about the foundations of classical equilibrium statistical mechanics see $[R]$.

The class of model systems the Gibbs states of which we are going to analyze consists of classical lattice systems:

These are systems on a cubic lattice $\mathbb{Z}^\nu$ with lattice constant $\delta \geq 0$; (unless otherwise stated $\delta = 1$). The observables of these systems are functions of classical "spin" random variables $\{\sigma_\alpha\}_{\alpha \in \mathbb{Z}^\nu}$: with each site $\alpha \in \mathbb{Z}^\nu$ there is associated a random variable σ_α with values in $\mathbb{R}^N$; $N = 1,2,\ldots$ is the number of components, and σ_α may be interpreted as a classical spin (for N=3) or as the

<u>position coordinates of some family of oscillators</u> attached
to site α; (i.e. we are dealing with classical spin systems, or, interpreted differently, with anharmonic crystals).

   With each finite set $B \subset Z^\nu$ we associate the σ-
algebra Σ_B of Borel sets in $\underset{\alpha \in B}{X} \; R^N_{(\alpha)}$, (where $R^N_{(\alpha)} \simeq R^N$),
and with $B = Z^\nu$ the σ-algebra Σ of Borel cylinder sets
of $\underset{\alpha \in Z^\nu}{X} \; R^N_{(\alpha)}$.

   The role of the a priori measure $d\{\sigma\}$ is played
by a probability measure on Σ:

   Given some Borel probability measure $d\tilde{\lambda}$ on R^N -
the "<u>single spin distribution</u>" - we set

$$d\{\sigma\} = \prod_{\alpha \in Z^\nu} d\tilde{\lambda} \, (\sigma_\alpha) \qquad\qquad (2.4)$$

In order to be able to specify the Hamilton function we
must introduce <u>periodic boundary conditions</u>:

   Let Λ be a finite rectangle in Z^ν;

$$\Lambda = \{\alpha \in Z^\nu : \alpha = \alpha_0 + l_1 \delta_1 + \ldots + l_\nu \delta_\nu\}, \qquad (2.5)$$

where α_0 is some fixed lattice vector, δ_i is the unit
vector with components δ_{ij}, $j=1,\ldots,\nu$, and the l_i's are
integers with $0 \le l_i \le L_i$, for some positive integers
L_i, $i=1,\ldots,\nu$.

   A point α of the form

$$\alpha = \alpha_0 + l_1 \delta_1 + \ldots + (L_j+1) \delta_j + \ldots + l_\nu \delta_\nu \text{ is indentified}$$

with

$$\hat{\alpha} = \alpha_0 + 1_1\delta_1 + \ldots + 0\delta_j + \ldots + 1_\nu\delta_\nu \in \Lambda,$$

in the sense that $\sigma_\alpha \equiv \sigma_{\hat{\alpha}}$.

Given α, we set $\alpha_+^i \equiv \alpha + \delta_i$. $\qquad\qquad$ (2.6)

The jth component of σ_α is denoted σ_α^j,

and $\partial^i \sigma_\alpha^j \equiv - (\sigma_\alpha^j - \sigma_{\alpha_+^i}^j)$,

$$F_\alpha^{ij} \equiv \partial^i \sigma_\alpha^j - \partial^j \sigma_\alpha^i \qquad\qquad (2.7)$$

With Λ we associate a cutoff Hamilton function

$$H_\Lambda^h(\{\sigma\}) = \frac{J}{2} \sum_{\alpha\in\Lambda} \sum_{i=1}^{\nu} (\partial^i \sigma_\alpha)^2 + h \cdot (\sum_{\alpha\in\Lambda} \sigma_\alpha) \qquad (2.8)$$

(In the analysis of lattice field theories involving vector fields one also encounters Hamilton functions of the form

$$H_\Lambda^F(\{\sigma\}) = \frac{J}{2} \sum_{\alpha\in\Lambda} F_\alpha^{ij} F_{ij,\alpha}). \qquad\qquad (2.9)$$

The constant $J > 0$ is the nearest neighbor-ferromagnetic coupling (related to the field strength in the case of lattice field theories [GRS]) and $h \in \mathbb{R}^N$ is the "external (magnetic) field".

The finite volume equilibrium state of the system defined by (2.4) and (2.8) with periodic b.c. at $\partial\Lambda$ (the boundary of Λ) at inverse temperature = 1 is given in terms of a probability measure $d\mu_\Lambda^h$ on Σ_Λ:

$$d\mu_\Lambda^h \left(\{\sigma\}\right) = Z_\Lambda^{-1} \exp\left[-H_\Lambda^h\{\sigma\}\right] \prod_{\alpha \in \Lambda} d\tilde{\lambda}\left(\sigma_\alpha\right)$$

$$(2.10)$$

$$= Z_\Lambda^{-1} \exp\left[\mathcal{J} \sum_{\alpha \in \Lambda} \sum_{i=1}^{\nu} \sigma_\alpha \cdot \sigma_{\alpha_+^i}\right] \prod_{\alpha \in \Lambda} d\lambda\left(\sigma_\alpha\right),$$

$$\left. \begin{array}{l} \text{where } d\lambda(\sigma) = e^{-\nu J \sigma^2} \quad d\tilde{\lambda}(\sigma), \\[2mm] \text{and} \quad Z_\Lambda = \int \exp\left[-H_\Lambda^h\{\sigma\}\right] \prod_{\alpha \in \Lambda} d\tilde{\lambda}\left(\sigma_\alpha\right) \end{array} \right\}$$

$$(2.11)$$

which is assumed to be __finite__ for all h; Z_Λ is the __partition function__ of the system.

The second expression for $d\mu_\Lambda^h$ in (2.10) clearly exhibits the ferromagnetic nature of the coupling between nearest neighbor spins.

The cutoff free energy density (in the following called __pressure__) is defined as

$$\alpha_\Lambda = \alpha_\Lambda(J,h) = \frac{1}{|\Lambda|} \log Z_\Lambda,$$

$$(2.12)$$

where $|\Lambda|$ is the number of points in Λ

Physical interpretation:

For $\nu=3$, $N=(1,2),3$ the measures $d\mu_\Lambda^h$ are the finite volume (cutoff) Gibbs states of somewhat naive models for classical ferromagnets or classical, anharmonic crystals in thermal equilibrium (at inverse temperature 1)

For suitable choices of the single spin distribution $d\tilde{\lambda}$ and variable lattice constants (possibly depending on the direction in the lattice) they describe the lattice approximations [GRS] to the space-time cutoff interacting measures of Bose quantum field theories in the

Euclidean description, see e.g. [E,Si] for extensive information about these theories, or of quantum crystals
in the imaginary time description, (see Section 8).

Examples:

1. Ising model:

$$N = 1, \quad \nu \geq 2, \quad d\lambda(\sigma) = \frac{1}{2}\{\delta(\sigma+1)+\delta(\sigma-1)\}d\sigma \qquad (2.13)$$

(or, more generally, $d\lambda(\sigma)$ some $\sigma \to -\sigma$ invariant probability
measure on $\mathbb{R}, \neq \delta(\sigma)d\sigma$).

2. Classical rotator:

$$N = 2, \quad \nu \geq 3, \quad d\lambda(\sigma) = \delta(|\sigma|-1)d^2\sigma \qquad (2.14)$$

3. Classical Heisenberg model:

$$N = 3, \quad \nu \geq 3, \quad d\lambda(\sigma) = \delta(|\sigma|-1)d^3\sigma \qquad (2.15)$$

4. $(\vec{\phi} \cdot \vec{\phi})^2_\nu$ -field theory:

$N = 1,2,3,\ldots, \quad \nu = 2$ or 3.

In this example it is assumed that the lattice constant $\delta > 0$
is variable. For conventional reasons we change our notation:

$\sigma \equiv \vec{\phi} = (\phi^1,\ldots,\phi^N)$. We redefine ∂^i, namely

$$\partial^i \phi^j_\alpha = -\frac{1}{\delta}(\phi^j_\alpha - \phi^j_{\alpha+i}), \qquad (2.16)$$

and we replace $\sum\limits_{\alpha \in \Lambda} \cdot$ by $\sum\limits_{\alpha \in \Lambda} \delta^3$.

We set J = 1 and

$$d\tilde{\lambda}(\vec{\phi}) = e^{-L_I(\vec{\phi})} d^N\phi, \text{ where}$$

(2.17)

$$L_I(\vec{\phi}) = \lambda(\vec{\phi}\cdot\vec{\phi})^2 - (M^2(\lambda,\delta,\nu)+\sigma) \vec{\phi}\cdot\vec{\phi}-\vec{h}\cdot\vec{\phi}+C$$

Here $\lambda > 0$, and $C = C(\lambda,\sigma,\vec{h},\delta,\nu)$ is chosen such that $d\tilde{\lambda}$ is a probability measure; the term $-M^2(\lambda,\delta,\nu) \vec{\phi}\cdot\vec{\phi}$ imple- ments <u>Wick ordering</u> (see e.g. [Si,Pa]) and, for $\nu=3$, a <u>mass renormalization</u> (see . [GJ1, Pa, SeSi]); it is chosen such that for $\nu=2$, [GRS], and $\nu=3$, [Pa], the family of measures $\{d\mu_\Lambda^{\vec{h}} \equiv d\mu_{\Lambda,\delta}^{\vec{h}}\}$ has a well defined limiting measure d $\nu_\Lambda^{\vec{h}}$, as $\delta\searrow 0$. Note that $M^2(\lambda,\delta,\nu)$ is <u>independent</u> of σ!

The moments of the limiting measure d $\nu_\Lambda^{\vec{h}}$ are the space-time cutoff Euclidean Green's- or Schwinger func- tions (EGF's) of the $(\vec{\phi}\cdot\vec{\phi})_\nu^2$ - quantum field model; [GRS]. For $\nu=2$ space-time dimensions the limit as $\delta\searrow 0$ has been proven to exist in [GRS]. For $\nu=3$ the proof of the same result is a rather formidable task and relies heavily on the extremely difficult constructions of [GJ1]; see [Pa].

5. sine - Gordon theory:

In this example the starting point is as in 4., but $\nu=2$, N=1, i.e. $\vec{\phi} = \phi$, and

$$L_I(\phi) = \lambda(\delta) \cdot \cos(\epsilon\phi+\theta),$$

(2.18)

with $\theta\in[0,2\pi]$, $0 < \epsilon < 2\sqrt{\pi}$, and $\lambda(\delta)\in \mathbb{R}$ some coupling constant that tends to ∞, as $\delta \searrow 0$ (implementing Wick ordering, [F1]). In this case the existence of a limit (known to be unique for $\varepsilon < 4 \cdot \pi^{-1/2}$) has been shown in [F1]. The moments of the limiting measure $d\nu_\Lambda(\phi)$ are the space-time cutoff EGF's of the sine-Gordon theory in two space-time dimensions, [F1,2,FSe].

6. Lattice vector field theory:

In this example $\nu = N = 4$; $\vec{\phi} = \vec{A} = (A^0, \ldots, A^3)$; $\delta > 0$; $H_\Lambda^F = H_\Lambda^F(\{A\})$ such as defined in (2.9);

$$d\tilde{\lambda}(\vec{A}) = e^{-L_I(\vec{A})} d^4A, \text{ with}$$

$$\left. \begin{array}{l} \text{e.g.} \\ L_I(\vec{A}) = \lambda(\vec{A} \cdot \vec{A})^2 - \sigma(\vec{A} \cdot \vec{A}), \ \lambda > 0 . \end{array} \right\} \tag{2.19}$$

Throughout the text some other models (classical lattice models and Euclidean field theories) which are related to models 1 - 6 will be mentioned and used to illustrate various concepts.

Very useful information on such models may be found in [R], [H], [E], [Si], [FSS].

We are now prepared to summarize some of the main results discussed in Part 1. All of them concern existence of the thermodynamic limit, new bounds on correlations ("Gaussian domination", [FSS]) and existence of phase transitions and symmetry breaking. (In the following sections we shall outline proofs for most of these results, omitting here and there some fine points of rigor which can be found in the references quoted).

In Section 3 we review the two main general methods for proving the existence of phase transitions in ferromagnetic systems:

1. Infrared (Gaussian) domination, [FSS]:

A direct approach to proving the existence of long range order, based on an analysis of two point correlation functions.

This method is applicable to multi-component spin systems with (or without any) discrete or continuous internal symmetries, provided the dimension of the underlying lattice (or space-time) is at least 3.

2. The Peierls argument, [Pe]:

This approach is based on estimating spin flip probabilities and the statistical weight of contours separating regions of opposite spin orientation.

The Peierls argument is applicable to one-component or anisotropic multi-component spin systems (with or without spin flip, i.e. $\sigma \rightarrow -\sigma$ symmetry), provided the dimension of the under lying lattice (or space-time) is at least 2.

Since method 1 fails in two dimensions, the most useful and efficient applications of the Peierls argument are to two dimensional systems, (such as the ϕ_2^4 - or pseudoscalar Yukawa$_2$ quantum field models; see Section 7, [GJS2], [F3]).

In Section 4 we discuss the existence of the thermodynamic limit by deriving new bounds on correlation functions of classical, nearest neighbor-ferromagnetic systems. In particular, we obtain a priori upper bounds on the connected (= truncated) correlation functions and the structure of their infrared singularities, e.g. an $O(k^{-2})$-bound on the connected two point function depending only on the size of the nearest neighbor-ferromagnetic coupling constant J; see [FSS]. These bounds supply the essential input to method 1 of proving occurence of phase transitions.

Definition:

Let K_h denote the class of all single spin distributions $d\tilde{\lambda}$ defined by

$$\{d\tilde{\lambda} : \exists \; \varepsilon > o \text{ such that } \int e^{\varepsilon|\sigma|+h\cdot\sigma} \; d\tilde{\lambda}(\sigma) < \infty\} \qquad (2.20)$$

Theorem A: (Existence of the thermodynamic limit)

For $\nu = 1,2,\ldots$, $N = 1,2\ldots$, a Hamilton function H_Λ^h given by (2.8) and some single spin distribution $d\tilde{\lambda} \in K_h$ there exists a sequence of rectangles (or cubes) $\{\Lambda_n\}_{n=o}^\infty$ such that

$$d\mu^h (\{\sigma\}) \equiv \lim_{n\to\infty} d\mu_{\Lambda_n}^h (\{\sigma\}) \qquad (2.21)$$

exists, in the sense that the characteristic functionals and all the moments of the measures $\{d\mu_{\Lambda_n}^h\}_{n=o}^\infty$ converge, as $n \to \infty$. The limiting measure $d\mu^h$ is a probability measure on Σ.

The expectation (state) determined by $d\mu^h$ is de-noted by $<->\;\equiv\;<->^{J,h}$. 　　　　　　　　　　(2.22)

Theorem B: (Gaussian domination, [FSS])

Under the assumptions of Theorem A

$$\left\langle e^{\alpha,i\; g_i(\alpha)\cdot\partial^i\sigma_\alpha}\right\rangle^{J,h}\;\leq\;e^{\frac{1}{2J}\;\alpha,i\;|\;g_i(\alpha)\;|^2}\;,\qquad (2.23)$$

__independently of__ $d\tilde{\lambda}$ __and__ h.

Theorem C: $(O(k^{-2})$ bound, [FSS])

Under the assumptions of Theorem A, the Fourier transform $d\omega(k)$ of the two point correlation function $<\sigma_\bullet\cdot\sigma_\alpha>$ has the form

$$\left.\begin{aligned} d\omega(k) &= \left[\alpha\;\delta_0\;(k)\;+\;F^C\;(k)\right]\;d^3k,\\[1mm] \text{where } 0 &\leq F^C(k)\;\leq\;\frac{\text{const.}}{J\cdot k^2}; \end{aligned}\right\}\qquad (2.24)$$

independently of $d\tilde{\lambda}$ and h; (α is the long range order).

Theorem C is a direct corollory of Theorem B. Although the proofs of Theorem B and C are elementary and based on a well known tool, the transfer matrix forma-lism, these results seem to appear for the first time in [FSS]. They form the technical core of method 1.

In __Section 5__ we combine __method 1__ with the bounds of Theorem C to prove occurence of phase transitions in

Models 2,3 and 6 and others.

In particular we prove:

Theorem D: (Phase transition in the classical N-vector
           models, [FSS])
Let $\nu \geq 3$, N = 1,2,3,... and

$$d\lambda(\sigma) \overset{e.g.}{=} \delta(|\sigma| - 1) \, d^N \sigma \qquad\qquad (2.25)$$

Then there exists some $J_c < \infty$ such that, for all $J > J_c$,
there is long range order (i.e. $\alpha > 0$). The state
$< - >^{J,0}$ is a mixture. There are at least S^{N-1} - many
pure phases, and the internal symmetry group O(N) is
broken in at least S^{N-1} - many pure phases (i.e. there
is spontaneous magnetization).

Remarks:

"S^{N-1} - many pure phases" is to be read as "as
many pure phases as there are points on the unit sphere
S^{N-1} in N dimensions". The group O(1) consists of the
two elements $\sigma \to \sigma$ and $\sigma \to -\sigma$.

Theorem D can be extended to all single spin dis-
tributions $d\lambda \in K_0$ invariant under O(N) and $\neq \delta_0(\sigma) d^N\sigma$,
and, for N = 1, to a class of single spin distributions
$d\lambda \in \underset{|h| < \infty}{\bigcup} K_h$ without any symmetry, at all, and to ferromagnets with many body and long range interactions and
impurities (which break translation invariance); see [FSS].

Finally we remark that in $\nu \leq 2$ dimensions spontaneous O(N) - breaking is impossible; (Mermin's theorem,

[M]). We conjecture however that, for $\nu = 2$, there exists
a critical point $J_c < \infty$ at which the correlation length
(the inverse of the "physical mass = exponential decay
rate of correlations") diverges. Moreover it is known
that, for $\nu = 2$, an <u>arbitrarily small anisotropy</u> suffices
to generate a phase transition for sufficiently large J.

A combination of Theorem C, Mermin's theorem and
correlation inequalities yields various bounds on critical exponents; see e.g. [LP],[FSS].

In <u>Section 6</u> we discuss the $(\vec{\phi}\cdot\vec{\phi})^2_3$ - quantum field
model (Model 4) in the continuum limit $\delta = 0$ (and properly
renormalized, [GJ1]). We prove

<u>Theorem E:</u> (Phase transitions for $(\vec{\phi}\cdot\vec{\phi})^2_3$, [FSS])

For $\nu = 3$, $N = 1,2,3$ $(4,5,\ldots)$, $\delta = 0$, $\lambda > 0$ fixed
and $\vec{h} = 0$

(1) there exists a finite constant $\sigma_c = \sigma_c(\lambda)$ such that
    for all $\sigma > \sigma_c$, the long range order α is positive
    and the physical vacuum of the theory is degenerate;

(2) for $N = 1,2,3$, all $\vec{h} = h\cdot\vec{e}$, where $\vec{e}$ is an arbitrary
    unit vector in $\Re^N$, $h \neq 0$, all Wightman axioms
    [StW,Jo,OS] are satisfied, including uniqueness of
    the vacuum[1], and

154

$$\lim_{h \to 0} \int d\mu^{\vec{h}}(\vec{\phi}) \; \vec{\phi} \; (o) = M^* \vec{e}, \; M^* > 0;$$

for N = 2,3 and h = O there exist N - 1 Goldstone bosons; (zero mass, scalar one particle states).[1]

In <u>Section 7</u> we give a simplified proof for the occurence of phase transitions in the ϕ_2^4 quantum field model (Model 4 for ν = 2 and N = 1), [GJS2]. The proof is based on the Peierls argument in the form of [GJS2]; see Section 3. Our techniques can be extended without difficulties to general P(ϕ)$_2$ models [E,Si], (where P is a positive, "almost even" but not necessarily even polynomial) the pseudoscalar Yukawa- [Y$_2$] and the sine-Gordon model (Model 5); see [F3] and Section 7.

What we shall show in Sections 6 and 7 is, in essence, that the following principle "governs" the occurence of phase transitions in super renormalizable field theories in two or three space-time dimensions:

---

[1] this result is due to [F4] based on results of [Fe O, Ma Sé].

Principle: [F3, FSS]

Let V be some superrenormalizable interaction term which depends on (possibly among others) a (pseudo-) scalar Bose field $\vec{\phi}$ and is "almost" invariant under a substitution transforming $\vec{\phi}$ into $-\vec{\phi}$.

Assume that, associated with the formal Lagrangean

$$V(\vec{\phi}) - \frac{\sigma}{2} : \vec{\phi} \cdot \vec{\phi} :,$$

and arbitrary $\sigma > 0$, there exists a theory satisfying all Wightman axioms with the possible exceptions of Lorentz covariance and uniqueness of the vacuum, (put differently, $V(\vec{\phi}) - \frac{\sigma}{2} : \vec{\phi} \cdot \vec{\phi} :$ is stable, or $V(\vec{\phi})$ stabilizes $- \frac{\sigma}{2} : \vec{\phi} \cdot \vec{\phi} :$, for all $\sigma > 0$). Then (a) for N=1 (i.e. $\vec{\phi}=\phi$) and σ large enough and (b) for N=1,2,3..., large enough and in three space-time dimensions, the physical vacuum is degenerate, i.e. there is a phase transition.

We will see in Part 2 that in two space-time dimensions this principle has an interesting consequence: the existence of a phase transition implies the existence of new, charged superselection ('soliton') sectors, [F7].

In Section 8 we mention some results for classical ferromagnetic lattice systems that follow from the Peierls argument, and we draw some general conclusions.

Section 3:

TWO GENERAL METHODS IN THE THEORY OF PHASE TRANSITIONS

The two methods of proving the existence of phase
transitions we explain in this section, Infrared Domination and the Peierls Argument, are of a general nature
(as opposed to methods based on deriving exact solutions,
etc.; see [PT]). In principle they are applicable to
general systems with infinitely many degrees of freedom
in a space (-time) represented by $\mathcal{R}^\nu$, or by some ν -
dimensional, regular lattice, typically $\mathbb{Z}^\nu$, and $\nu \geq 3$,
$\nu \geq 2$, respectively. In practice, however, successful
applications have thus far been limited to (classical),
ferromagnetic systems, (a class which is larger than it
may perhaps seem).

Our systems are described in terms of some C^*-algebra
$\mathcal{A}$ of local observables and a state $< - >$ on $\mathcal{A}$[1]; ($\mathcal{A}$
may be an algebra of bounded functions of some fields,
an abelian algebra of random variables, etc. In statistical mechanics the state $< - >$ is typically a Gibbs
equilibrium state, in quantum field theory e.g. a Euclidean vacuum expectation value).

In the following we may imagine that the under-
lying space is $\mathcal{R}^\nu$, (since a ν-dimensional lattice can
be embedded in $\mathcal{R}^\nu$). We assume that $\mathcal{A}$ has a local
structure:

---

[1] The following general remarks are not to be read
with too much emphasis on precision.

With each bounded, open subset $B \subset \mathcal{R}^{\nu}$ there is associated a local algebra $\mathcal{A}(B)$, and if B_1 and B_2 are disjoint - (in local, relativistic field theory, if B_1 and
B_2 are space-like separated; a case not considered in
the following) - all the elements of $\mathcal{A}(B_1)$ commute
with all the elements of $\mathcal{A}(B_2)$; $\mathcal{A}$ is assumed to be the
norm closure of $\bigcup_{B \subset \mathcal{R}^{\nu}} \mathcal{A}(B)$.

From $\mathcal{A}$ and $< - >$ we obtain a Hilbert space $\mathcal{H}$, a
cyclic vector Ω, and a representation π of $\mathcal{A}$ on $\mathcal{H}$, by
the G,N.S. construction; see e.g. $[R,H]$.

Let $\overline{\mathcal{A}}(\sim B)$ denote the __weak closure__ (on $\mathcal{H}$) of
$\bigcup_{B_1 \subset \sim B} \pi(\mathcal{A}(B_1))$. Furthermore, let

$$\overline{\mathcal{A}}_{\infty} = \bigcap_{B \subset \mathcal{R}^{\nu}} \overline{\mathcal{A}}(\sim B) , \tag{3.1}$$

and $\mathcal{H}_{\infty} = \{A\Omega : A \in \overline{\mathcal{A}}_{\infty}\}^{-}$ \hfill (3.2)

(where $\{\}^{-}$ denotes the strong closure of $\{\}$).
It follows from the cyclicity of Ω for $\pi(\mathcal{A})$ and from
the definition (3.1) of $\overline{\mathcal{A}}_{\infty}$ that

$$\overline{\mathcal{A}}_{\infty} = \{\lambda I : \lambda \in \mathbb{C}\} <=> \mathcal{H}_{\infty} = \{\lambda \Omega : \lambda \in \mathbb{C}\} \tag{3.3}$$

__Definition 3.1:__

We say that $< - >$ is a __pure phase state__ if and only
if dim $\mathcal{H}_{\infty} = 1$ (i.e. if $\overline{\mathcal{A}}_{\infty}$ consists only of multiples
of the identify I). If dim $\mathcal{H}_{\infty} > 1$ then $< - >$ is not a
pure phase state and "there is a phase transition"

See [Fö] for a discussion of these notions in classical statistical mechanics and [F5] for the case of Euclidean field theory.

Whether definition 3.1 is reasonable under very general circumstances is beyond the authors knowledge. In the framework of the models we are going to consider it is.

Let P denote the selfadjoint projection onto the orthogonal complement $\mathcal{H}_\infty^\perp$. We define the truncated expectation of the product A^*A, $A \in \mathcal{O}$, in the state $< - >$ by

$$<A^*A>^T = <A^*PA> \equiv (\pi(A)\Omega, \, P\pi(A)\Omega) \qquad (3.4)$$

In a pure phase, i.e. if $< - >$ is a pure phase state,

$$<A^*A>^T = <A^*A> - |<A>|^2 . \qquad (3.5)$$

Hence if, for some $A \in \mathcal{O}$,

$$\bar{\alpha}_A \equiv <A^*A> - |<A>|^2 - <A^*A>^T > 0 \qquad (3.6)$$

then $< - >$ is not a pure phase (state), i.e. there is a phase transition. The number $\bar{\alpha}_A$ is called the long-range order (associated with A and $< - >$).

For a proof that there is a phase transition it thus suffices to choose a suitable local observable A (in the models A is typically a classical spin or some function of the Euclidean field) and to derive

(I) an <u>upper bound</u>

$$<A^*A>^T \leq C_1$$

(II) <u>a lower bound</u>

$$<A^*A> - |<A>|^2 \geq C_2$$

(3.7)

such that $C_2 - C_1 > 0$.

Then $\bar{\alpha}_A \geq C_2 - C_1 > 0$.   (3.8)

_____

So far we have merely discussed some generalities about phase transitions. Next we want to explain general methods for deriving the crucial inequality (3.8). It turns out that, in general, the <u>hard estimate</u> in the proof of (3.8) is (I).

In order to make these considerations more concrete we now assume that <u>space translations</u> (in the case of a lattice theory translations by lattice vectors) are represented by a group $\{\tau_x\}$ of * automorphisms of $\mathcal{O}$ with the property that

$$\tau_x(\mathcal{O}(B)) = \mathcal{O}(B_x),$$ (3.9)

where B_x is the translate of the region B by the vector x. We also assume in general that the state $< - >$ is <u>translation invariant</u>:

$$<\tau_x(A)> = <A>, \text{ for all } A \in \mathcal{O}$$ (3.10)

It follows from (3.10) that there exists a unitary group $\{U_x\}$ on $\mathcal{H}$ implementing $\{\tau_x\}$, i.e.

$$\tau_x(A) = U_x^* A U_x, \text{ for all } A \in \mathcal{A}. \tag{3.11}$$

We let $\mathcal{H}_I$ be the closed subspace of $\mathcal{H}$ consisting of vectors that are <u>invariant</u> under $\{U_x\}$. By a very straight forward argument it is seen that $\mathcal{H}_I$ is a <u>closed sub-space</u> of $\mathcal{H}_\infty$. Let P_I denote the selfadjoint projection on to $\mathcal{H}_I^\perp$. We set $A_x = \tau_x(A)$, $A = A_O$, $A \in \mathcal{A}$. The <u>connected expectation</u> of the product $A_y^* A_x$ is defined by

$$<A_y^* A_x>^C = <A_y^* P_I A_x> \equiv (U_y^* \pi(A)\Omega,\ P_I\ U_x^*\ \pi(A)\Omega). \tag{3.12}$$

Since $\mathcal{H}_I \subseteq \mathcal{H}_\infty$, $P_I \geq P$, so that

$$<A^*A>^C \geq <A^*A>^T \tag{3.13}$$

Thus, if

$$\alpha_A \equiv <A^*A> - |<A>|^2 - <A^*A>^C > 0, \tag{3.14}$$

then dim $\mathcal{H}_I > 1$, dim $\mathcal{H}_\infty \geq$ dim $\mathcal{H}_I > 1$, and $\bar{\alpha}_A > 0$.

See [R,H] for more information and references.

3.1 Infared domination, [FSS]:

This is the first of the announced two methods for proving the existence of phase transitions. Without loss of generality we may choose an observable A that can be approximated in norm by strictly local observables

"arbitrarily rapidly" and such that the Fourier transform $d\omega(k)$ of $<A^*A_x>$ has support in some compact set B.
(In the case of a theory on a lattice this is automatically satisfied; B is the first Brillouin zone. In the
continuum case, let $\overset{o}{A}$ be a strictly local observable,
h a Schwartz test function, the Fourier transform of
which has support in B, and define $A^{\#} \equiv \int h(x) \overset{o}{A}{}^{\#}_x$).
Obviously

$$<A^*_y A_x> = <A^*_y A_x>^C + <A^*_y (I-P_I) A_x>.$$

By the definition of P_I, $<A^*_y (I-P_I) A_x>$ is <u>independent</u>
of x and y.

Therefore

$$<\text{grad } A^*_y \cdot \text{grad } A_x> = <\text{grad } A^*_y \cdot \text{grad } A_x>^C \qquad (3.15)$$

(where, for a lattice theory, "grad" is the finite
difference gradient).

In the case of a relativistic field theory the results
of [AHR] prove that

$$<\text{grad } A^*(h) \cdot \text{grad } A(h)> \leq C \ ||h||^2_2 \qquad (3.16)$$

where $A^{\#}(h) = \int h(x) A^{\#}_x$, $||h||^2_2 = \int |h(x)|^2$, and, in the
case of a <u>canonical</u> Bose field theory, the constant C
is finite and <u>independent</u> of the specific model. The
analogy between canonical Bose field theories in the
Euclidean description and classical ferromagnets [GRS,
SG, DN] suggests that, for a suitable choice of A, inequality (3.16) holds in classical spin systems with

nearest neighbor-ferromagnetic couplings, with a constant C only depending on these couplings, but independent of the single spin distribution $d\tilde{\lambda}$. This is proven
in Section 4; see [FSS] for the original results.

We now assume inequality (3.16) and discuss its
implications. It is asserted that it immediately gives
an upper bound on $<A^*A>^C$, provided the dimension ν of
space is at least 3:

From (3.16) we obtain

$$d\omega(k) = \left[\gamma \delta_0(k) + F^C(k)\right]d^\nu k,$$

where $\gamma = |<A>|^2 + \alpha_A$, and   (3.17)

$$0 \leq F^C(k) \leq C k^{-2}$$

The proof is immediate, But from (3.17) it follows that

$$<A^*A>^C \leq C \int_B \frac{d^\nu k}{k^2}$$ (3.18)

which is finite for $\nu \geq 3!$ (For $\nu \leq 2$, k^{-2} is not integrable at $k = 0$).

In order to prove that $\alpha_A > 0$ it now suffices to show that

$$<A^*A> - |<A>|^2 > C \int \frac{d^\nu k}{k^2}$$ (3.19)

This estimate always depends on the characteristic
features of a specific model, in contrast to (3.16) so
that no general methods are available.

In a pure phase

$$|<A>|^2 = <A^*A> - <A^*A>^C$$

$$\geq <A^*A> - C \int \frac{d^\nu k}{k^2} \tag{3.20}$$

This inequality is often very useful for proving <u>dis-continuity</u> of <A>, e.g. "spontaneous magnetization", as some coupling constants of a given model are varied, which also proves the existence of a phase transition. We note that (for $A = A^*$) inequality (3.16) follows from

$$|<e^Z \text{ grad } A(h)_{>}| \leq e^{C'|Z|^2} ||h||_2^2 \tag{3.16'}$$

This is a consequence of analyticity in Z and the Cauchy estimate; see Section 4 and [FSS].

Finally we remark that the method for proving the existence of phase transitions consisting of inequalities (3.16) or (3.16') and (3.19) does not explicitly depend on the internal symmetries of the system described by $\mathcal{O}l$, < - >.

As phase transitions are often accompanied by the spontaneous breaking of such symmetries it is now clear - if not already obvious from (3.18) - that this method does in general not apply to two dimensional systems , as in two dimensions continuous symmetries cannot be broken; [M,E Sw].

———————

3.2 The Peierls argument, [Pe, GJS2]:

This is the second general method for proving occurence of phase transitions we propose to review. Let the algebra $\mathcal{O}$ and the state $< - >$ be as discussed above. We assume that the G.N.S. Hilbert space $\mathcal{H}$ reconstructed from $\mathcal{O}$ and $< - >$ is <u>separable</u>.

We cover $\mathcal{R}^\nu$ with a grid of mesh 1. Let $\mathcal{C}$ denote the family of all <u>unit cubes</u> of this grid, and let $\mathcal{B}$ be the collection of <u>all faces</u> of all unit cubes in $\mathcal{C}$.

As described above, with each $\square \in \mathcal{C}$ (or each finite union of cubes in $\mathcal{C}$) there is associated a local algebra $\mathcal{O}(\square)$, and

$$\tau_x(\mathcal{O}(\square)) = \mathcal{O}(\square_x), \text{ for all } x \in \mathbf{Z}^\nu, \tag{3.21}$$

where $\square_x$ is the translate of $\square$ by the vector x. Let $P_1, P_2, \ldots, P_m$ be $m \geq 2$ <u>commuting</u>, <u>selfadjoint</u> projections in $\mathcal{O}(\square_o)$ with

$$P_1 + P_2 + \ldots + P_m = I \tag{3.22}$$

and $<P_i(x)> \geq \varepsilon > 0$, $(\varepsilon \leq \frac{1}{m})$, $\tag{3.23}$

for all i = 1,....,m, and all x in $\mathbf{Z}^\nu$. Here $P_i(x) \equiv \tau_x(P_i) \in \mathcal{O}(\square_x)$. (If $< - >$ is translation invariant it suffices to assume (3.23) for one x).

Next, suppose we are able to prove that, for all (sufficiently large) x and y and all $i \neq j$

$$< P_i(x) \; P_j(y) > \; \leq \; \delta \; \epsilon^2 \qquad\qquad (3.24)$$

for some constant $\delta < 1$.

Then, for any sequence $\{x_n\}_{n=0}^{\infty}$ of points diverging to ∞
for which

$$w - \lim_{n \to \infty} P_i(x_n) \text{ exists, for some } i,$$

$$w - \lim_{n \to \infty} P_i(x_n) \neq \{\lim_{n \to \infty} \; < P_i(x_n) >\} \; I, \qquad (3.25)$$

an immediate consequence of (3.23) and (3.24).

Since $\mathcal{H}$ is <u>separable</u>, we can always choose a sequence
$\{x_n\}_{n=0}^{\infty}$ diverging to ∞ such that

$$\pi_i \equiv w\text{-}\lim_{n \to \infty} P_i(x_n) \qquad\qquad (3.26)$$

exists, for all $i = 1, \ldots, m$.

Clearly $\pi_i \in \bar{\mathcal{O}}_\infty$, so that $\psi_i \equiv \pi_i \Omega \in \mathcal{H}_\infty$. We therefore
conclude from (3.25) that the state $< - >$ is <u>not</u> a pure
phase in the sense that dim $\mathcal{H}_\infty > 1$.

From (3.23) and (3.26) we have that

$$(\Omega, \psi_i) \; = \; < \pi_i > \; \geq \; \epsilon, \text{ for all } i,$$

so that $(\psi_i, \psi_i) \geq \epsilon^2$, $\qquad\qquad (3.27)$

in particular $\psi_i \neq \vec{0}$, for all i.

Furthermore for all $i \neq j$,

$$(\psi_i, \psi_j) = \lim_{n \to \infty} \left[\lim_{k \to \infty} <P_i(x_n) P_j(x_k)> \right] \qquad (3.28)$$

$$\leq \delta \, \varepsilon^2, \text{ by } (3.24)$$

and, since

$$<P_i(x_n) \, P_j(x_k)> = <P_i(x_n)^* \, P_j(x_k) \, P_i(x_n)> \geq 0,$$

$$(\psi_i, \psi_j) \geq 0.$$

A little elementary geometry shows that (3.27) and (3.28) yield the following

Proposition 3.1:

   If the constant δ is smaller than some constant δ_m (of purely geometric origin) then the m vectors $\{\psi_i\}_{i=1}^m$ are linearly independent and dim $\mathcal{H}_\infty \geq m$.

   This shows that, for $\delta < \delta_m$, the state $< - >$ is a mixture (convex combination) of at least m pure phases.

   The Peierls argument supplies a systematic way of estimating $<P_i(x) P_j(y)>$, $i \neq j$, which we now explain:

   Let $\Lambda_{(x,y)}$ be some rectangle of unit cubes containing the smallest rectangle of unit cubes that contains $\square_x$ and $\square_y$.

   A configuration C is a function on $\mathbb{Z}^\nu \cap \Lambda_{(x,y)}$ with values in $\{1, \ldots, m\}$ such that $C(x) = i$ and $C(y) = j$.

$$(3.29)$$

Since $\sum\limits_{\ell=1}^{m} P_\ell (x) = I$, for all x,

$$\langle P_i (x) \ P_j (y) \rangle = \langle P_i (x) \ P_j (y) \prod_{\substack{z \in \mathbf{Z}^\nu \cap \Lambda_{(x,y)} \\ z \neq x,y}} (\sum_{\ell=1}^{m} P_\ell (z))) \rangle$$

$$= \sum_C \langle \prod_{z \in \mathbf{Z}^\nu \cap \Lambda_{(x,y)}} P_{C(z)} (z) \rangle \qquad (3.30)$$

Given $\Lambda_{(x,y)}$, we define a family $\Gamma_{(x,y)}$ of <u>connected</u> surfaces γ in $\mathcal{B}$ (=all faces of unit cubes in $\mathcal{C}$) with the properties that

(1) $\gamma \subset \Lambda_{(x,y)}$, $\gamma \cap \partial \Lambda_{(x,y)} = \emptyset$

$\left.\begin{array}{l} \\ \\ \\ \\ \\ \\ \end{array}\right\} (3.31)$

(2) γ decomposes $\Lambda_{(x,y)}$ into precisely two

disjoint subsets $B_x (\gamma) \supset \Box_x$ and $B_y (\gamma) \supset \Box_y$.

An element of $\Gamma_{(x,y)}$ is called a <u>contour</u>.

If γ is a contour in $\Gamma_{(x,y)}$ we define $N(\gamma)$ to be the set of <u>all</u> nearest neighbor cubes in $\mathcal{C}$ with <u>one common face</u> in γ.

Recalling that in (3.30) $i \neq j$, we conclude that, given a configuration C, there exists a contour $\gamma(C) \in \Gamma_{(x,y)}$ with the properties that

(1) for all $(\square_z, \square_{z'}) \in N(\gamma(C))$ with

$\quad \square_z \subset B_x (\gamma(C))$

$\quad\quad C(z) = i, \quad C(z') \neq i$

(2) There exists a connected set $B_C \subset B_x (\gamma(C))$

$\quad$ such that $\square_x \subset B_C$, and $C(z) = i$, for all

$\quad \square_z \subset B_C$, and $\partial B_C \supset \gamma(C)$.

$\hspace{9cm} (3.32)$

(See also Fig. 3 below).

We may now derive an upper bound for the r.h.s. of (3.30):

$$\langle P_i(x)\, P_j(y) \rangle = \sum_C \left\langle \prod_{z \in \mathbb{Z}^\nu \cap \Lambda}^{\checkmark} {}_{(x,y)} P_{C(z)}(z) \right\rangle$$

$$= \sum_{\gamma \in \Gamma_{(x,y)}} \sum_{\{C : \gamma(C) = \gamma\}} \left\langle \prod_{z \in \mathbb{Z}^\nu \cap \Lambda}{}_{(x,y)} P_{C(z)}(z) \right\rangle$$

$$\leq \sum_{\gamma \in \Gamma_{(x,y)}} \sum_{C(z') \neq i} \left\langle \prod_{(\square_z, \square_{z'}) \in N(\gamma)} P_i(z) P_{C(z')}(z') \right\rangle .$$

$\hspace{9cm} (3.33)$

For the <u>proof</u> of (3.33) we apply the definition of γ(C),
use the fact that

$$\{P_\ell(z) : z \in \mathbf{Z}^\nu \cap \Lambda_{(x,y)}, \; \ell=1,\ldots,m\}$$

is a family of commuting, positive, selfadjoint
operators with $0 \leq P_\ell(z) \leq I$, and finally recall that

$$A \, B \leq A' \, B'$$

if A,B, A' and B' are commuting, positive bounded
operators with $0 \leq A \leq A'$ and $0 \leq B \leq B'$. Let $|\gamma|$ denote
the total surface of γ, (i.e. the number of faces in γ).

<u>Theorem 3.2:</u> (see [GJS2])

Assume that, for all $\Lambda_{(x,y)}$ such as defined above and
all contours $\gamma \in \Gamma_{(x,y)}$, and with

$$C(z) = i \neq C(z'), \text{ for all } (\Box_z, \Box_{z'}) \in N(\gamma)$$

$$\left\langle \prod_{(\Box_z, \Box_{z'}) \in N(\gamma)} P_i(z) \, P_{C(z')}(z') \right\rangle \leq e^{-K|\gamma|} \qquad (3.34)$$

for some constant $K \geq K(\nu,m,\varepsilon,\delta)$, where $K(\nu,m,\varepsilon,\delta)$ is
a fixed constant <u>only depending</u> on the dimension ν of
space, the number m of projections and two positive
numbers ε and δ (see (3.23) and (3.24)).

Then $\left\langle P_i(x) \, P_j(y) \right\rangle \leq \delta \, \varepsilon^2$, for all $j \neq i$. \qquad (3.35)

For $K > K(\nu,m,\varepsilon,\delta_m)$ the state < - > is a mixture of at least m pure phases; (see Proposition 3.1).

Remarks:

For the purposes of these lectures it suffices to prove Theorem 3.1 for the case $\nu = m = 2$ considered in [GJS2]. We remark however that the analysis of the general case is perhaps a little more than an academic exercise: It is important in the analysis of multi-dimensional systems which are expected to have triple- or m-tuple points (i.e., for certain choices of the parameters of the system, it is expected to have at least three, or m, pure phases).

We also note that method 3.1 (infrared domination) generally only proves existence of phase transitions without proving more than obvious information (derived from the structure of the internal symmetries) about the manifold of pure phases. This is not so in method 3.2 (the Peierls argument), as one learns from Proposition 3.1 and Theorem 3.2. Sometimes it is necessary to combine (among others) both methods to get the desired information on the structure of the pure phases of a system.

Proof of Theorem 3.2 (for $\nu=m=2$):

For the purpose of good intuition we write $P_1 = P_+$, $P_2 = P_-$ ("spin up-down").

We now pick two arbitrary points x and y in $\mathbb{Z}^2$. It is then to be proved that (3.34) implies (3.35). The family $\mathcal{C}$ is now a family of unit squares, the set $\mathcal{B}$

of their faces a family of <u>bonds</u>. Let $R_{(x,y)}$ be the smallest rectangle of unit squares containing the squares $\square_x$ and $\square_{y'}$ and let $|\partial R_{(x,y)}|$ be the length of its boundary, (i.e. the number of bonds in $\partial R_{(x,y)}$).

Let now $\Lambda_{(x,y)}$ be an arbitrary rectangle of unit squares with the property that <u>all</u> contours $\gamma \in \Gamma_{(x,y)}$, where $\Gamma_{(x,y)}$ has been defined in (3.31), which have distance 0 from $\partial \Lambda_{(x,y)}$ (i.e. touch $\partial \Lambda_{(x,y)}$) have <u>at least</u> length $\frac{1}{2} |\partial R_{(x,y)}|$.

There are two adjacent faces f_1, f_2 of $R_{(x,y)}$ (of total length $\frac{1}{2} |\partial R_{(x,y)}|$) such that each contour $\gamma \in \Gamma_{(x,y)}$ contains at least one bond $b_i(\gamma)$ (a "bond of entrance into $R_{(x,y)}$") contained in the interior of $R_{(x,y)}$ that touches $f_1 \cup f_2$ but <u>none</u> of the other two faces of $\partial R_{(x,y)}$.

Figure 3 below shows two sites x and y, the corresponding rectangles $R_{(x,y)}$ (the shaded region) and $\Lambda_{(x,y)}$ the faces f_1, f_2 and three contours in $\Gamma_{(x,y)}$ with some bonds of entrance

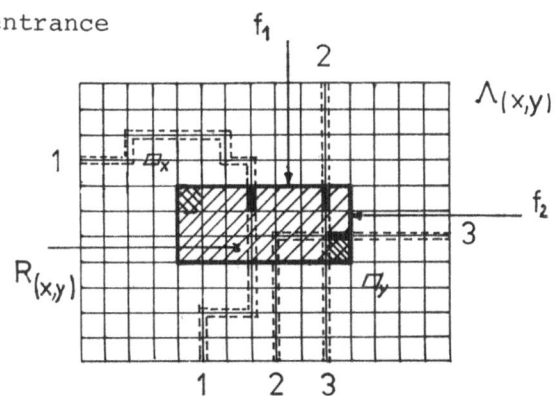

Fig. 3

172

There are $\frac{1}{2} |\partial R_{(x,y)}|$ different bonds of entrance
for the contours in $\Gamma_{(x,y)}$. This follows from the definition (3.31) of $\Gamma_{(x,y)}$ and the definition of the class
of bonds of entrance. For the class of contours in $\Gamma_{(x,y)}$
of length n there are no more than

$$\min \{n, \frac{1}{2} |\partial R_{(x,y)}|\} \tag{3.36}$$

different bonds of entrance, a consequence of (3.31),
the definitions of $R_{(x,y)}$ and of the bonds of entrance.
Using now the definitions of $\Lambda_{(x,y)}$ and $\Gamma_{(x,y)}$ we conclude from (3.36) that, for fixed n, there are at most
n different bonds of entrance into $R_{(x,y)}$, for the class
of all contours in $\Gamma_{(x,y)}$ of length n.

If we now choose the bond of entrance as the first
bond of a contour in $\Gamma_{(x,y)}$ and apply a standard

── argument we conclude that, for fixed n,
there are at most

$$n \; 3^{n-1} \tag{3.37}$$

contours $\gamma \in \Gamma_{(x,y)}$ with $|\gamma| = n$.

Finally we note that the length of a <u>closed</u> contour
$\gamma \in \Gamma_{(x,y)}$ is <u>even</u>. For <u>closed</u> contours (3.37)   (3.38)
may obviously be replaced by $n \; 3^{n-2}$.   (3.37')

Lemma 3.3:

Let K be the constant introduced in Theorem 3.2, (3.34), and assume that $K > \log 3$.
Then

$$\langle P_+(x)\, P_-(y) \rangle \leq \sum_{\gamma \in \Gamma_{(x,y)}} \left\langle \prod_{(\Box_z,\, \Box_{z'})} P_+(z)\, P_-(z') \right\rangle$$

$$\leq \sum_{n=2}^{\infty} 2n\, 3^{2n-2}\, e^{-K \cdot 2n} < \infty \qquad (3.39)$$

Proof:

The first inequality in Lemma 3.3 is inequality (3.33). Now

$$\sum_{\gamma \in \Gamma_{(x,y)}} \left\langle \prod_{(\Box_z,\, \Box_{z'}) \in N(\gamma)} P_+(z)\, P_-(z') \right\rangle$$

$$= \sum_{n} \sum_{\{\gamma \in \Gamma_{(x,y)}\,;\, |\gamma| = n\}} \left\langle \prod_{(\Box_z,\, \Box_{z'}) \in N(\gamma)} P_+(z)\, P_-(z') \right\rangle$$

$$= \sum_{n} \sum_{\{\gamma \in \Gamma_{(x,y)}\,;\, \gamma \text{ closed}\}}^{|\gamma| = n} \left\langle \prod_{(\Box_z,\, \Box_{z'}) \in N(\gamma)} P_+(z)\, P_-(z') \right\rangle$$

$$+ E(\Lambda_{(x,y)}),$$

where $E(\Lambda_{(x,y)})$ represents the sum over __all__ terms labelled by contours $\gamma \in \Gamma_{(x,y)}$ that are __not__ closed. But, by assumption (3.34) and (3.37), $E(\Lambda_{(x,y)}) \searrow 0$, as $\Lambda_{(x,y)} \nearrow \mathcal{R}^2$.

The proof now follows immediately from (3.37), (3.38) and the assumed inequality (3.34).

$$\text{Q.E.D.}$$

Remark:

For $\nu > 2$ and m arbitrary (3.39) is replaced by

$$\left\langle P_i(x) P_j(y) \right\rangle \leq \sum_{n=\nu}^{\infty} C_n (m-1)^{2n} e^{-K \cdot 2n}, \qquad (3.39')$$

where C_n is the number of closed contours in $\Gamma_{(x,y)}$ of surface 2n, and $C_n \propto e^{\text{const.}n}$.

It is obvious that, as $K \nearrow \infty$, the r.h.s. of (3.39) and (3.39') tend to 0. This completes the proof of Theorem 3.2.

$$\text{Q.E.D.}$$

———————

Summary of Section 3:

We have proven that the existence of a phase transition in a system described by $(\mathcal{O}\!l, <->)$, in the sense that $<->$ is a mixture of at least two pure phases (dim $\mathcal{H}_\infty \geq 2$) follows from

3.1 inequalities (3.16) and (3.19), or

3.2 inequalities (3.23) and (3.34),

and that information on the number of pure phases can be obtained from combining 3.1 and 3.2 with the structure of the internal symmetries.

In Sections 4-8 we prove these inequalities for various model systems (Models 1-6), thus establishing the results announced in Section 2.

Section 4:

INFRARED (GAUSSIAN) DOMINATION
AND THERMODYNAMIC LIMIT

In this section we prove inequalities (3.16) and (3.16') in the form of Theorem B, (2.23), for the correlation functions of the classical, ferromagnetic systems introduced in Section 2, (2.8)-(2.11), and for a class of Euclidean field theory models. These estimates, in particular Theorem B are basic for

(1) taking the thermodynamic limit $\Lambda \nearrow \mathbf{Z}^{\nu}$,
(2) applying method 3.1 to proving the occurence of phase transitions, see [FSS];
(3) the proof of the basic inequality (3.34), Theorem 3.2, in method 3.2 (the Peierls argument).

The inequality we are going to prove asserts that the correlation functions of such systems are dominated by <u>Gaussian correlations</u> with covariance (= two point function) in momentum space $\sim O(k^{-2})$. It is related to and motivated by the grad Φ- bounds of canonical quantum field theory, [GJ2, He].

4.0 The $\partial^i \sigma - (\text{grad } \phi-)$ bounds: Theorem B

Let $H_\Lambda^h(\{\sigma\})$ (Hamilton function) and $d\mu_\Lambda^h(\{\sigma\})$ (Gibbs state) be as defined in (2.8),(2.10).

We pick some direction $\mathbf{z}^\nu$, hence forth called T-direction (the direction of transfer). Without loss of generality this may be the 1-direction. The hyper plane in $\mathbf{z}^\nu$ perpendicular to T and passing through the origin is denoted $T^\perp$. Let the space cutoff Λ be a rectangle of the form

$$\Lambda = [-L,L] \times \Lambda_1 \tag{4.1}$$

with $\Lambda_1 = [-L_2,L_2] \times \ldots \times [-L_\nu,L_\nu] \subset T^\perp$.

The point $\alpha = (\ell_1,\ldots, L_i+1,\ldots,\ell_\nu)$ is

identified with $\hat{\alpha} = (\ell_1,\ldots,-L_i,\ldots,\ell_\nu)$, $\tag{4.2}$

(corresponding to <u>periodic</u> boundary conditions).

Let $\langle-\rangle_\Lambda^J \equiv \langle-\rangle_\Lambda^{J,h}$ denote expectation in $d\mu_\Lambda^h$.

<u>Theorem 4.1:</u> ($\equiv$ Theorem B, [FSS])

For $\nu=1,2,\ldots,$ $N=1,2,\ldots$ and all single spin dis-tributions $d\tilde{\lambda}$ in the class K_h defined in (2.20)

$$\left\langle e^{\sum_{\alpha,\ell} h_\ell(\alpha)\cdot\partial^\ell \sigma_\alpha} \right\rangle_\Lambda^J \leq e^{\frac{1}{2J}\sum_{\alpha,\ell} |h_\ell(\alpha)|^2}, \tag{4.3}$$

independently of d$\tilde{\lambda}$,h, and Λ.

The proof of Theorem 3.1 is in three steps. The basic tool is

4.1 The Transfer-Matrix formalism, $[T]$

Without loss of generality we may first assume that

$$d\tilde{\lambda}(\sigma) = g(\sigma) d^N\sigma, \tag{4.4}$$

for some non-negative function g of integral 1 and compact support X. The general case follows later by a limiting argument.

We set $\alpha = (i,I)$, $i \in T$. $I \in T^{\ell}$, and define

$$\vec{x}_i = \{\sigma_{(i,I)}\}_{I \in \Lambda_1} = \bigoplus_{I \in \Lambda_1} \sigma_{(i,I)}; \tag{4.5}$$

clearly $\vec{x}_i \in \mathcal{R}^p$, with $p = N \cdot |\Lambda_1|$; we write

$$\Omega_1 = X^{\times |\Lambda_1|}, \tag{4.6}$$

$$G(\vec{x}_i) = \prod_{I \in \Lambda_1} g(\sigma_{(i,I)}) e^{h \cdot \sigma_{(i,I)}}$$

$$\times \exp\left[-\frac{J}{2} \sum_{I \in \Lambda_1} \sum_{\ell=2}^{\gamma} (\partial^{\ell} \sigma_{(i,I)})^2\right], \tag{4.7}$$

and define

$$F(\vec{x}) = G(\vec{x})^{\frac{1}{2}}. \tag{4.8}$$

In these new notations

$$Z_\Lambda d\mu_\Lambda^h(\{\sigma\}) = \prod_{i=-L}^{L} G(\vec{x}_i) \, e^{-J/2 \, (\vec{x}_i - \vec{x}_{i+1})^2} \, d^p x_i \quad (4.9)$$

Let T denote the operator on $L^2(\Omega_1) \equiv L^2(\Omega_1, d^p x)$ defined by the kernel

$$K(\vec{x}, \vec{x}') = F(\vec{x}) \, e^{-(J/2)(\vec{x}-\vec{x}')^2} F(\vec{x}') \qquad (4.10)$$

T is called the __transfer matrix__.

For the purposes of taking the __thermodynamic limit__ and proving the existence of phase transitions we must consider perturbed transfer matrices.

Let $F_1(\vec{x})$, $F_2(\vec{x})$ be two functions in $L^2(\Omega_1)$ and $\vec{a}$ some vector in $\mathcal{R}^p$.

We define $T_{\vec{a}}(F_1, F_2)$ to be the operator on $L^2(\Omega_1)$ with kernel

$$K_{\vec{a}}^{(F_1, F_2)}(\vec{x}, \vec{x}') = e^{-\frac{|\vec{a}|^2}{2J}} \, \overline{F_1(\vec{x})} \, e^{-(J/2)(\vec{x}-\vec{x}')^2} e^{\vec{a} \cdot (\vec{x}-\vec{x}')}$$

$$\times F_2(\vec{x}')$$

$$= \overline{F_1(\vec{x})} \, e^{-(J/2)(\vec{x}-\vec{x}' - \frac{\vec{a}}{2J})^2} F_2(\vec{x}), \quad (4.11)$$

and $T(F_1) = T_{\vec{0}}(F_1, F_1)$. $\qquad (4.12)$

Lemma 4.2:

(1) T is a positive, selfadjoint, trace-class, positivity improving operator on $L^2(\Omega_1)$; $T_{\vec{a}}$ (F_1,F_2) is trace-class.

(2) $||T||$ is an eigenvalue of T of multiplicity 1, and the corresponding eigenvector ϕ_0 can be chosen to be non-negative with $||\phi_0||_2 = 1$.

Proof:

(1) We first show that $T_{\vec{a}}$ (F_1,F_2) is trace-class, for all $\vec{a} \in \mathfrak{R}^P$, F_1,F_2 in $L^2(\Omega_1)$. Since

$$T = T_{\vec{o}}(F,F) = T(F),$$

this also yields that T is trace-class. For this purpose we must show that

$$\text{Tr}(|T_{\vec{a}}(F_1,F_2)|)$$

is finite, where

$$|T_{\vec{a}}(F_1,F_2)| = \sqrt{T_{\vec{a}}(F_1,F_2)^* \, T_{\vec{a}}(F_1,F_2)}$$

is the absolute value of $T_{\vec{a}}$ (F_1,F_2).

Let $\{h_n\}_{n=o}^{\infty}$ be some complete, orthonormal system for $L^2(\Omega_1)$. Then

$$\text{Tr}(|T_{\vec{a}}(F_1,F_2)|) = \sum_{n=o}^{\infty} (h_n, |T_{\vec{a}}(F_1,F_2)|h_n) \qquad (4.13)$$

We now estimate each term in the series on the r.h.s. of (4.13):

Let ψ and ϕ be arbitrary vectors in $L^2(\Omega_1)$. Then

$$|(\psi, T_{\vec{a}}(F_1,F_2)\phi)|$$

$$= |\int\int d^P x\, d^P x'\, (F_1\psi)(\vec{x})\, e^{-\frac{J}{2}(\vec{x}-\vec{x}'-\frac{\vec{a}}{J})^2}\, (F_2\phi)(\vec{x}')|$$

$$= |(\frac{2\pi}{J})^{P/2} \int d^P k\, \overline{(F_1\psi)(\vec{k})}\, e^{-\frac{\vec{k}^2}{2J}}\, e^{i(\frac{\vec{k}\cdot\vec{a}}{J})}\, (F_2\phi)(\vec{k})|$$

$$\leq |(\frac{2\pi}{J})^{P/2} \int d^P k\, |(F_1\psi)(\vec{k})|^2\, e^{-\frac{\vec{k}^2}{2J}}\, |^{1/2}$$

$$\times |(\frac{2\pi}{J})^{P/2} \int d^P k\, |(F_2\phi)(\vec{k})|^2\, e^{-\frac{\vec{k}^2}{2J}}\, |^{1/2} \tag{4.14}$$

(by the Schwarz inequality)

$$= (\psi, T(F_1)\psi)^{1/2}\, (\phi, T(F_2)\phi)^{1/2} \tag{4.15}$$

<u>Polar decomposition</u> tells us that

$$T_{\vec{a}}(F_1,F_2) = U|T_{\vec{a}}(F_1,F_2)|, \tag{4.16}$$

for some <u>unitary</u> operator U on $L^2(\Omega_1)$.
Combining (4.15) with (4.16) we obtain

$$|(\phi, |T_{\vec{a}}(F_1,F_2)|\psi)| = |(U\phi, T_{\vec{a}}(F_1,F_2)\psi)|$$

$$\leq (\phi, T_U(F_1)\phi)^{1/2}\, (\psi, T(F_2)\psi)^{1/2}, \tag{4.17}$$

where $T_U(F_1) \equiv U^* T(F_1)\, U$ \tag{4.18}

Thus

$$(h_n, |T_{\vec{a}}(F_1,F_2)|h_n) \le (Uh_n, T(F_1) Uh_n)^{1/2}$$

$$\times (h_n, T(F_2) h_n)^{1/2},$$

and, by the Schwarz inequality for sequences and (4.13),

$$Tr(|T_{\vec{a}}(F_1,F_2)|) \le [\sum_{n=o}^{\infty} (Uh_n, T(F_1) Uh_n)]^{1/2}$$

$$\times [\sum_{n=o}^{\infty} (h_n, T(F_2) h_n)]^{1/2} \qquad (4.19)$$

Since U is unitary and $\{h_n\}_{n=o}^{\infty}$ is a complete ortho-normal system, so is $\{Uh_n\}_{n=o}^{\infty}$. Thus, the r.h.s. of (4.19) is equal to

$$\{Tr(T(F_1))\}^{1/2} \{Tr(T(F_2))\}^{1/2} \qquad (4.20)$$

Next, we remark that $T(F_1)$ is <u>positive</u>:
From the definition (4.12) it is obvious that $T(F_1)$ is symmetric, and positivity follows from the fact that $e^{-\frac{J}{2}\vec{x}^2}$ is a function of positive type. If A is a trace-class operator on $L^2(\Omega_1)$ with a kernel $C(\vec{x},\vec{x}')$ that is bounded and jointly continuous in $\vec{x}$ and $\vec{x}'$ then

$$Tr A = \int d^p x \; C(\vec{x},\vec{x}) \qquad (4.21)$$

From (4.21) and a limiting argument we now conclude that, for i=1,2,

$$\text{Tr } (T(F_i)) = \int d^Px \ |F_i(\vec{x})|^2 = ||F_i||_2^2 \qquad (4.22)$$

Thus we obtain from (4.19), (4.20) and (4.22) that

$$\text{Tr}(|T_{\vec{a}}(F_1,F_2)|) \leq ||F_1||_2 \cdot ||F_2||_2 < \infty.$$

Since $T = T(F)$, T is positive and trace-class. This completes the proof of (1).

Part (2) follows directly from the (generalized) Perron-Froboenius theorem.

$$\text{Q.E.D.}$$

Let $H_{-L}, \ldots, H_L$ be arbitrary functions on Ω_1 with the property that $H_j \ G$ is integrable, for all j, where G is defined in (4.7), and set $F_j \equiv (H_j G)^{1/2}$

Let $\alpha = (i,I)$ label the points of Λ, and set

$$\vec{a}_i = \{h(i,I)\}_{I\in\Lambda_1} \equiv \bigoplus_{I\in\Lambda_1} h(i,I), \qquad (4.23)$$

where $h(i,I) \in \mathcal{R}^N$, for all $(i,I) \in \Lambda$.

Finally $\langle \cdot \rangle_\Lambda^J$ denotes expectation with respect to $d\mu_\Lambda^h$; see (4.9).

Lemma 4.3: ("Feynman-Kac formula").

$$\left\langle (\prod_{j=-L}^{L} H_j) \ e^{-\Sigma_{\alpha\in\Lambda} h(\alpha) \partial^1 \sigma_\alpha} \right\rangle_\Lambda^J$$

$$= e^{\frac{1}{2J} \sum_{\alpha \in \Lambda} |h(\alpha)|^2} \quad \frac{\text{Tr} \ (\prod_{j=-L}^{L} T_{\vec{a}_j} (F_j, F_{j+1}))}{\text{Tr} \ (T^{2L+1})} \qquad (4.24)$$

Proof:

   If we integrate (4.9) and use (4.10) and (4.21) we obtain

$$Z_\Lambda = \text{Tr} \ (T^{2L+1}) \qquad (4.25)$$

Writing out the numerator on the r.h.s. of (4.24) ex-plicitly in terms of integrals, using the fact that $\prod_{j=-L}^{L} T_{\vec{a}_j} (F_j, F_{j+1})$ is trace-class, by Lemma 4.2, (1) and applying (4.21), next writing out the l.h.s. of (4.24) explicitly, using (4.9), finally applying (4.25) we immediately see that (4.24) holds.

Q.E.D.

Remark:

   If we combine Lemmata 4.2, (2) and 4.3 we see that the cutoff can always be removed in one direction ; $(L \to \infty)$. We also note that

$$||T|| = \lim_{L \to \infty} \{\text{Tr} \ (T^{2L+1})\}^{1/2L+1} \qquad (4.26)$$

These observations are sometimes useful in the analysis of two dimensional systems, as is well known.

4.2 A trace inequality

The following is one of the most basic estimates of Section 4.

__Lemma 4.4:__ (see [FSS])

$$|\text{Tr} \; (\prod_{j=-L}^{L} T_{\vec{a}_j} \; (F_j , F_{j+1}))|$$

$$\leq \prod_{i=-L}^{L} \{\text{Tr}(T(F_i)^{2L+1})\}^{\frac{1}{2L+1}} \qquad (4.27)$$

__Proof:__

By __Hölder's inequality__ for traces; see e.g. [DS]

$$|\text{Tr} \; (\prod_{j=-L}^{L} T_{\vec{a}_j} \; (F_j , \; F_{j+1}))|$$

$$\leq \prod_{j=-L}^{L} \{\text{Tr} \; (|T_{\vec{a}_j} \; (F_j , F_{j+1})|^{2L+1})\}^{\frac{1}{2L+1}}$$

Lemma 4.4 therefore follows from the inequality

$$\text{Tr}(|T_{\vec{a}} \; (F_1 , F_2)|^{2L+1})$$

$$\leq \{\text{Tr} \; (T(F_1)^{2L+1})\}^{1/2} \{\text{Tr}(T(F_2)|^{2L+1}\}^{1/2}$$

Let $A = |T_{\vec{a}} \; (F_1 , F_2)|$, $B = T_U \; (F_1)$; see (4.18), and $C = T \; (F_2)$

Using the definition of the trace, see (4.13), inequality (4.17) and the Schwarz inequality for sequences we get

$$\text{Tr } (A^{2L+1}) \leq \{\text{Tr } (A^L B A^L)\}^{1/2} \{\text{Tr } (A^L C A^L)\}^{1/2}$$

Using again Hölder's inequality we obtain

$$\{\text{Tr } (A^L B A^L)\}^{1/2} \leq \{\text{Tr}(B^{2L+1})\}^{\frac{1}{2(2L+1)}} \{\text{Tr}(A^{2L+1})\}^{\frac{L}{2L+1}}$$

and similarly for $\{\text{Tr}(A^L C A^L\}^{1/2}$.

Hence

$$0 \leq \text{Tr}(A^{2L+1}) \leq \{\text{Tr}(A^{2L+1})\}^{\frac{2L}{2L+1}} \{\text{Tr}(B^{2L+1})\}^{\frac{1}{2(2L+1)}}$$

$$\times \{\text{Tr}(C^{2L+1})\}^{\frac{1}{2(2L+1)}}$$

If we divide this inequality by $\{\text{Tr}(A^{2L+1})\}^{\frac{2L}{2L+1}}$ and raise it then to the $(2L+1)^{\text{st}}$ power we arrive at

$$\text{Tr}(A^{2L+1}) \leq \{\text{Tr}(B^{2L+1})\}^{1/2} \{\text{Tr}(C^{2L+1})\}^{1/2}$$

$$= \{\text{Tr}(T_U(F_1)^{2L+1})\}^{1/2}\{\text{Tr}(T(F_2)^{2L+1})\}^{1/2}$$

$$= \{\text{Tr}(T(F_1)^{2L+1})\}^{1/2} \{\text{Tr}(T(F_2)^{2L+1})\}^{1/2},$$

and we have used the unitary invariance of the trace.

Q.E.D.

4.3 "Nelson's symmetry"

We now assume that the functions $H_{-L}, \ldots, H_L$ introduced above, see Lemma 4.3, are of the form

$$H_i(\vec{x}_i) = \prod_{I \in \Lambda_1} f_{(i,I)} \, (\sigma_{(i,I)})$$

$$\times \exp\left[\sum_{I \in \Lambda_1} \sum_{\ell=2}^{\nu} h_\ell(i,I) \, \partial^\ell \sigma_{(i,I)} \right] \qquad (4.28)$$

for some non-negative functions $f_{(i,I)}$ on $\mathcal{R}^N$ compatible with the requirement that $H_i G$ be integrable, for all $i = -L, \ldots, L$.

We want to estimate

$$\left\langle \prod_{\alpha \in \Lambda} f_\alpha(\sigma_\alpha) \, \exp\left[-\sum_{\alpha \in \Lambda} \sum_{\ell=1}^{\nu} h_\ell(\alpha) \, \partial^\ell \sigma_\alpha \right] \right\rangle_\Lambda^J$$

$$= \left\langle \left(\prod_{i=-L}^{L} H_i \right) \exp\left[-\sum_{\alpha \in \Lambda} h_1(\alpha) \, \partial^1 \sigma_\alpha \right] \right\rangle_\Lambda^J \qquad (4.29)$$

By Lemma 4.3 this is equal to

$$e^{\frac{1}{2J} \sum_{\alpha \in \Lambda} |h_1(\alpha)|^2} \frac{\mathrm{Tr}\left(\prod_{j=-L}^{L} T_{\vec{a}_j} (F_j, F_{j+1}) \right)}{\mathrm{Tr}\, (T^{2L+1})} \qquad (4.30)$$

which is non-negative. From Lemma 4.4 we know that (4.30) is bounded above by

$$e^{\frac{1}{2J} \sum_{\alpha \in \Lambda} |h_1(\alpha)|^2} \prod_{i=-L}^{L} \left\{ \frac{\mathrm{Tr}\, (T(F_i)^{2L+1})}{\mathrm{Tr}\, (T^{2L+1})} \right\}^{\frac{1}{2L+1}} \qquad (4.31)$$

Since all directions in $\mathbf{Z}^\nu$ (parallel to a face of Λ)

can be chosen as <u>directions of transfer</u>, we may choose
next the 2-direction as our new direction of transfer.
A minute of reflection which involves rewriting Lemma 4.3
and (4.31) in terms of transfer matrices in the 2-direc-
tion shows that

$$e^{-\frac{1}{2J} \sum_{\alpha \in \Lambda} |h_2(\alpha)|^2} \frac{\text{Tr}(T(F_i)^{2L+1})}{\text{Tr}(T^{2L+1})}$$

is precisely of the form of the l.h.s. of (4.27) (with
transfer matrices in the 2-direction) and can therefore
be estimated by using Lemma 4.4; (see also [FSS]; in
Euclidean field theory this observation is called
"Nelson's symmetry", since Nelson first applied it there.
See [Si], [Gu]).

Repeating these observations another $(\nu-2)$ times we
arrive at

<u>Lemma 4.5:</u>   (from [FSS] $\cup$ [FS])

If we define

$$\alpha_\Lambda = \frac{1}{|\Lambda|} \log Z_\Lambda, \text{ and, for f a non-negative}$$

function on $\mathcal{R}^N$,   (4.32)

$$\alpha_\Lambda(f) = \frac{1}{|\Lambda|} \log \left\{ Z_\Lambda \left\langle \prod_{\alpha \in \Lambda} f(\sigma_\alpha) \right\rangle_\Lambda^J \right\}$$

then

$$\left\langle \prod_{\alpha \in \Lambda} f_\alpha(\sigma_\alpha) \exp\left[-\sum_{\alpha \in \Lambda} \sum_{\ell=1}^{\nu} h_\ell(\alpha) \partial^\ell \sigma_\alpha \right] \right\rangle^J_\Lambda$$

$$\leq e^{\frac{1}{2J} \sum_{\alpha \in \Lambda} \sum_{\ell=1}^{\nu} |h_\ell(\alpha)|^2} e^{\sum_{\alpha \in \Lambda} [\alpha_\Lambda(f_\alpha) - \alpha_\Lambda]} \tag{4.33}$$

This beautiful inequality is a more general form of Theorem 4.1, so that we have now completed the proof of the latter. Following [FS], we will later refer to (4.33) as "chess board estimate". Of course this estimate extends to complex functions $f_\alpha(\sigma_\alpha)$ if we take on the l.h.s. of (4.33) the absolute value and, on the r.h.s. of (4.33), replace f_α by $|f_\alpha|$.

4.4 The termodynamic limit

Next we want to apply the chessboard estimate (4.33) to prove existence of the thermodynamic limit. For this purpose we first derive bounds on the "pressures" α_Λ and $\alpha_\Lambda(f)$.

For $\Lambda = [-L_1, L_1] \times \ldots \times [-L_\nu, L_\nu]$ we write

$$\alpha_\Lambda(f) = \alpha_{L_1, \ldots, L_\nu}(f).$$

Lemma 4.6:

(1) Let f be any non-negative function. Then $\alpha_\Lambda(f)$ is monotone decreasing in Λ, in particular

$$\alpha_\Lambda(f) \leq \alpha_{1, \ldots, 1}(f),$$

where $\alpha_{1,\ldots,1}(f) = \log \int f(\sigma) \, e^{h \cdot \sigma} \, d\tilde{\lambda}(\sigma)$

(2) $\alpha_\Lambda(f) > 2^\nu \, \alpha_{2,\ldots,2}(f) - (2^\nu - 1) \alpha_{1,\ldots,1}(f)$,

for $L_i \geq 2$, all $i = 1, \ldots, \nu$.

Proof:

We set $H(\vec{x}_i) = \prod_{I \in \Lambda_1} f(\sigma_{(i,I)}) \, G(\vec{x}_i)$,

where G is as in (4.7), and $\hat{F}(\vec{x}) = \sqrt{H(\vec{x})}$.

We then note that

$$\alpha_\Lambda(f) = \frac{1}{|\Lambda_1|} \log \{ \mathrm{Tr}(T(\hat{F})^{2L_1 + 1}) \}^{\frac{1}{2L_1 + 1}}$$

a consequence of (4.25) and (4.32).

Using the definition of the trace (with $\{h_n\}_{n=0}^\infty$ all the orthogonal eigenvectors of the positive operator $T(F)$) and the fact that $x^{1/p}$ is concave on $\{x \geq 0\}$, for all $p > 1$, we see that

$$\left. \begin{array}{l} \{ \mathrm{Tr}(T(\hat{F})^p) \}^{\frac{1}{p}} \\ \text{is monotone decreasing in } p. \end{array} \right\} \qquad (4.34)$$

Since the logarithm is monotone, and $|\Lambda_1|$ is independent of L_1, $\alpha_{L_1,\ldots,L_\nu}(f)$ is monotone decreasing in L_1. Part (1) now follows from Nelson's symmetry.

Proof of (2):

By Hölder's inequality for the trace

$$0 \leq \mathrm{Tr}(T(\hat{F})^2) \leq \{\mathrm{Tr}(T(\hat{F})^{2L_1}\}^{\frac{2L_1+1}{2L_1}} \;\; \{\mathrm{Tr}(T(\hat{F})^{2L_1+1})\}^{\frac{1}{2L_1+1}}$$

By taking logarithms, dividing by $|\Lambda_1|$ and using (4.34) we get

$$\alpha_{L_1,\ldots,L_\nu}(f) \geq 2\,\alpha_{2,L_2,\ldots,L_\nu}(f) - \alpha_{1,L_2,\ldots,L_\nu}(f)$$

$$(4.35)$$

From Lemma 4.6, (1) we have

$$\alpha_{1,L_2,\ldots,L_\nu}(f) \leq \alpha_{1,1,\ldots,L_\nu}(f).$$

Repeating now (4.35) we get

$$\alpha_{L_1,\ldots,L_\nu}(f) \geq 4\,\alpha_{2,2,L_3,\ldots,L_\nu}(f) - 3\alpha_{1,1,L_3,\ldots,L_\nu}(f)$$

$$\geq 8\,\alpha_{2,2,2,\ldots,L_\nu}(f) - 4\,\alpha_{2,2,1,\ldots,L_\nu}(f)$$

$$- 3\,\alpha_{1,1,1,\ldots,L_\nu}(f)$$

$$\geq 8\,\alpha_{2,2,2,\ldots,L}(f) - 7\,\alpha_{1,1,1,\ldots,L_\nu}(f)$$

and we have used (4.34) again,

$$\geq \ \ldots\ldots$$

$$\geq 2^{\nu}\alpha_{2,\ldots,2}\,(f)-(2^{\nu}-1)\,\alpha_{1,\ldots,1}(f)\,.$$

This completes the proof of (2).

Q.E.D.

The following inequality is sometimes more convenient (see Section 6): Suppose that, for all $L_1,\ldots,L_\nu$

$$\alpha_{L_1,\ldots,L_\nu}(f) \leq \bar{\alpha}(f)$$

Then

$$\alpha_{L_1,\ldots,L_\nu}(f) \geq 2^{\nu}\alpha_{2,\ldots,2}\,(f) - (2^{\nu}-1)\bar{\alpha}(f) \qquad (4.36)$$

Estimating $\alpha_{2,\ldots,2}(f)$ for our lattice systems is a straightforward task which we leave to the reader. An immediate consequence of Lemma 4.6 is

Corollary 4.7:

If $\int f(\sigma)\,e^{h\cdot\sigma}d\tilde{\lambda}(\sigma) > 0$

$$\lim_{\Lambda\nearrow\mathbf{Z}^\nu}\alpha_\Lambda(f) = \inf_\Lambda\ \alpha_\Lambda(f) = \alpha_\infty\,(f)\ \text{exists.}$$

We are now prepared to extend Theorem 4.1 and Lemma 4.5 to arbitrary single spin distributions in the class:

$$K_h = \{d\tilde{\lambda}: \exists\,\varepsilon > 0 \text{ such that } \int e^{\varepsilon|\sigma|}\,e^{h\cdot\sigma}\,d\tilde{\lambda}(\sigma) < \infty\,\}$$

and then to prove the existence of a thermodynamic limit for the correlation functions.

We recall that, so far, Theorem 4.1 and Lemma 4.5 have only been proven for single spin distributions

$$d\tilde{\lambda}(\sigma) = g(\sigma) \, d^N\sigma,$$

where g is some non-negative, measurable function of integral 1 and compact support.

The extension of our estimates to K_h is a straightforward consequence of the chessboard estimate (4.33), Lemma 4.6 and Corollary 4.7:

Let $f_\alpha(\sigma) = e^{g(\alpha) \cdot \sigma_\alpha}$

for some function g on Λ with values in $\mathcal{R}^N$. Let B denote the support of g; $(B \subseteq \Lambda)$. Then the chessborad estimate (4.33) yields

$$\left\langle \exp\left[\sum_{\alpha \in \Lambda} g(\alpha) \cdot \sigma_\alpha \right] \right\rangle_\Lambda^J \leq e^{\sum_{\alpha \in B} [\alpha_\Lambda (\exp g(\alpha) \cdot \sigma) - \alpha_\Lambda]} \qquad (4.37)$$

By Lemma 4.6 the r.h.s. of (4.37) is bounded by

$$\exp\left[\sum_{\alpha \in B} \{\alpha_{1,\ldots,1}(\exp g(\alpha) \cdot \sigma) - 2^\nu \alpha_{2,\ldots,2} + (2^\nu - 1)\alpha_{1,\ldots,1}\} \right]$$

so that we obtain, using Lemma 4.6, (1)

$$\left\langle \exp\left[\sum_{\alpha \in \Lambda} g(\alpha) \cdot \sigma_\alpha \right] \right\rangle_\Lambda^J \leq$$

$$\leq \prod_{\alpha \in B} \; [\; \int e^{g(\alpha) \cdot \sigma} e^{h \cdot \sigma} \; d\tilde{\lambda}(\sigma)]$$

$$\times \; [\; \int e^{h \cdot \sigma} \; d\tilde{\lambda}(\sigma)] \quad (2^{\nu}-1)|B| \cdot e^{-2^{\nu}|B|} \alpha_2, \ldots, 2 \qquad (4.38)$$

The following two observations are important:

1. The r.h.s. of (4.38) is independent of Λ;
2. for $|g(\alpha)| < \epsilon$, $\alpha \in \Lambda$, the r.h.s. of (4.38) is uniformly bounded on the class of <u>all</u> single spin distributions $d\tilde{\lambda}$ for which

$$\int e^{\epsilon |\sigma|} \; e^{h \cdot \sigma} \; d\tilde{\lambda}(\sigma) \leq \text{const.}$$

Next, let $d\tilde{\lambda}_{\infty}(\sigma)$ be some fixed single spin distribution in K_h.

Then there exists some $\epsilon_{\infty} > 0$ such that

$$\int e^{\epsilon_{\infty}|\sigma|} \; e^{h \cdot \sigma} \; d\tilde{\lambda}_{\infty}(\sigma) \; < \; \infty$$

Using now (4.38) it is seen by a simple limiting argument that there exists a sequence $\{g_n\}_{n=0}^{\infty}$ of non-negative, measurable functions of compact support (and integral 1) such that, for

$$d\tilde{\lambda}_n(\sigma) \equiv g_n(\sigma) \; d^N\sigma,$$

and for all functions $g(\alpha)$ with $|g(\alpha)| < \epsilon_{\infty}$, for all $\alpha \in \Lambda$, the expectations of $\exp\left[\sum_{\alpha} g(\alpha) \cdot \sigma_{\alpha}\right]$ in the Gibbs

measures corresponding to $(\Lambda, \{d\tilde{\lambda}_n\})$ converge to the one
of $\exp[\Sigma g(\alpha) \cdot \sigma_\alpha]$ in the Gibbs measure corresponding to
$(\Lambda, d\tilde{\lambda}_\infty)$. Moreover the pressures introduced in (4.32)
corresponding to $(\Lambda, \{d\tilde{\lambda}_n\})$ converge to the ones corresponding to $(\Lambda, d\tilde{\lambda}_\infty)$.   Therefore Theorem 4.1
and Lemma 4.5 immediately extend to any expectation
$<->_\Lambda^J$ corresponding to a single spin distribution $d\tilde{\lambda}_\infty$ in
K_h. We may now pass to the thermodynamic limit: From
the above considerations, (4.37) and Corollary 4.7 we
conclude that, for g some function of compact support
B bounded in modulus by ε_∞ and single spin distribution
$d\tilde{\lambda}_\infty$,

$$\lim_{\Lambda \nearrow \mathbf{Z}^\nu} \sup \left\langle \exp\left[\sum_{\alpha \in \Lambda} g(\alpha) \cdot \sigma_\alpha \right] \right\rangle_\Lambda^J$$

$$\leq e^{\sum_{\alpha \in B} \left[\alpha_\infty (\exp g(\alpha) \cdot \sigma) - \alpha_\infty \right]}$$ (4.39)

From definition (4.32), Hölder's inequality and Corollary
4.7 we conclude that

α_∞ (exp g·σ) is convex in g.

Therefore the r.h.s. of (4.39) is finite and continuous
on the class of all functions g on $\mathbf{Z}^\nu$ of rapid decrease
bounded in modulus by $\varepsilon_\infty/2$.

If we combine this with (4.38) and apply a standard
compactness argument (Cantor's diagonal procedure) we
obtain

Theorem 4.8:

There exists a sequence $\{\Lambda_n\}_{n=0}^{\infty}$ of rectangles in-creasing to $\mathbb{Z}^{\nu}$ such that, for all $d\tilde{\lambda} \in K_h$,

$$d\mu^h(\{\sigma\}) \equiv \lim_{n\to\infty} d\mu_{\Lambda_n}^h (\{\sigma\}) \text{ exists,}$$

in the sense that all the moments of the measures $\{d\mu_{\Lambda_n}^h\}_{n=0}^{\infty}$ (i.e. the correlation functions) converge to the moments of some probability measure, denoted $d\mu^h$, on Σ.

Theorem 4.1 and the chessboard estimate, Lemma 4.5. hold, mutatis mutandis , for the expectation $\langle-\rangle^J \equiv \langle-\rangle^{J,h}$ determined by $d\mu^h$.

_____

We note that if $\langle-\rangle^J$ is a pure phase state (if not decompose it into pure phase states, [Fö]) then

$$\left\langle e^{\sum_{\alpha} g(\alpha)\sigma_{\alpha}'} \right\rangle^J \leq e^{1/2J \sum_{\alpha} g(\alpha)\left[(-\Delta)^{-1}g\right](\alpha)} \tag{4.40}$$

where $\sigma_{\alpha}' \equiv \sigma_{\alpha} - \left\langle \sigma_{\alpha} \right\rangle^J$, and Δ is the finite difference Laplacean; (4.40) follows directly from Theorem 4.1; see [FSS]. It justifies the term: "Gaussian domination".

Theorem 4.9: (see Theorem C; [FSS])

$$\left\langle (\sum_{\alpha,\ell} g_{\ell}(\alpha)\partial^{\ell}\sigma_{\alpha})^2 \right\rangle^J \leq \frac{1}{J} \sum_{\alpha,\ell} |g_{\ell}(\alpha)|^2$$

Proof:

This form of Theorem 4.9 is proven in $[FSS]$. If we replace $\frac{1}{J}$ by $\frac{5}{3}\frac{\nu}{J}$ then Theorem 4.9 follows directly from Theorem 4.1:

$$\left| \left\langle e^{z \sum_{\alpha,\ell} g_\ell(\alpha) \partial^\ell \sigma_\alpha} \right\rangle^J \right| \leq e^{1/2J |Rez|^2 \sum_{\alpha,\ell} |g_\ell(\alpha)|^2}$$

and the fact that $\left\langle e^{z \sum_{\alpha,\ell} g_\ell(\alpha)\partial^\ell \sigma_\alpha} \right\rangle$ is an entire function of z, by the Cauchy estimate

Q.E.D.

Theorem B, Section 2, is an immediate consequence of Theorem 4.9 and (3.16), (3.17).

The final issue of this section is to discuss the dependence of all our estimates on the <u>lattice spacing</u> δ.

If the underlying cubic lattice of the system has lattice constant δ we make the following substitutions:

We redefine ∂^ℓ,

$$\left.\begin{aligned} & \partial^\ell \sigma_\alpha = -\frac{1}{\delta}[\sigma_\alpha - \sigma_{\alpha_+^\ell}]; \text{ see } (2.7), (2.16) \\ & \sum_{\alpha,\ell} \rightarrow \sum_{\alpha,\ell} \delta^\nu, \text{ at \underline{all} places} \\ & \text{As } \delta \searrow 0, \sum_{\alpha,\ell} \delta^\nu \rightarrow \sum_\ell \int d^\nu \alpha \end{aligned}\right\} \qquad (4.41)$$

We leave it to the reader to check that if we take into account (4.41) <u>all</u> our estimates, Theorem 4.1 and 4.9, and Lemmata 4.5, (4.6) are <u>uniform</u> in δ.

These estimates form the technical core of the remainder of Part I and make the following sections reasonably short.

4.5 <u>The decomposition into pure phases</u>
(This subsection may be skipped)

The final issue of Section 4 is to establish the connection between the general framework introduced in Section 3 and the models considered in Sections 2 and 4.

Given a finite set $B \subset \mathbf{Z}^\nu$, let Σ_B denote the σ-algebra of Borel sets on $\prod_{\alpha \in B} \mathcal{R}^N_{(\alpha)}$.

An interesting consequence of the <u>chessboard estimate</u> Lemma 4.5 (in a slightly more general form), and Lemma 4.6 is the following

<u>Theorem:</u>

For the measures constructed in Theorem 4.9 $d\mu^h/_{\Sigma_B}$, the restriction of $d\mu^h$ to Σ_B, is absolutely con-

tinuous with respect to $\prod_{\alpha \in B} \cdot d\tilde\lambda(\sigma_\alpha)$ $\qquad$ (4.42)

An analogue of this theorem for the $P(\phi)_2$ models in the continuum limit $\delta = 0$, [E,Si], has been obtained in [FS]. The case considered here follows from the same arguments if one takes into account Lemmata 4.5 <u>and</u> 4.6.

This theorem permits us to define

$$\mathcal{O}(B) = L^\infty(\underset{\alpha \in B}{\times} \mathcal{R}^N_{(\alpha)}, \prod_{\alpha \in B} d\tilde\lambda(\sigma_\alpha)) \qquad (4.43)$$

The algebra $\mathcal{O}$ is then defined as in Section 3. Given a measure $d\mu^h$ of the type constructed in Theorem 4.9, we may then define $\overline{\mathcal{O}}$ ($\sim$B) and $\overline{\mathcal{O}}_\infty$ as in Section 3, (3.1). (In this case the G.N.S. Hilbert space $\mathcal{H}$ is $= L^2(X_{\mathcal{R}^N_{(\alpha)}, \alpha \in \mathbb{Z}^\nu}, \Sigma, d\mu^h))$. We define Σ_∞ to be the smallest σ-algebra such that all functions in $\overline{\mathcal{O}}_\infty$ are Σ_∞-measurable. Then $d\mu_\infty$ is the restriction of $d\mu^h$ to Σ_∞. Let X be some convenient model for the support of $d\mu_\infty$.

We now recall a general result concerning the <u>decomposition</u> of a state $<\to>^J$ of the type constructed in Theorem 4.9 <u>into pure phases</u>.

.

<u>Theorem 4.10:</u> [Fö]

For all $A \in \mathcal{O}$

$$\langle A \rangle^J = \int_X d\mu_\infty(\chi) \langle A \rangle^J_\chi$$

and, for $d\mu_\infty$-almost all $\chi \in X$, $<\to>^J_\chi$ is a <u>pure phase state</u>.

The measures corresponding to different pure phase states are <u>mutually singular</u>.

If dim $\mathcal{H}_\infty = m$ there exist m different pure phase states, (a consequence of the definition of $\overline{\mathcal{O}}_\infty$ and (3.2))

<u>Remark:</u>

For the interested reader we remark that, for $d\mu_\infty$-almost all $\chi \in X$, the states $<\to>^J_\chi$ and $<\to>^J$ satisfy the

same "Dobrushin-Lanford-Ruelle equations"; [R,GRS].
Theorem 4.10 has been extended to the continuum limit
($\delta=0$; $\mathbb{Z}^\nu \to \mathcal{R}^\nu$), i.e. the case of (Nelson-Symanzik
positive) Euclidean field theory. More precisely, let
$d\mu$ be some probability measure defined on the σ-algebra
generated by the Borel cylinder sets of $\mathcal{S}'_{real}$ ($\mathcal{R}^\nu$)
which satisfies Osterwalder-Schrader positivity [OS] in
the form of [F5] and the moments of which are Euclidean
Green's functions (Wightman functions restricted to the
Euclidean points) of some relativistic quantum field
theory. Such a measure is called a quantum measure, [F5].

Let Σ_I be the smallest σ-algebra such that all
functions in $\mathcal{H}_I$ (see (3.11)-(3.12)) are Σ_I-measurable,
$d\mu_I = d\mu|_{\Sigma_I}$. Let X be a convenient model for the support
of $d\mu_I$.

Theorem 4.10': [F5]

For all $\Delta \in \Sigma$

$$\mu(\Delta) = \int_X d\mu_I(\chi) \; \mu_\chi(\Delta),$$

where, for $d\mu_I$-almost all $\chi \in X$, μ_χ is a Euclidean invariant quantum measure ergodic under the action of the
translation group. Different "pure phase" quantum
measures are mutually singular.

Theorems 4.10 and 4.10' will be used freely in the
subsequent sections without explicit reference. The
statement "there are $n \leq \infty$ pure phases" is to be understood in the sense of the integral decompositions of

Theorems 4.10 and 4.10', with the precision that dim $\mathcal{H}_\infty$ = n, (dim $\mathcal{H}_I$=n), i.e. the support of $d\mu_\infty$ ($d\mu_I$) contains n points.

Section 5:

PHASE TRANSITIONS AND SPONTANEOUS SYMMETRY BREAKING FOR THE CLASSICAL N-VECTOR MODELS (MODELS 1-3,6)

In this section we combine method 3.1 (Infrared Domination, see inequalities (3.16) and (3.19)) with Theorems 4.1 and 4.9 to derive the existence of phase transitions for some class of the states $<->^J$ constructed in Theorem 4.8, in particular for Models 1-3 with $v \geq 3$ and Model 6 introduced in Section 2.

Theorem 5.1 (= Theorem D, [FSS])

Let $v \geq 3$ and $N = 1,2,3,\ldots,$. Assume that $d\lambda$ is an O(N) invariant single spin distribution in the class K_o defined in (2.20), and

$$d\lambda(\sigma) \neq \delta_o(\sigma)d^N\sigma.$$

Let the Hamilton function be given by

$$H_\Lambda = J \sum_{\alpha \in \Lambda} \sum_{i=1}^{v} \sigma_\alpha \cdot \sigma_{\alpha+i}; \text{ see (2.10).}$$

Then there exists some finite $J_c > 0$ such that, for

$J > J_C$ the state $\longleftrightarrow^J$ (constructed in Theorem 4.8) is <u>not</u> a pure phase state; (see 4.5). The long range order α_{σ_o}, (Section 3), is <u>strictly positive</u>.

There are at least S^{N-1} many pure phases, and for a set of pure phases of non-vanishing μ_∞-measure $\langle \sigma_o \rangle^J_\chi \neq 0$, (i.e. $\longleftrightarrow^J_\chi$ is "spontaneously magnetized").

Proof:

We prove this theorem for the special case where there exists some $\delta > 0$ such that

$$\lambda(\{\sigma : |\sigma| \leq \delta\}) = 0 \tag{5.1}$$

The general case is proven in $[FSS]$ by the same methods but involves some more detailed estimates on the pressure

$$\alpha_\infty(J) = \lim_{\Lambda \nearrow \mathbb{Z}^\nu} \frac{1}{|\Lambda|} \log Z_\Lambda, \text{ as a function of } J.$$

We apply the strategy of Section 3.1:
For the observable A we choose σ_o.
Then (5.1) gives

$$\langle A^* A \rangle = \langle \sigma_o \cdot \sigma_o \rangle^J \geq \delta^2 \tag{5.2}$$

Let $d\omega(k)$ be the Fourier transform of $\langle \sigma_o \, \sigma_\alpha \rangle^J$. From Theorem 4.9 we obtain, (see also Theorem C, Section 2 and (3.17))

$$\left.\begin{aligned}
& d\omega(k) = [\gamma \delta_o(k) + F^C(k)]d^\nu k, \\
& \text{where } \gamma = |\langle \sigma_o \rangle^J|^2 + \alpha_{\sigma_o}, \text{ and} \\
& 0 \leq F^C(k) \leq \frac{1}{J}[2\nu - 2\{\sum_{\ell=1}^\nu \cos k^\ell\}]^{-1}
\end{aligned}\right\} \tag{5.3}$$

Integration of (5.3) and use of (5.2) gives

$$\delta^2 \leq \gamma + \frac{1}{J} \int_B d^\nu k \left[2\nu - 2 \left\{ \sum_{\ell=1}^{\nu} \cos k^\ell \right\} \right]^{-1}$$

$$= \gamma + \text{const.} \frac{1}{J} \tag{5.4}$$

Here B is the first Brillouin zone and the constant on the r.h.s. of (5.4) is finite, provided $\nu \geq 3$; (for $\nu = 2$ the integral diverges logarithmically at k=0).

Next we note that $\langle \sigma_o \rangle^J_{J_o} = 0$. This follows from Theorem 4.9, since $\langle \sigma_o \rangle^J_\Lambda = 0$, for all finite Λ.

Thus, using (5.3) and (5.4)

$$\alpha_{\sigma_o} \geq \delta^2 - \text{const.} \frac{1}{J} \tag{5.5}$$

This is obviously positive for large enough J. We may now apply Theorem 4.10:

In a pure phase there is no long range order. Thus

$$\langle \sigma_o \cdot \sigma_o \rangle^J - \alpha_{\sigma_o} = \langle \sigma_o \cdot \sigma_o \rangle^{J,T}$$

$$= \int_X d\mu_\infty(\chi) \left\{ \langle \sigma_o \cdot \sigma_o \rangle^J_\chi - |\langle \sigma_o \rangle^J_\chi|^2 \right\}$$

i.e. $\int_X d\mu_\infty(\chi) |\langle \sigma_o \rangle^J_\chi|^2 = \alpha_{\sigma_o} > 0,$

for $J > J_C$ Q.E.D.

Theorem 5.2:

Let $\nu = N = 4$ and let H_Λ^F, $d\tilde{\lambda}$, λ and σ be as in Section 2, Model 6; see (2.9), (2.19). Let $\mathopen{<}-\mathclose{>}^\sigma$ be the corresponding state constructed in Theorem 4.8.

e.g.
Then, for $J = 1$ and each fixed $\lambda > 0$, there exists some finite $\sigma_c(\lambda)$ such that, for $\sigma > \sigma_c(\lambda)$, $\mathopen{<}-\mathclose{>}^\sigma$ is <u>not</u> a pure phase state.

Proof:

For the observalbe A (method 3.1) we may e.g. choose A^O.

Let $d\omega(k)$ be the Fourier transform of $\left\langle A_o^O\, A_\alpha^O \right\rangle^\sigma$.

Then Theorem 4.9 (= Theorem C) adapted to this model: $\sum_\ell g_\ell(\alpha)\, \partial^\ell\, \sigma_\alpha \;\rightarrow\; \sum_{\ell=1}^{3} g_\ell(\alpha)\, \partial^\ell A^O$, shows that

$$d\omega(k) = \left[\gamma \delta_o(k) + F^C(k)\right] d^4 k,$$

$$\text{where } \gamma = \left|\left\langle A^O \right\rangle^\sigma\right|^2 + \alpha_{AO}, \text{ and}$$

$$0 \le F^C(k) \le \frac{1}{J}\left[3 - \sum_{\ell=1}^{3} \cos k^\ell\right]^{-1}$$

$$\left.\begin{array}{c}\\[3em]\end{array}\right\} \qquad (5.6)$$

As in the proof of Theorem 5.1 we show that

$$\left\langle A^O \right\rangle^\sigma = 0$$

By integrating (5.6) we therefore find

$$\left\langle (A_o^o)^2 \right\rangle^\sigma = \alpha_{Ao} + C, \tag{5.7}$$

where the constant C is independent of λ and σ; and J = 1. Next we note that

$$\left\langle (A_o^o)^2 \right\rangle^\sigma = \frac{1}{4} \left\langle \vec{A}_o \cdot \vec{A}_o \right\rangle, \tag{5.8}$$

by Nelson's symmetry.

Let $\alpha_\infty(J,\lambda,\sigma)$ be the pressure of the system:

$$
\alpha_\infty(J,\lambda,\sigma) = \lim_{\Lambda \nearrow \mathbb{Z}^4} \frac{1}{|\Lambda|} \log \int e^{-H_\Lambda^F(\{A\})}
$$

$$
\times \prod_{\alpha \in \Lambda} e^{-\lambda(\vec{A}_\alpha \cdot \vec{A}_\alpha)^2 + \sigma \vec{A}_\alpha \cdot \vec{A}_\alpha} d^4 A_\alpha \tag{5.9}
$$

From Jensen's inequality and the chess board estimate (Lemma 4.5 and Theorem 4.8) we get

$$
e^{-2\sigma \left\langle \vec{A}_o \cdot \vec{A}_o \right\rangle^\sigma} \leq \left\langle e^{-2\sigma \vec{A}_o \cdot \vec{A}_o} \right\rangle^\sigma
$$

$$
\leq e^{\alpha_\infty(1,\lambda,-\sigma) - \alpha_\infty(1,\lambda,\sigma)} \tag{5.10}
$$

From (5.9) and (2.9) we get

(1) $\alpha_\infty(J,\lambda,-1) \leq \alpha_\infty(0,0,-1) < \infty$

(2) $\alpha_\infty(J,\lambda,1) \geq$ Const. λ^{-1}, as $\lambda \searrow 0$, uniformly in $J \in [0,1]$.

(3) $\alpha_\infty(1,\lambda,\sigma) = \alpha_\infty(\frac{1}{\sigma}, \frac{\lambda}{\sigma^2}, 1) + \frac{1}{\sigma^2}$

Combining (1) - (3) we get (for $\sigma \geq 1$)

$$\alpha_\infty(1,\lambda,-\sigma) - \alpha_\infty(1,\lambda,\sigma)$$

$$\leq \alpha_\infty(0,0,-1) - \alpha_\infty(\tfrac{1}{\sigma}, \tfrac{\lambda}{\sigma^2}, 1)$$

$$\leq - \text{Const.} \ \sigma^2, \ \text{as} \ \sigma \nearrow \infty \tag{5.11}$$

Taking the logarithm of (5.10) and using (5.11) we conclude:

$$\langle \vec{A}_o \cdot \vec{A}_o \rangle \geq O(\sigma), \ \text{as} \ \sigma \nearrow \infty \tag{5.12}$$

Theorem 5.2 now follows from (5.7), (5.8) and (5.12).

$$\text{Q.E.D.}$$

Remark:

Of course, (5.12) agrees with the prediction of the
naive Goldstone picture, [Co]. A physically more interesting result of the type of Theorem 5.2 is discussed
in [F3].

Theorem 5.3: ([FSS])

Let $\nu \geq 3$, N=1 and $d\lambda$ some Borel probability, measure
on $\mathcal{R}$ with the properties that

1. $d\lambda$ has no symmetries
2. $d\lambda \in \bigcap_h K_h$.

3. There exist positive numbers ε and δ such that
   $\lambda((-\infty,-\delta]) \geq \varepsilon, \ \lambda([\delta,\infty)) \geq \varepsilon$.

Let

$$H_\Lambda(\{\sigma\}) = J \sum_{\alpha \in \Lambda} \{ \sum_{i=1}^{\nu} \sigma_\alpha \sigma_{\alpha_+^i} + h \sigma_\alpha \}; \quad (\text{see } (2.10)).$$

Then there exists some finite J_C such that, for all $J > J_C$, $\langle \sigma_o \rangle^{h,J}$ is <u>discontinuous</u> in h at some $h=h_{crit.}$ (J), for at least one $h_{crit.}$ (J), and there is a phase transition at (J, $h_{crit.}$ (J)) which is <u>not</u> accompanied by the spontaneous breaking of any symmetry.

<u>Proof:</u>

We prove Theorem 5.3 for the special case, where

$$\lambda([-\delta,\delta]) = 0 \tag{5.13}$$

For the general case see [FSS].
Cleary estimates (5.3) and (5.4) apply to this case, thus using (5.13) and (5.4),

$$\delta^2 \leq \langle \sigma_o^2 \rangle^{h,J} = |\langle \sigma_o \rangle^{h,J}|^2 + \alpha_{\sigma_o} + \text{const.} \frac{1}{J}$$

As a consequence of the FKG inequalities the state $\langle \text{-->}^{h,J}$ may always be constructed in such a way that $\alpha_{\sigma_o} \equiv 0$; see [FS,FSS].

Thus

$$|\langle \sigma_o \rangle^{h,J}| \geq \sqrt{\delta^2 - \text{const.} \frac{1}{J}} \tag{5.14}$$

Given some positive $\varepsilon' < \delta$, there exists therefore a finite $J_C(\varepsilon')$ such that

$$| \langle \sigma_0 \rangle^{h,J} | \geq \varepsilon', \text{ for all } J \geq J_c(\varepsilon') \tag{5.15}$$

Using now hypotheses 2. and 3. we see by a very simple argument that

$$\lim_{h \nearrow \infty} \langle \sigma_0 \rangle^{h,J} > 0, \lim_{h \searrow -\infty} \langle \sigma_0 \rangle^{h,J} < 0 \tag{5.16}$$

Combination of (5.15) and (5.16) proves that $\langle \sigma_0 \rangle^{h,J}$ has at least one discontinuity in h. Thus , for some $h = h_{crit.}$ (J), there exist at least two different pure phase states.

By hypothesis 1. this phase transition is <u>not</u> accompanied by spontaneous symmetry breaking.

Q.E.D.

Remark:

Theorem 5.3 also holds in $\nu=2$ dimensions. The proof is then based on the Peierls argument. See Section 8.

Remarks:

The results proven in Theorems 5.1 and 5.3 are both very satisfactory and insufficient: They are very satis-factory, because they show that in $\nu \geq 3$ dimensions <u>spontaneous breaking of continuous internal symmetries does occur</u> and that <u>phase transitions are not always accompanied by spontaneous symmetry breaking</u>, (i.e. the concept of phase transitions is more fundamental than and independent of the one of symmetry breaking. This can also be shown in 2 dimensions; see Section 8).

They are insufficient, because no more than trivial information, derived from the structure of the internal symmetries, about the manifold of pure phases has been achieved.

By choosing suitable single spin distributions $d\lambda$ essentially any manifold of pure phases compatible with the internal symmetries can be achieved. This weakness of method 3.1 can often be cured by combining it with the Peierls argument, (method 3.2); see Section 8.

The relevance of results of the form of Theorem 5.2 must be discussed elsewhere.

Section 6:

THE $(\vec{\phi} \cdot \vec{\phi})_3^2$ - QUANTUM FIELD MODEL

In this section we prove the existence of a phase transition and the spontaneous breaking of the internal $O(N)$-symmetry for the $(\vec{\phi} \cdot \vec{\phi})_3^2$ - quantum field model $(\vec{\phi} = (\phi^1, \ldots, \phi^N)$ is an N-tuple of real, scalar Bose fields). For $N = 2,3$ we show that there are $N - 1$ Goldstone bosons, (zero mass one particle states). This establishes Theorem E, Section 2.

6.1 Existence and Wightman axioms

We briefly summarize the basic results concerning

existence and Wightman axioms for the $(\vec{\phi} \cdot \vec{\phi})_3^2$ - models. These results are very deep and difficult to prove, so that all proofs will be omitted.

First we recall the definition of these models, in the case of a positive lattice constant δ. The single spin distribution is given by

$$d\tilde{\lambda}(\vec{\phi}) = e^{-L_I(\vec{\phi})} d^N\phi, \text{ where}$$

(6.1)

$$L_I(\vec{\phi}) = \lambda(\vec{\phi} \cdot \vec{\phi})^2 - (M^2(\lambda,\delta) + \sigma)\vec{\phi} \cdot \vec{\phi} + \vec{h} \cdot \vec{\phi}$$

We consider a sequence of lattice constants converging to 0, e.g.

$$\delta_n = 2^{-n}, \quad n = 0, 1, 2, \ldots .$$

(6.2)

Let Λ be a rectangle in $\mathbb{Z}_{\delta_n}^3$ of the form

$$[-L_1, L_1] \times [-L_2, L_2] \times [-L_3, L_3],$$

where L_i is an arbitrary, fixed positive number (e.g. an integer) independent of n, for $i = 1, 2, 3$.

We redefine $|\Lambda|$: $|\Lambda| = 8 L_1 L_2 L_3$ (6.3)

The finite volume "Gibbs measure" is then given by

$$d\mu_\Lambda^n(\vec{\phi}) = Z_\Lambda(\lambda,\sigma,\vec{h},\delta_n)^{-1} e^{-\frac{1}{2} \sum_{x \in \Lambda} \delta_n^3 \sum_{\ell=1}^{3} (\partial^\ell \vec{\phi}_x)^2}$$

(6.4)

$$e^{-\sum_{x \in \Lambda} \delta_n^3 [L_I(\vec{\phi}_x) - C_1(\lambda,\sigma,\delta_n)]}$$

$$\times \ e^{-C_2(\lambda,\sigma,\delta_n)} \ [L_1 \ L_2 + L_1 \ L_3 + L_2 \ L_3] \tag{6.4}$$

where ∂^{ℓ} is defined as in (2.16) and $Z_{\Lambda}(\lambda,\sigma,\vec{h},\delta_n)$ is chosen such that $d\mu_{\Lambda}^n$ is a probability measure.

Theorem 6.1:

For each fixed, bounded rectangle Λ and a proper choice of $\{M^2(\lambda,\delta_n)\}_{n=0}^{\infty}$, $\{C_i(\lambda,\sigma,\delta_n)\}_{n=0}^{\infty}$, $i = 1,2$, specified in [GJ1, Pa, Se Si]

$$d\mu_{\Lambda}^{\lambda,\sigma}(\vec{\phi}) = \lim_{n\to\infty} d\mu_{\Lambda}^n(\vec{\phi})$$

exists, in the sense that all the moments and the characteristic functionals of the measures $\{d\mu_{\Lambda}^n\}_{n=0}^{\infty}$ converge, as $n\to\infty$.

Remarks:

A convenient model for the measure space on which $d\mu_{\Lambda}^{\lambda\sigma}$ is defined is $\mathcal{S}'_{real}(\mathcal{R}^3)^{\times N}$ equipped with the σ-algebra generated by all Borel cylinder sets; points in this space are denoted $\vec{\phi}(.)$, and the "observables" (see Section 3) may be built up from the functions

$$\{e^{i\vec{\phi}(\vec{f})} : \vec{f} \in \mathcal{S}'_{real}(\mathcal{R}^3)^{\times N}, \ \text{supp} \ \vec{f} \subseteq \Lambda\}$$

Theorem 6.1 is due to Park; see [Pa] and refs. given there. In his proof Park uses the fundamental techniques and results of [GJ1] and of [Fe].

We let $<->_\Lambda^{\lambda,\sigma}$ denote the expectation determined by $d\mu_\Lambda^{\lambda,\sigma}$. We then define the finite volume pressure (vacuum energy density)

$$\alpha_\Lambda(\lambda,\sigma) = \frac{1}{|\Lambda|} \log \left\langle e^{\sigma:\vec{\phi}\cdot\vec{\phi}:(\chi_\Lambda)} \right\rangle_\Lambda^{\lambda,0} \tag{6.5}$$

where $:\vec{\phi}\cdot\vec{\phi}: (\chi_\Lambda) = \int_\Lambda d^3x :\vec{\phi}\cdot\vec{\phi}:(x)$, and

$$:\vec{\phi}\cdot\vec{\phi}:(x) = \underset{|\varepsilon|\to o}{\text{w-lim}} \{\vec{\phi}(x)\cdot\vec{\phi}(x+\varepsilon) - \frac{N}{4\pi|\varepsilon|}\}; \tag{6.6}$$

The precise meaning of (6.6) is explained in [FSS].

Theorem 6.2:

(1) For $N = 1,2,3$ and arbitrary, fixed $\lambda > o$, $\sigma \geq o$ and all $\vec{h} \neq o$

$$d\mu^{\lambda,\sigma,\vec{h}}(\vec{\phi}) = \underset{\Lambda \nearrow \mathfrak{R}^3}{\lim} e^{-|\Lambda|\alpha_\Lambda(\lambda,\sigma)} e^{\sigma:\vec{\phi}\cdot\vec{\phi}:(\chi_\Lambda)} d\mu_\Lambda^{\lambda,o}(\vec{\phi})$$

exists, in the sense specified in Theorem 6.1. There exists a sequence $\{h_n\}_{n=o}^\infty$ converging to O such that for all unit vectors $\vec{e} \in \mathfrak{R}^3$

$$d\mu^{\lambda,\sigma,\vec{e}}(\vec{\phi}) = \underset{n\to\infty}{\lim} d\mu^{\lambda,\sigma,h_n\vec{e}}(\vec{\phi})$$

exists, in the same sense. Finally

$$\alpha_\infty(\lambda,\sigma) = \underset{\Lambda \nearrow \mathfrak{R}^3}{\lim} \alpha_\Lambda(\lambda,\sigma) \text{ exists,}$$

(for all $\vec{h}$, and, for $\vec{h} = o$, is independent of $\vec{e}$).

(2) The moments of the measures constructed in part (1)
are the Euclidean Green's functions of some unique
relativistic quantum field theory satisfying all
Wightman-Osterwalder-Schrader axioms.

Remarks:

In the form stated here part (1) has been proven
in [F6]. The difficult portions of the proof are however
due to [MS] and [FeO 1,2]. The basic results of these
refs. are combined in [F6] with the Lee-Yang theorems
of [SG, DN] (or, for N=1,2, with correlation inequalities
of [GRS, DN]) to derive (1). For N=1 independent proofs
have been given in [Pa,Fe O2].

In [MS] and [Fe O1] it is also shown that, for
$\sigma = o$, $\{M^2(\lambda,\delta_n)\}_{n=o}^{\infty}$ can be chosen such that $d\mu^{\lambda,o,\vec{e}}$ is
independent of $\vec{e}$ and continuous in λ in a weak sense,
even at $\lambda = o+$; the weak limit of $\{d\mu^{\lambda,o,\vec{e}}\}$, as $\lambda \searrow o$, is
the Gaussian measure $d\mu_o$ describing a free, massless
field.   (6.7)

For $\vec{h} \neq o$ the physical mass of the theory constructed
in Theorem 6.2, (1) is strictly positive; see [F6] and
[Fe O2].

6.2 Phase transitions and Goldstone bosons

Let $\langle\text{-}\rangle^{\lambda,\sigma,\vec{e}}$ denote the expectation with respect to $d\mu^{\lambda,\sigma,\vec{e}}$.

Theorem 6.3: ($\simeq$ Theorem E, $[\text{FSS}]$)

For $N=1,2,3$ and each $\lambda>0$, there exists some finite $\sigma_C(\lambda)$ such that, for $\sigma>\sigma_C(\lambda)$,

$$\langle\vec{\phi}(o)\rangle^{\lambda,\sigma,\vec{e}} = M^*\vec{e},$$

with $M^* \neq 0$; the internal $O(N)$ symmetry of the Lagrangean is spontaneously broken by $\langle\text{-}\rangle^{\lambda,\sigma,\vec{e}}$, and there exist $N-1$ Goldstone bosons.

Proof:

From $[\text{F6}]$ and $[\text{FSS}]$ we know that

$$\langle:\vec{\phi}\cdot\vec{\phi}:(o)\rangle^{\lambda,\sigma} = \lim_{|x|\to o} \{ \langle\vec{\phi}(o)\cdot\vec{\phi}(x)\rangle^{\lambda,\sigma} - \frac{N}{4\pi|x|} \} \qquad (6.8)$$

exists and is <u>finite</u>, for <u>all</u> finite σ.

Next

$$\langle\vec{\phi}(o)\cdot\vec{\phi}(x)\rangle^{\lambda,\sigma,\vec{e}} = (M^*)^2 + \frac{1}{(2\pi)^3}\int d^3k \, e^{ikx} \, F^C(k)$$

a consequence of Theorem 4.9 (see also (3.17)) and Theorem 6.2, (2); (in particular, the cluster property of $\langle\text{-}\rangle^{\lambda,\sigma,\vec{e}}$).

Theorem 4.9 (in the limit $\delta=o$, see (4.41)) gives

$0 \leq (k^i)^2 \, F^C(k) \leq N$, for all $i=1,2,3$.

From this one can conclude that

$$0 \leq F^C(k) \leq \frac{N}{k^2} \; , \quad \text{(see [FSS])} \tag{6.9}$$

(This can also be concluded from (6.8) and the <u>Källen-Lehmann representation</u>)

Thus

$$\langle :\vec{\phi}\cdot\vec{\phi}:(o) \rangle^{\lambda,\sigma,\vec{e}} = (M^*)^2 + \frac{1}{(2\pi)^3} \int d^3k \, [F^C(k) - \frac{N}{k^2}]$$

$$\leq (M^*)^2 \tag{6.10}$$

It now suffices to show that $\langle :\vec{\phi}\cdot\vec{\phi}:(o) \rangle^{\lambda,\sigma,\vec{e}}$ is positive, for σ sufficiently large. Then $M^* > o$ and, since $<->^{\lambda,\sigma,\vec{e}}$ is invariant under the substitution $\vec{e} \wedge \vec{\phi} \rightarrow -\vec{e} \wedge \vec{\phi}$, by construction,

$$\langle \vec{\phi}(o) \rangle^{\lambda,\sigma,\vec{e}} = M^* \, \vec{e}.$$

Let $\square$ be the characteristic function of some <u>unit</u> cube. By translation invariance

$$\langle :\vec{\phi}\cdot\vec{\phi}:(o) \rangle^{\lambda,\sigma,\vec{e}} = \langle :\vec{\phi}\cdot\vec{\phi}:(\square) \rangle^{\lambda,\sigma,\vec{e}}$$

<u>Jensen's inequality</u> and the <u>chessboard estimate</u> (Lemma 4.5, Theorem 4.8) give

$$e^{-\sigma \langle :\vec{\phi}\cdot\vec{\phi}:(\square) \rangle^{\lambda,\sigma,\vec{e}}} \leq \langle e^{-\sigma :\vec{\phi}\cdot\vec{\phi}:(\square)} \rangle^{\lambda,\sigma,\vec{e}}$$

$$\leq e^{\alpha_\infty(\lambda,o) - \alpha_\infty(\lambda,\sigma)}$$

Hence

$$\left\langle :\vec{\phi}\cdot\vec{\phi}:(\square)\right\rangle^{\lambda,\sigma,\vec{e}} \geq \alpha_\infty(\lambda,\sigma)-\alpha_\infty(\lambda,o) , \qquad (6.11)$$

and it suffices now to prove positivity of

$\alpha_\infty(\lambda,\sigma) - \alpha_\infty(\lambda,o)$, for sufficiently large σ. By (6.5) $\alpha_\infty(\lambda,o) = 0$.

Using (6.5), Theorem 6.2, (1) and the chess board estimate we conclude that

$$e^{|B|\alpha_\infty(\lambda,\sigma)} \geq \left\langle e^{\sigma:\vec{\phi}\cdot\vec{\phi}:(\chi_B)}\right\rangle^{\lambda,o} \qquad (6.12)$$

for all bounded cubes $B \subset \mathcal{R}^3$.

By scaling lengths, λ and σ it is easily seen that it now suffices to show that, for fixed $\sigma>o$ and some bounded cube B,

$$\left\langle e^{\sigma:\vec{\phi}\cdot\vec{\phi}:(\chi_B)}\right\rangle^{\lambda,o} > 1 \qquad (6.13)$$

for sufficiently small λ.
By the chess board estimate

$$\left\langle e^{-\sigma:\vec{\phi}\cdot\vec{\phi}:(\chi_B)}\right\rangle^{\lambda,o} \leq e^{|B|\ \alpha_\infty(\lambda,-\sigma)} \qquad (6.14)$$

and the r.h.s. is <u>uniformly bounded</u> in $\lambda \in [o,\lambda_o]$, for each finite λ_o; by [GJ1].

From [MS,Fe 01] we know that

$$\left\langle [:\vec{\phi}\cdot\vec{\phi}:(\chi_B)]^m\right\rangle^{\lambda,o} \text{ converges to}$$

$$\left\langle \left[:\vec{\phi}\cdot\vec{\phi}:(x_B) \right]^m \right\rangle_o , \quad \text{as } \lambda \searrow o, \text{ for all } m < \infty.$$

Here $\langle\text{-->}\rangle_o$ is the Euclidean (Gaussian) expectation of a free massless field; see (6.7).

Finally $\displaystyle\sum_{m=o}^{M} \frac{\sigma^{2m}}{(2m)!} \left\langle \left[:\vec{\phi}\cdot\vec{\phi}:(x_B) \right]^{2m} \right\rangle_o$ \hfill (6.15)

diverges to $+\infty$, as $M\to\infty$, for large enough $|B|$, as one varifies by an explicit calculation. Combining these facts we conclude that

$$\left\langle e^{\sigma:\vec{\phi}\cdot\vec{\phi}:(x_B)} \right\rangle^{\lambda,o} =$$

$$2 \sum_{m=o}^{\infty} \frac{\sigma^{2m}}{(2m)!} \left\langle \left[:\vec{\phi}\cdot\vec{\phi}:(x_B) \right]^{2m} \right\rangle^{\lambda,o} - \left\langle e^{-\sigma:\vec{\phi}\cdot\vec{\phi}:(x_B)} \right\rangle^{\lambda,o}$$

which diverges to $+\infty$, as $\lambda \searrow o$, by (6.14) and (6.15). This proves (6.13); (more details can be found in [FSS]).

The existence of N-1 Goldstone bosons, (massless one particle states coupled to the physical vacuum by the field $\vec{e}\wedge\vec{\phi}$), follows essentially from the Goldstone theorem in the form of [ESw].

$$\text{Q.E.D.}$$

Remarks:

(1) For N>3 the occurence of phase transitions can still be proven (by almost identical arguments), but Lorentz covariance of the theory is unknown.

(2) For N=2,3 and $\sigma>\sigma_C(\lambda)$, the physical mass of the

theory tends to 0, as $h \searrow 0$, <u>linearly</u> in h; see [F6, LP].

(3) It is shown in [DN] by means of correlation inequalities that

$$\left\langle (\vec{e} \cdot \vec{\phi})(0)(\vec{e} \cdot \vec{\phi})(x) \right\rangle^{\lambda, \sigma, e} = O((\left\langle (\vec{e}_\wedge \vec{\phi})(0) \cdot$$

$$(\vec{e}_\wedge \vec{\phi})(x) \right\rangle^{\lambda, \sigma, \vec{e}})^2),$$

as $|x| \to \infty$.

<u>Section 7:</u>

<u>PHASE TRANSITIONS IN TWO DIMENSIONAL</u>
<u>QUANTUM FIELD MODELS</u>

In this section we consider the $P(\phi)_2-$, [E,Si], $\lambda \cos(\epsilon\phi)_2-$, [F1, FSe], and the pseudoscalar Yukawa$_2$ quantum field models, [Y$_2$]. We prove that, for certain values of the bare couplings, these models have phase transitions which are sometimes, but not necessarily, accompanied by the breaking of an internal symmetry transforming ϕ into $-\phi$.

We consider the $P(\phi)_2$ models in some detail, whereas for $\lambda \cos(\epsilon\phi)_2$ (Model 5) and the pseudo scalar Yukawa model we just state the results; see [F3] for detailed proofs. Our methods are inspired by [GJS2] and [FSS].

We assume that the reader is somewhat familiar with the $P(\phi)_2$ Euclidean field theory; see $[E,Si]$. We just recall the most important definitions:

Let $\Lambda = [-L_1,L_1] \times [-L_2,L_2] \subset \mathcal{R}^2$,

where L_1,L_2 are positive integers.

The free field Gaussian measure on $\mathcal{S}'_{real}(\mathcal{R}^2)$ with mean 0, bare mass = 1 and periodic boundary conditions at $\partial\Lambda$ is denoted $d\mu^o_\Lambda$, and $<->^o_\Lambda$ is the expectation determined by $d\mu^o_\Lambda$. The colons $:-:$ denote Wick ordering of polynomials in the (pseudo-)scalar Euclidean field ϕ with respect to bare mass = 1, (i.e. with respect to $<->^o \equiv <->^o_{\mathcal{R}^2}$; $[Si]$). If Q is some polynomial and $B \subset \mathcal{R}^2$ some bounded rectangle then

$$:Q(\phi):(B) \equiv \int_B :Q(\phi):(x)d^2x \tag{7.1}$$

The Euclidean action of the $P(\phi)_2$ theory is given by

$$U_\Lambda(P) \equiv \int_\Lambda :P(\phi):(x)\ d^2x, \tag{7.2}$$

where P is some __positive__ polynomial.

We define

$$d\mu^P_\Lambda(\phi) \equiv \{ \left\langle e^{-U_\Lambda(P)} \right\rangle^o_\Lambda \}^{-1}\ e^{-U_\Lambda(P)}\ d\mu^o_\Lambda(\phi), \tag{7.3}$$

(the cutoff interacting measure; see $[E,Si]$),

$$\alpha^P_\Lambda(h) \equiv \frac{1}{|\Lambda|}\ \log \left\langle e^{-U_\Lambda(P)+h\phi(\Lambda)} \right\rangle^o_\Lambda, \tag{7.4}$$

and, for fixed, positive R,

$$\alpha_\Lambda^P(\lambda,\sigma) \equiv \frac{1}{|\Lambda|} \log \left\langle e^{-\lambda U_\Lambda(R) +\frac{\sigma}{2}:\phi^2:(\Lambda)} \right\rangle_\Lambda^o \qquad (7.5)$$

(the finite volume vacuum energy densities; [GRS]) It
is known that, for α_Λ as in (7.4), (7.5), $\alpha_\infty = \lim_{\Lambda \nearrow \mathcal{R}^2} \alpha_\Lambda$
exists, see [GRS]; $\alpha_\infty^P(h)$ is a convex function of h and
hence it is continuously differentiable in h except
possibly at countably many values of h.

We now state without proof the basic result concerning existence of the infinite volume limit.

Theorem 7.1:

(1) If $\alpha_\infty^P(h)$ is continuously differentiable in h at
some point h_o then

$$d\mu^{P+h_o x}(\phi) = \lim_{\Lambda \nearrow \mathcal{R}^2} d\mu_\Lambda^{P+h_o x}(\phi)$$

exists, in the sense of Section 6, (and the limiting
measure is independent of "classical boundary conditions"; see [FS]).

(2) $d\mu_\pm^P(\phi) = \lim_{\substack{h \searrow o \\ h \nearrow o}} d\mu^{P+hx}(\phi)$

exists, and

$$d\mu_+^P = d\mu_-^P \text{ iff } \alpha_\infty^P(h)$$

is continuously differentiable at h = o.

(3) The moments of the infinite volume measures constructed in (1) and (2) are the Euclidean Green's
functions of a unique, relativistic quantum field
theory satisfying all the axioms of Wightman and
Osterwalder-Schrader.

Remarks:

In this form Theorem 7.1 is due to [FS]. However
the results and methods of this ref. are based in an
essential way on earlier results; see [E,Si] and refs.
given there.

Theorem 7.1 not only asserts existence of the infinite volume limit, but it also shows that, in a very
strong sense (see (1) and [FS]), phase transitions do
not occur if $\alpha_\infty^P(h)$ is continuously differentiable in h
at h=o (see (2), (3)).

We now want to prove that, for certain choices of P

$$d\mu_+^P \neq d\mu_-^P$$

$$(<\!\!=\!\!> \quad \frac{d\alpha_\infty^P(h)}{dh} \quad \text{is discontinuous at h=o} \qquad\qquad (7.6)$$

$$<\!\!=\!\!> \text{ there is a phase transition}).$$

Our proof is based on the Peierls argument in the form
of [GJS2]; see Section 3, method 3.2. Some knowledge of
Section 3.2 is henceforth assumed.

Theorem 7.2:

For an arbitrary, <u>even</u>, positive polynomial R and a fixed $\sigma \in (1, 2\pi^2)$ there exists a <u>positive</u> λ_c such that, for

$$P(x) \equiv \lambda R(x) - \frac{\sigma}{2} x^2, \text{ and } 0 < \lambda < \lambda_c,$$

there is a phase transition, in the sense of (7.6), and the states (expectations) <->$_\pm$ determined by $d\mu_\pm^P$ break the $\phi \to -\phi$ symmetry of P.

Remarks:

1. Using scaling and re-Wick ordering it is seen that this theorem extends to polynomials
   $P(x) = \lambda R(x)$, $\lambda > \lambda_{critical}$, $(\sigma = 0)$; see $[GJS2]$.

2. Theorem 7.2 is due to $[GJS2]$. A simplified proof and a generalization to <u>non-even</u> P's, as well as extensions to $\lambda \cos (\epsilon\phi)_2$ and pseudo-scalar Yukawa$_2$ are presented in $[F3]$.

Proof:

Cover $\mathcal{R}^2$ with a grid of mesh 1. Let $\square$ be some unit square of this grid. According to method 3.2 we must define m mutually orthogonal projections on $\mathcal{H} = L^2(\mathcal{S}'_{real}(\mathcal{R}^2), d\mu_\pm^P)$ which are "supported" on $\square$, i.e. $\Sigma_\square$ - measurable functions on $\mathcal{S}'_{real}(\mathcal{R}^2)$, and the sum of which is the identity:

We let $\chi_\pm$ $(\square)$ be the characteristic functions of the measurable sets

$$\{\phi \in \text{supp } d\mu^o : \phi(\square) \underset{\leq}{\overset{\geq}{}} 0\} \subset \mathscr{S}'_{\text{real}} \ (\mathscr{R}^2). \qquad (7.7)$$

In the notation of 3.2 we have $m = 2$, $P_1 = \chi_+(\square)$, $P_2 = \chi_-(\square)$; obviously

$\chi_+(\square) + \chi_-(\square) = 1$.

Suppose now that, for all $x \in \mathbb{Z}^2$,

$$\left\langle \chi_+(\square_o) \ \chi_-(\square_x) \right\rangle_\pm \leq \frac{1}{4} - \delta, \qquad (7.8)$$

for some $\delta > 0$.

Since $<->_\pm$ is a pure phase state, by Theorem 7.1, (3),

$$\left\langle \chi_+(\square_o) \ \chi_-(\square_x) \right\rangle_\pm \ \rightarrow \ \left\langle \chi_+(\square_o) \right\rangle_\pm \ \left\langle \chi_-(\square_o) \right\rangle_\pm,$$

as $|x| \rightarrow \infty$. $\qquad (7.9)$

But $\left\langle \chi_-(\square_o) \right\rangle_\pm = 1 - \left\langle \chi_+(\square_o) \right\rangle_\pm$

We set $\left\langle \chi_+(\square_o) \right\rangle_\pm = M_\pm^* \qquad (7.10)$

By construction of the state $<->_+$, see Theorem 7.1, (2), and the assumed evenness of P

$$M_+^* \geq 1/2 \qquad (7.11)$$

Combining (7.8)-(7.11) we conclude

$$M_+^* - (M_+^*)^2 < \frac{1}{4} - \delta, \ M_+^* \geq 1/2, \text{ hence } M_+^* \geq \frac{1}{2} + \sqrt{\delta} \qquad (7.12)$$

Furthermore $M_-^* = 1 - M_+^* \leq \frac{1}{2} - \sqrt{\delta}$ (7.13)

which proves the phase transition, and, since, for a state <-> invariant under $\phi \to -\phi$, $\langle x_+(\square) \rangle = 1/2$, this also shows that the states <->$_\pm$ break the $\phi \to -\phi$ symmetry of the dynamics. (The chain of arguments (7.8) - (7.13) is taken from [GJS2]).

We are now left with proving (7.8):
By Theorem 3.2, inequalities (3.34), (3.35), (7.8) follows from the inequality

$$\left\langle \prod_{(\square,\square') \in N(\gamma)} x_+(\square) \, x_-(\square') \right\rangle_\pm \leq e^{-K|\gamma|} \qquad (7.14)$$

for all contours $\gamma \in \Gamma_{(x,y)}$ (see Section 3) and some sufficiently large constant K. (By Lemma 3.3, $K \gtrsim \log 3.5$ yields (7.8)). Let J be some positive number to be chosen later. We let $x_\pm^1(\square)$ be the characteristic functions of

$$\{\phi : \phi(\square) \begin{array}{c} \geq J \\ \leq -J \end{array}\},$$

and $x_\pm^2(\square)$ the ones of

$$\{\phi : \phi(\square) \in [0, \pm J]\},$$

Then $x_\pm = x_\pm^1 + x_\pm^2$ (7.15)

We insert (7.15) into the l.h.s. of (7.14) and expand. This yields

$$\left\langle \prod_{(\square,\,\square')\in N(\gamma)} x_+(\square)\; x_-(\square') \right\rangle_\pm \tag{7.16}$$

$$= \sum_{\substack{\epsilon(\square)=1,2 \\ \epsilon(\square')=1,2}} \left\langle \prod_{(\square,\square')\in N(\gamma)} x_+^{\epsilon(\square)}(\square)\, x_-^{\epsilon(\square')}(\square') \right\rangle_\pm$$

Let $N_1(\gamma)$ be some maximal subset of $N(\gamma)$ with the property that if $(\square_1,\square_1')$ and $(\square_2,\square_2')$ are two different elements of $N_1(\gamma)$ then $\square_i \neq \square_i'$, for $i=1,2$. Obviously

$$|N_1(\gamma)| \geq \tfrac{1}{4}\,|N(\gamma)| = \frac{|\gamma|}{4}\,, \tag{7.17}$$

where $|N_1(\gamma)|$ is the number of pairs $(\square,\square')$ in $N_1(\gamma)$.

We now label the pairs $(\square,\square')$ in $N_1(\gamma)$ by some index $i \in \{1,2,\ldots,|N_1(\gamma)|\} \equiv I_\gamma$.
We then define

$$x_1(i) = x_+^1(\square)\, x_-^1(\square'),\quad x_2(i) = x_+^2(\square) \tag{7.18}$$

$$x_3(i) = x_-^2(\square'),\quad x_4(i) = x_+^2(\square)\, x_-^2(\square')$$

Since $0 \leq x_\pm^\epsilon(\square) \leq 1$, for all $\square$ and $\epsilon = 1,2$, we now get from (7.16)

$$\left\langle \prod_{(\square,\,\square')\in N(\gamma)} x_+(\square)\,x_-(\square') \right\rangle_\pm$$

$$\leq \sum_{\{p(i)=1,\ldots,4\}} \left\langle \prod_{i\in I_\gamma} x_{p(i)}(i) \right\rangle_\pm \tag{7.19}$$

Next we note that, for i = $(\square, \square')$,

$$0 \le x_1(i) \le F_1(i) \equiv e^{-2J} \, e^{\phi(\square)-\phi(\square')}$$

$$0 \le x_2(i) \le F_2(i) \equiv e^{\frac{\sigma J^2}{2} \, (1-\frac{1}{J^2} \, \phi(\square)^2)}$$

$$0 \le x_3(i) \le F_3(i) \equiv e^{\frac{\sigma J^2}{2} \, (1-\frac{1}{J^2} \, \phi(\square')^2)}$$

$$0 \le x_4(i) \le F_4(i) \equiv F_2(i) \, F_3(i) \, ,$$

$$(7.20)$$

an immediate consequence of (7.18); see also [GJS2].
Combining (7.19) and (7.20) we get

$$\left\langle \prod_{(\square,\square') \in N(\gamma)} x_+(\square) \, x_-(\square') \right\rangle_\pm$$

$$\le \sum_{\{p(i)=1,\ldots,4\}} \left\langle \prod_{i \in I_\gamma} F_{p(i)}(i) \right\rangle_\pm \qquad (7.21)$$

The idea is now to estimate the r.h.s. of (7.21) by
means of "Gaussian domination" and the chess board
estimate (Lemma 4.5, Theorem 4.8, (4.41)).

Let $h_i(x)$ be the function on $\mathcal{R}^2$ with support in
$\square \cup \square'$ the graph of which is given by Fig. 4:

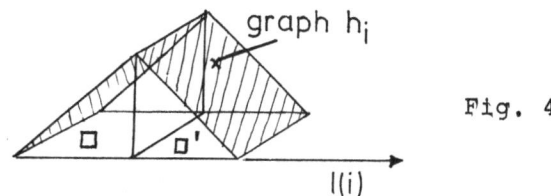

graph h_i

Fig. 4

l(i)

$\ell(i)$ is the direction perpendicular to the common face of $\Box$ and $\Box'$. Then

$$\phi(\Box)-\phi(\Box') = -\int h_i(x)\ \partial^{\ell(i)}\phi(x)\ d^2x,$$

(7.22)

where $\partial^{\ell(i)}\phi(x) \equiv (\dfrac{\partial}{\partial x^{\ell(i)}}\ \phi)(x)$.

$$\left.\begin{array}{l} \text{Obviously } ||h_i||_2^2 = \frac{2}{3}\ ,\ \text{and} \\[2em] \text{supp } h_i \cap \text{supp } h_{i'} = \phi,\ \text{for } i \neq i' \end{array}\right\}$$

(7.23)

By (7.20) and (7.22)

$$F_1(i) = e^{-2J}\ e^{\partial^{\ell(i)}\phi(h_i)}$$

(7.24)

Next we define

$$\hat{\alpha}_\Lambda(\lambda,\sigma) \equiv$$

(7.25)

$$\frac{1}{|\Lambda|}\ \log\ \left\langle e^{-\lambda U_\Lambda(R)}\ e^{\frac{\sigma}{2}\left[:\phi^2:(\Lambda)-\sum_{\Box\subset\Lambda}\ :\phi(\Box)^2:\right]} \right\rangle_\Lambda^o\ +\alpha_\Lambda^o$$

where α_Λ^o is the thermodynamic pressure per unit volume of the free field of bare mass 1 in a periodic box of length $2L_1$ at inverse temperature $2L_2$. From Corollary 4.7 we infer existence of $\hat{\alpha}_\infty(\lambda,\sigma)$ and from Lemma 4.6.(1)

$$\hat{\alpha}_\infty(\lambda,\sigma) \leq \hat{\alpha}_{\Lambda_o}(\lambda,\sigma),\ \Lambda_o = [-1,1]\times[-1,1]$$

(7.26)

We now define weights

$$W_1 = 2J - \frac{1}{3} ,$$

$$W_2 = W_3 = \frac{1}{2} W_4 = - \frac{\sigma J^2}{2} + \frac{\sigma}{2} \left\langle \phi(\square)^2 \right\rangle^{\,o}$$

$$+ \alpha_\infty(\lambda,\sigma) - \hat{\alpha}_\infty(\lambda,\sigma)$$

Lemma 7.3:

$$\left\langle \prod_{i \in I_\gamma} F_{p(i)}^{(i)} \right\rangle_\pm \le e^{-i \sum\limits_{i \in I_\gamma} W_{p(i)}} \tag{7.27}$$

Proof:

We note that

$$F_2(i) = e^{\frac{\sigma J^2}{2} - \frac{\sigma}{2} \left\langle \phi(\square)^2 \right\rangle^{\,o}} e^{- \frac{\sigma}{2} : \phi(\square)^2 :}$$

similarly for $F_3(i)$, (replace $\square$ by $\square'$),
and $F_4(i) = F_2(i) \, F_3(i)$

$$\left. \right\}\ \ \ (7.28)$$

If we inspect (7.24) and (7.28) and use the fact that $i \in I_\gamma$ labels a subset of $N(\gamma)$ consisting of disjoint pairs, (i.e. pairs with mutually disjoint supports), see (7.17), we immediately conclude that the <u>chess board</u> estimate can be applied to the l.h.s. of (7.27); (see also [GJS2, FS, F3]). But the chess board estimate and the definition of the weights $W_1, \ldots, W_4$ yield (7.27). This completes the proof.

Finally we propose to prove that, for each finite J, and some fixed $\sigma \in (1, 2\pi^2)$

$W_i \nearrow \infty$, as $\lambda \searrow 0$, $\qquad\qquad\qquad\qquad\qquad\qquad$ (7.29)

for $i = 2,3,4$.

This follows from

Lemma 7.4:

Let σ be some fixed number in the open internal $(1, 2\pi^2)$. Then, for each finite $\lambda_o > 0$,

(1) $\hat{\alpha}_\infty(\lambda, \sigma)$ is bounded uniformly in $\lambda \in [0, \lambda_o]$, and

(2) $\alpha_\infty(\lambda, \sigma) \nearrow \infty$, as $\lambda \searrow 0$.

Proof:

By (7.26) and (7.25) (and conditioning, [GRS])

$$e^{\hat{\alpha}_\infty(\lambda,\sigma)} \leq e^{\hat{\alpha}_{\Lambda_o}(\lambda,\sigma)}$$

$$= \left(\left\langle e^{-\lambda U_{\Lambda_o}(P)} \; e^{\frac{\sigma}{2}\left[:\phi^2:(\Lambda_o) - \sum_{\Box \in \Lambda_o} :\phi(\Box)^2:\right]} \right\rangle^o_{\Lambda_o}\right)^{1/4} e^{\alpha \overset{o}{\Lambda_o}}$$

$$\leq \left(\left\langle e^{-2\lambda U_{\Lambda_o}(P)} \right\rangle^o_{\Lambda_o}\right)^{1/8}$$

$$\left(\left\langle e^{\sigma\left[:\phi^2:(\Box) - :\phi(\Box)^2:\right]} \right\rangle^o_{\Box}\right)^{1/2} e^{\alpha \overset{o}{\Lambda_o}} \qquad (7.30)$$

It is well known that the first factor on the r.h.s. of (7.30) is bounded unformly in $\lambda \in [0, \lambda_o]$. The second factor can be calculated explicitly:

$$\left\langle e^{\sigma\left[:\phi^2:(\Box) - :\phi(\Box)^2:\right]} \right\rangle^o_{\Box}$$

$$= \exp\left\{ \sum_{n=2}^\infty \frac{(2\sigma)^n}{n} \sum_{o \neq k \in \Gamma_\Box} [k^2 + 1]^{-n} \right\} \qquad (7.31)$$

where $\Gamma_\square$ is the momentum space lattice dual to $\square$. If $k \in \Gamma_\square$ and $k \neq 0$ then, obviously, $k^2 \geq 4\pi^2$, so that

$$\{(2\sigma)^n \sum_{0 \neq k \in \Gamma_\square} [k^2+1]^{-n}\}^{1/n} \to \frac{2\sigma}{4\pi^2+1} < 1,$$

as $n \to \infty$, which proves that the series in the exponent of the r.h.s. of (7.31) converges. This proves (1). Part (2) follows from exactly the same arguments used in Section 6, (6.13)-(6.15). This completes the proof of Lemma 7.4. If we now combine (7.21), (7.27) and (7.29) we see that there exists some $\lambda_{crit.} > 0$ such that, for all $\lambda \in (0, \lambda_{crit.})$, inequality (7.14) is true, for some $K \geq \log 3.5$. This implies (7.8) and completes the proof of the theorem.   Q.E.D.

Remarks

(1)   The phase transition for the pseudoscalar Y_2 theory (which has a symmetry taking ϕ to $-\phi$) follows by the same proof. One just must verify that, in a periodic box, grad $\phi-$ bounds and the chess board estimate apply.[1] This is straightforward, but see [F3]. The arguments proving that $\alpha(\lambda,\sigma) - \hat{\alpha}(\lambda,\sigma) \nearrow \infty$, as $\lambda \searrow 0$ (or $\lambda \nearrow 0$), for fixed $\sigma > \sigma_{crit}$ are as in the case of the $P(\phi)_2$ models, (Lemma 7.4 & Section 6).

(2)   For the $\lambda \cos(\epsilon\phi)_2$-theory (i.e. Model 5 of Section 2) the proof of existence of a phase transition is based on the same strategy as the one of Theorem 7.2, i.e. the Peierls argument, grad ϕ-bounds and the chessboard esti-

---

[1] i.e. the analogue of Lemma 4.5, Theorem 4.8.

mate. These estimates follow directly from Section 4
and the convergence of the lattice approximation proven
for $0 < \varepsilon^2 < \frac{16}{\pi}$ in [F1]. The estimates on the statistical
weight of contours (inequ. (7.14)) are then obtained as
in (7.20), (7.21) and (7.27), but in (7.20) we must set

$J = \frac{\pi}{4\varepsilon}$ (this is a convenient but not the only possible
choice), and the functions $F_2(i)$, $F_3(i)$ (see (7.20)) are
replaced by

$$e^{-\frac{\tilde{\lambda}}{\sqrt{2}}} \ e^{\lambda:\cos\varepsilon\phi\,(\Box^{(i)})}:\ ,$$

where $\tilde{\lambda} = \lambda\left[\left\langle\cos\varepsilon\phi\,(\Box\,)\right\rangle^{\,o}\right]^{-1}$

$$= \tilde{\lambda}\ \exp\left[\frac{\varepsilon^2}{2}\ (\Box\,,\,(-\Delta+1)^{-1}\Box)\right],$$

and <u>Wick ordering</u> is always done with respect to a <u>bare</u>
<u>mass = 1</u>.

The required estimates on the vacuum energy density
$\alpha_\infty(\lambda,\varepsilon,m_o)$ and on

$\hat{\alpha}_\infty(\lambda,\varepsilon,m_o) \leq \hat{\alpha}_{\Lambda_o}(\lambda,\varepsilon,m_o)$

$$= \frac{1}{4}\ \log\left\langle e^{\lambda\left[:\cos\varepsilon\phi:(\Lambda_o)-\sum\limits_{\Box\,\dot\subset\,\Lambda_o}:\cos\varepsilon\phi\,(\Box\,):\right]}\right\rangle^{\,o}_{\Lambda_o,m_o} + \alpha^o_{\Lambda_o}(m_o)$$

where m_o is the bare mass in the free Lagrangean and the
Gaussian measure with expectation $< - >^o_{\Lambda_o,m_o}$, are however
more delicate; but see [F3], also [F1].
We obtain

<u>Theorem 7.5:</u>

For the $\lambda\cos(\varepsilon\phi)_2$-theory, there exists some $\lambda_{crit.} > 0$

such that, for all $\lambda > \lambda_{crit.}$, all $\varepsilon^2 < \varepsilon^2_{crit.}$ $(\lambda) < \frac{16}{\pi}$ and

for $m_o \leq C \cdot \varepsilon$ (for some positive constant C and some $\varepsilon^2_{crit.}$

$(\lambda) > o)$ there is a phase transition, and the $\phi \rightarrow -\phi$

symmetry of the Lagrangean is spontaneously broken.

(3) Let $R(x)$, λ and σ be defined as in Theorem 7.2. Let Q be some arbitrary, real, not necessarily even poly-nimial with deg Q $(\leq)$ deg R. We define

$$P(x) = \lambda (R(x) + \varepsilon Q(x)) - \frac{\sigma}{2} x^2 + hx,$$

(7.32)

where $\sigma \in (1, 2\pi^2)$, $0 < \lambda < \lambda_c$,

with λ_c such as in Theorem 7.2, and $h \in [1,1]$. Then we have

Theorem 7.6:

There exists some $\varepsilon_{crit.} > 0$ such that, for all $\varepsilon \in (-\varepsilon_{crit.}, \varepsilon_{crit.})$ there are at least two disjoint $P(\phi)_2$-states (see Theorem 7.1), i.e. a phase transition, for at least one $h = h_{crit.}$ $(\lambda, \sigma, \varepsilon) \in [-1, 1]$.

Remark:

This is a field theoretic version of Theorem 5.3 and shows that phase transitions may occur without being accompanied by symmetry breaking.

The proof of Theorem 7.6 follows by first noting that all estimates established in the proof of Theorem 7.2, in particular inequality (7.8), hold uniformly in

ε and h, for $-\varepsilon_0 \le \varepsilon \le \varepsilon_0$, $-h_0 \le h \le h_0$ and some sufficiently small, positive ε_0, h_0. From Theorem 7.1, (2)
(cluster property for $<->_\pm$!) and inequality (7.8) we
now conclude that

$$| \langle x_+(\square) - x_- (\square) \rangle_\pm^{\varepsilon,h} |^2$$

$$= \lim_{|x| \to \infty} \langle (x_+(\square_0) - x_- (\square_0))(x_+(\square_x) - x_- (\square_x)) \rangle_\pm^{\varepsilon,h}$$

$$\ge \min_{y \in [0,1]} (y^2 + (1-y)^2) - 2 (\tfrac{1}{4} - \delta)$$

$$= 2\delta , \qquad\qquad\qquad (7.33)$$

where $<->_\pm^{\varepsilon,h}$ are the clustering states corresponding to
the polynomial P defined in (7.32) which have been constructed in Theorem 7.1, (2).

Next, Theorem 5.2 of ref. [FS] tells us that, for
all but countably many $h \ne 0$,

$$\lim_{|\varepsilon| \to 0} \langle x_+(\square) - x_- (\square) \rangle_\pm^{\varepsilon,h} = \langle x_+(\square) - x_- (\square) \rangle_\pm^{0,h}$$

$$(7.34)$$

and the r.h.s. of (7.34) is strictly positive, for $h>0$,
and strictly negative, for $h<0$, as a consequence of
Theorem 7.2, inequality (7.12) and the FKG-inequalities
that give monotonicity of $\langle x_+(\square) - x_- (\square) \rangle_\pm^{0,h}$ in h.

Combining this fact with (7.33) and (7.34) we con-
clude that, for sufficiently small $|\varepsilon| \ne 0$, the function
$\langle x_+(\square) - x_- (\square) \rangle_\pm^{\varepsilon,h}$ has a discontinuity in h at at

least one $h_{crit.} \in [-h_o, h_o]$, whence

$$<\!-\!>_+^{\varepsilon,h} \neq <\!-\!>_-^{\varepsilon,h} \,, \text{ for } h = h_{crit.} \qquad\qquad Q.E.D.$$

Conjecture:

For **all** positive polynomials R and fixed $\sigma \in (1, 2\pi^2)$ there exists some $\lambda_C = \lambda_C(R, \sigma) > 0$ with the property that for $0 < \lambda < \lambda_C$, there is an h_C such that, for

$$P(x) = \lambda R(x) - \frac{\sigma}{2} x^2 + h_C\, x,$$

$$<\!-\!>_+^P \neq <\!-\!>_-^P \,.$$

Our present estimates on α_∞ and $\hat{\alpha}_\infty$ are not sharp enough to provide a proof of this conjecture.

Remark:

The results of Section 7 are the basic input for our construction of <u>soliton sectors</u> in Part 2.

Section 8:

<u>QUANTUM CRYSTALS AND MORE</u>
<u>ABOUT THE PEIERLS ARGUMENT</u>

8.1 <u>Quantum Crystals</u> (a straightforward application of
[FSS])

Here we briefly sketch an extension of our results of Sections 4 and 5 to systems which may be interpreted as simple models of anharmonic (quantum) crystals. We consider a cubic lattice

$$
\mathbf{Z}_{\underline{\delta}}^{\nu} = \mathbf{Z}_{\delta_1} \times \dots \times \mathbf{Z}_{\delta_\nu} \text{, where}
$$

$$
\mathbf{Z}_{\delta} = \{x : x = \delta \cdot m; \, m \in \mathbf{Z}\}, \tag{8.1}
$$

i.e. δ_i is the lattice constant in the direction i. We set

$$
-\partial^i \sigma_\alpha = \frac{1}{\delta_i} (\sigma_\alpha - \sigma_{\alpha_+^i}),
$$

with α_+^i, ∂^i as in (2.6), (2.7).

We abbrviate $\prod_{i=1}^{\nu} \delta_i$ by $\pi\underline{\delta}$; \qquad (8.2)

Λ denotes a rectangle in $\mathbf{Z}_{\underline{\delta}}^{\nu}$.

The Hamilton function associated with Λ is of the form

$$
H_\Lambda^h (\{\sigma\}) = \frac{1}{2} \sum_{\alpha \in \Lambda} \pi\underline{\delta} \, \{ \sum_{i=1}^{\nu} J_i \, (\partial^i \sigma_\alpha)^2 + h\sigma_\alpha \} \tag{8.3}
$$

with periodic boundary conditions at $\partial\Lambda$.

Let $d\tilde{\lambda}$ be some single spin distribution in the class K_h introduced in (2.20), and let $<->_\Lambda$ denote the expectation associated with H_Λ^h and $d\tilde{\lambda}$. Then, for all $\Lambda \subseteq \mathbf{Z}_{\underline{\delta}}^{\nu}$, the following slightly more general version of <u>Theorem 4.1</u> holds:

$$\left\langle e^{\sum\limits_{\alpha,i} \pi\underline{\delta}\ g_i(\alpha)\cdot\partial^i\sigma_\alpha}\right\rangle_\Lambda \leq e^{\frac{1}{2}\sum\limits_{\alpha,i} \pi\underline{\delta}\ \frac{1}{J_i}\ |g_i(\alpha)|^2} \tag{8.4}$$

The proof is almost as in Section 4 (Steps 4.1-4.3), details are left to the reader.

An interesting example for this somewhat more general situation is an <u>anharmonic quantum crystal</u>:

Here $\nu=4$, the number N of components of σ is arbitrary; (σ is now interpreted as an N tuple of oscillator coordinates).

We let J_1, J_2, J_3 be arbitrary positive numbers; (for simplicity we may suppose that $J_1=J_2=J_3=J>0$), and e.g.

$$\delta_1 = \delta_2 = \delta_3 = 1.$$

Furthermore

$$J_4 = \frac{1}{M},\ \delta_4 = \frac{\beta}{2L+1},\ L = 1,2,3,\ldots, \tag{8.5}$$

for some positive constants M and β.

The rectangle Λ has the form

$$\Lambda = \Lambda_1 \times [0,\beta] \tag{8.6}$$

with Λ_1 a rectangle $\subseteq \mathbb{Z}^3$.

The single spin distribution is given by

$$d\tilde{\lambda}(\sigma) = e^{-\delta_4 V(\sigma)} d^N\sigma \tag{8.7}$$

where V is some real, measurable function on $\mathcal{R}^N$ with

$$\int e^{-\beta V(\sigma)} d^N\sigma < \infty. \tag{8.8}$$

The equilibrium measure corresponding to

H_Λ^o - see (8.3) - $\delta_4 = \dfrac{\beta}{2L+1}$ - see (8.5) - and $d\tilde{\lambda}$ is

denoted $d\mu_{\Lambda_1,L}^\beta$

Then, for $|\Lambda_1| < \infty$, one can prove:

$$\lim_{L\to\infty} \int d\mu_{\Lambda_1,L}^\beta (\{\sigma\}) \; e^{i \sum_\alpha \frac{\beta}{2L+1} g(\alpha)\sigma_\alpha}$$

$$= \int d\mu_{\Lambda_1,\infty}^\beta (\{\sigma\}) \; e^{i\sum_{I\in\Lambda_1} \int_0^\beta dt\, g(t,I)\sigma_I(t)} \tag{8.9}$$

for all bounded, periodic functions g(t,I) on $[o,\beta] \times (\Lambda_1\cap\mathbf{Z}^3)$ which are continuous in t. Here $d\mu_{\Lambda_1,\infty}^\beta$ is a perturbed Wiener measure with $N|\Lambda_1|$ dimensional state space and periodic boundary conditions at $t=0,\beta$. The proof of convergence in (8.9) follows by combining the Trotter product formula and the Golden-Thompson inequality (using (8.8)) with the Feynman-Kac formula.

The measure $d\mu_{\Lambda_1,\infty}^\beta$ is the __equilibrium path space measure__ at inverse temperature β of a quantum crystal with __quantum mechanical Hamiltonian__

$$\mathcal{H}_{\Lambda_1} = \sum_{I\in\Lambda_1} \{-\frac{1}{2M}\Delta_{\sigma_I} + V(\sigma_I) + \sum_{\{I':|I-I'|=1\}} \frac{J}{4}(\sigma_I-\sigma_{I'})^2\}$$

$$\tag{8.10}$$

where Δ_{σ_I} is the Laplacean in the variables σ_I.

The moments of $d\mu^{\beta}_{\cdot,\infty}$ are the imaginary time ("tempera-ture-ordered") Green's functions of the system.

The point we want to make is that inequality (8.4) combined with a lower bound on the second moment of $d\mu^{\beta}_{\mathbf{Z}^3,\infty}$

$$\int d\mu^{\beta}_{\mathbf{Z}^3,\infty} (\{\sigma\}) \ \sigma_o \cdot \sigma_o \geq \text{const.} > 0, \qquad (8.11)$$

uniformly in β, (which must be derived from the proper-ties of V; see e.g. Section 5, Theorem 5.2) yields existence of phase transitions for sufficiently large β, for many (physically interesting) choices of V; (note that the truncated second moment is bounded above by $O(\beta^{-1})$ by (8.4)).

We leave it to the reader to make a list of poten-tials V for which (8.11) is valid and to exploit con-sequences of the Lee-Yang theorems of [SG,DN], for the case where V is a quartic polynomial.

8.2 Some more consequences of the Peierls argument

(a) Theorem 5.3 remains true in $\nu=2$ dimensions:

This is an extension of a recent result of Sylvester and van Bejeren [SvB] to single spin distributions that are not necessarily invariant under $\sigma \to -\sigma$.

The (somewhat complicated) proof follows from the Peierls argument and is very similar in spirit to the proofs of Theorems 7.2 and 7.6.

Rather than proving discontinuity of $<\sigma_0>^{h,J}$ at some $h = h_{crit.}$ (as in Theorem 5.3) one proves discontinuity of $<\chi_+(o) - \chi_-(o)>^h$ at some $h = h_{crit.}$ (see Theorem 7.6). Details are contained in some unpublished work of the author.

(b) <u>Anisotropic N-vector models</u>; [Ma,Ku]

Let $d\lambda(\sigma) \neq \delta_0(\sigma) d^N\sigma$ be some $O(N)$-invariant Borel probability measure on $\mathscr{R}^N$.

We define $d\lambda_J(\sigma) = d\lambda(\frac{\sigma}{\sqrt{J}})$, $\qquad\qquad$ (8.12)

and $\nu \geq 2$.

Let a finite volume Gibbs measure be given by

$d\mu_\Lambda^{\varepsilon,J}(\{\sigma\}) =$

$Z_\Lambda(\varepsilon,J)^{-1} \exp[-\frac{1}{4} \sum_{\substack{\alpha \in \Lambda \\ |\alpha'-\alpha|=1}} \{(\sigma_\alpha - \sigma_{\alpha'})^2 + \varepsilon(\sigma_\alpha^1 - \sigma_{\alpha'}^1)^2\}]$

$\times \prod_{\alpha \in \Lambda} e^{W_\varepsilon(\sigma_\alpha)} d\lambda_J(\sigma_\alpha)$, $\qquad\qquad$ (8.13)

where $W_\varepsilon(\sigma) = 2\varepsilon(\sigma^1)^2 - \varepsilon\sigma\cdot\sigma$, and let $d\mu^{\varepsilon,J}$ be some infinite volume limit of $\{d\mu_\Lambda^{\varepsilon,J}\}$.

Theorem:

For arbitrarily small, fixed $\varepsilon > 0$ (or for $\varepsilon = e\ J^{-1}$ log J, with $e > \frac{1}{2}(N-1)$) there exists some $J_C < \infty$ such that, for all $J > J_C$, there is a <u>phase transition</u>: $d\mu^{\varepsilon,J}$ is a superposition of at least two pure phase (ergodic) measures some of which are spontaneously magnetized.

Remark:

This theorem extends work of Malishew [Ma] and Kunz [Ku]. The proof is again based on the Peierls argument and is contained in (unpublished) work of E. Lieb and the author.

(c) <u>Triple points</u>

Let $\nu = 2,3,\ldots$, $N = 1$, and

$$d\tilde{\lambda}(\sigma) \stackrel{e.g.}{=} \cdot \frac{1}{2W_J+1}\ \{W_J\ \delta(\sigma-J)+\delta(\sigma) + W_J\ \delta(\sigma+J)\}d\sigma, \quad (8.14)$$

where W_J is e.g. a monotone increasing function with the property that $W_J = 1$, for $J = J_{crit.} \geq J_\nu$, (for a certain $J_\nu < \infty$). Let H_Λ be the usual Hamilton function and $d\mu^J$ some infinite volume Gibbs measure corresponding to H_Λ and $d\tilde{\lambda}$ (as in (8.14)).

Then
(c1) for sufficiently small $J > 0$ $d\mu^J$ is <u>ergodic</u> and <u>unique</u> (in the sense that it is independent of boundary conditions used in the construction of the thermodynamic limit). There is no spontaneous magnetization. The proof

is by means of a "high temperature expansion".

(c2) for $J=J_{crit.}$ $d\mu^J$ is the superposition of at least three different pure phase (ergodic) measures, at least two of which have spontaneous magnetization.

Proof:

This follows by combining the results of Section 3.2 (Proposition 3.1 and Theorem 3.2, with m=3 orthogonal projections at each site: onto $\sigma=\pm J$, $\sigma=o$), the lattice version of the first inequality in (7.20) (see also (7.24)) and Gaussian domination (Theorem 4.1).

$$\text{Q.E.D.}$$

(c3) Suppose e.g. that $W_J \nearrow \infty$, as $J \nearrow \infty$. Then we conjecture that, for $J \gg J_{crit.}$ there are precisely two different pure phase equilibrium states with non-vanishing, opposite spontaneous magnetization. (Presumably this follows from low temperature methods. It is known that under these conditions there are at least two different pure phases.)

Remark:

Examples of the type of (c) with an arbitrary degree of sophistication may be invented; (also for the case where $N>1$; e.g. $d\lambda(\sigma)$ a convex combination of $\{\delta(|\sigma|-j_\ell J)d^N\sigma\}_{\ell=1}^n$, $0 \le j_1 < \ldots < j_n$, $0<J$). The point is that the Peierls argument in the form of method 3.2 and Theorem 4.1 combined with (7.20), (7.21), (7.24) supply a systematic approach to their analysis.

The interest in such examples depends on our ability
of inventing and analyzing models that may describe certain aspects of real physical systems. One interesting
open problem of the type discussed here is to prove or
disprove the existence of triple points in $P(\phi)_2$ quantum
field models, (with deg $P \geq 6$).

Part 2:

Section 9:

PHASE TRANSITIONS AND THE SPONTANEOUS
OCCURRENCE OF CHARGED SUPER-SELECTION (SOLITON-) SECTORS

In this section we explain a basic connection between phase transitions and the existence of non-trivial
super-selection rules ("topological charges") in two
dimensional quantum field models.

We show that (under suitable hypotheses specified
in [F7]) the following holds.

Theorem 9.1: [F7]

In two space-time dimensions, any quantum field
theory with an n-fold degenerate physical vacuum has
n(n-1) charged super-selection (soliton) sectors (disjoint from the vacuum sectors). These sectors are
labelled by the elements of a finite group, (the so
called soliton-group).

We shall not present a general proof of this theorem
at this place. This has been done in $[F7, \S 6]$; see also
$[Ro]$. We rather illustrate the general theory of soliton-
sectors in two dimensions by studiing one specific
model, (the anisotropic $(\vec{\phi} \cdot \vec{\phi})_2^2$-model). But we remark
that Theorem 9.1 applies to all the models analyzed in
Section 7 and proves the existence of non-trivial topological charges and soliton states for those models; $[F7]$.
Particularly clear interpretations of the _physical_ mechanism responsible for the existence of soliton states
can be given in the context of the sine-Gordon theory,
with or without mass term; see $[Co\ 2,3,\ F1,\ F\ Se,\ F7]$;
Perhaps the most interesting model is the pseudo scalar
Yukawa model. Mathematically, the $P(\phi)_2$-models (with
non-even P) are most fascinating from the point of view
of solitons: For these models the existence of soliton-
states is still _unknown_.

Here we propose to study the anisotropic $(\vec{\phi} \cdot \vec{\phi})_2^2$ -
model, with $\vec{\phi} = (\phi_1, \phi_2)$ a pair of neutral, scalar fields,
and with the continuous O(2)-symmetry explicitly broken
by a _mass term_. This because, for this model, our results are most complete, and the presentation of the
analysis of solitons is simplest; but see $[F7]$. We
briefly define this model:

The _bare mass_ is fixed to be =1, and Wick ordering
is always done with respect to bare mass = 1. The interaction Lagrangean of the model is given by

$$\lambda : (\vec{\phi} \cdot \vec{\phi})^2 : - \frac{\sigma}{2} : \phi_1^2 : , \quad \lambda > 0, \quad \sigma \in (1, 2\pi^2), \qquad (9.1)$$

whence the Euclidean action

$$U_\Lambda = \int_\Lambda [\lambda : (\vec{\phi} \cdot \vec{\phi})^2 : (x) - \frac{\sigma}{2} : \phi_1^2 : (x)] d^2x \qquad (9.2)$$

Existence of this model in the infinite volume limit has been proven in $[F4,6]$. With ϕ, $\chi_+(\square)$,.... replaced by ϕ_1, χ_+ $(\phi_1(\square))$,... Theorem 7.2 immediately extends to this model and proves the existence of a phase transition and spontaneous $\phi_1 \to -\phi_1$ symmetry breaking, for all $0 < \lambda < \lambda_c$, for some finite λ_c; (see Theorem 7.2, Lemma 7.4).

We now fix $\sigma > 1$ and $\lambda < \lambda_c$. Then there are two <u>different</u>, pure phase Euclidean vacuum expectations

$$<->_+ \neq <->_- \qquad (9.3)$$

satisfying all the axioms of Osterwalder and Schrader $[OS]$, and

$$\langle \phi_1 \rangle_+ = - \langle \phi_1 \rangle_- = \Phi_c > 0 \qquad (9.4)$$

The Osterwalder-Schrader reconstruction gives two pure Wightman states ω_+ and ω_- with the property that

$$\omega_\pm (\prod_{i=1}^n \vec{\phi}(\vec{f}_i)) = W_\pm^{(n)} (\vec{f}_1, \dots, \vec{f}_n) \qquad (9.5)$$

Here $\vec{\phi}(\vec{f}) = \phi_1(f^1) + \phi_2(f^2)$ is the smeared <u>relativistic</u> quantum field, $f_i^j \in \mathscr{S}(\mathscr{R}^2)$, for all i,j and $W_\pm^{(n)}$ are the (pure phase) n-point Wightman distributions (obtained as a boundary value of the analytic continuation in the time arguments of the n point Euclidean Green's functions, i.e. the nth moments of $<->_\pm$).

In the following we propose to construct two states ω_s and $\omega_{\bar{s}}$ interpolating between ω_+ and ω_- (the soliton-, anti-soliton state, resp.).

Our construction is best explained in the frame-work of the theory of algebras of local observables [HK] adapted to the two dimensional models; see [GJ 2,3]

Let $\vec{\phi}(\vec{x}) = \vec{\phi}(\vec{x},o)$ denote the time 0-quantum field, and

$$\vec{\pi}(\vec{x}) = (\frac{\partial}{\partial t} \vec{\phi})(\vec{x},o) \tag{9.6}$$

As the $(\vec{\phi} \cdot \vec{\phi})^2_2$ - model is a __canonical__ quantum field theory, π_1, π_2 are the momenta canonically conjugate to ϕ_1, ϕ_2, respectively. Let 0 be some finite union of compact intervals on space. We define a real, local Sobolev space of distributions supported in 0 by

$$\mathcal{H}_{-1/2}(0) = \{f : f \text{ real-valued, supp } f \subseteq 0, ||f||_{-1/2} < \infty\},$$

$$\tag{9.7}$$

where $||f||^2_{-1/2} \equiv (f, (\frac{d^2}{d\vec{x}^2} + 1)^{-1/2} f)_{L^2}$, and

$$\mathcal{H}_0 = \mathcal{H}_{-1/2}(0) \times \mathcal{H}_{-1/2}(0).$$

A __local von Neumann algebra__ $\mathcal{O}(0)$ is defined as the weak closure of the operators

$$\{e^{i(\vec{\phi}(\vec{f}) + \vec{\pi}(\vec{g}))} : \vec{f}, \vec{g} \in \mathcal{H}_0\}$$

on the Fock space $\mathcal{F}$ of the __free__ field $\vec{\phi}_{j}(\lambda = \sigma = 0)$, with bare mass 1.

A C^*-algebra $\mathcal{A}$ is now obtained by taking the closure of $\bigcup_{O \subset \mathcal{R}} \mathcal{A}(O)$ in the operator norm.

The following results, due to Glimm and Jaffe [GJ2,3], are known for the $(\vec{\phi} \cdot \vec{\phi})_2^2$ - theory:

F1) The restrictions of the states $\omega_\pm$ to the algebras $\mathcal{A}(O)$ are ultraweakly continuous, for all bounded O. This is the so called locally Fock property established in [GJ2].
As a consequence, the vacua ω_+, ω_- are well defined states on the algebra $\mathcal{A}$.

F2) The Hilbert spaces $\mathcal{H}_\pm$ obtained from the Wightman distributions (9.5) by means of Wightman reconstruction coincide with the Hilbert spaces $\mathcal{H}_\pm^{G.N.S.}$ obtained from $(\omega_\pm, \mathcal{A})$ by means of the Gel'fand-Naimark-Segal construction; see [GJ2,3]. (Note that, a priori, $\mathcal{H}_\pm$ could be a proper subspace of $\mathcal{H}_\pm^{G.N.S.}$).

F3) The representations of $\mathcal{A}$ on $\mathcal{H}_+$, $\mathcal{H}_-$ are irreducible (a consequence of the fact that ω_+ and ω_- are pure phase vacuum states). They are disjoint from each other.

F4) Let $\bar{O}$ be the union of diamonds (intersection of a forward and a backward light cone) with base O, and let $\bar{O}_{(\Lambda,a)}$ be the space-time region obtained from $\bar{O}$ by a Poincaré transformation (Λ,a).

Since one knows the quantum fields $\vec{\phi}(\vec{f})$, $\vec{f} \in \mathcal{J}_{real}(\mathbb{R}^2)^{\times 2}$ are selfadjoint, one can define local algebras $\mathcal{O}(\bar{0})$ in the obvious manner. Glimm and Jaffe prove [GJ2] that

$$\mathcal{O}(0) = \mathcal{O}(\bar{0}) \tag{9.8}$$

The Poincaré automorphisms of $\mathcal{O}$, formally defined by

$$(\tau_{(\Lambda,a)}\vec{\phi})(\vec{x},t) = \vec{\phi}(\Lambda^{-1}(\vec{x}-\vec{a}_1, t-a_o)) \tag{9.9}$$

have the property that

$$\tau_{(\Lambda,a)}(\mathcal{O}(\bar{0})) = \mathcal{O}(\bar{0}_{(\Lambda,a)}). \tag{9.10}$$

They are implemented on $\mathcal{H}_\pm$ by the unitaries of the representation U of the Poincaré group on $\mathcal{H}_\pm$ obtained by Wightman reconstruction; see e.g. [CJ, McB] and refs. given there.

Definition:

Given some bounded diamond $\bar{0}$ we let $\sim\bar{0}$ denote its causal complement. In two space-time dimensions $\sim\bar{0} = \bar{0}_L \cup \bar{0}_R$, where $\bar{0}_{L(R)}$ is a wedge region opening to the left (the right).

Construction of soliton states:

We are now prepared to explain the construction of the states ω_s and $\omega_{\bar{s}}$, (a C^*-algebra construction). We want these states to have the following two basic properties; (the reader is advised to consult Section 1 for heuristic motivation):

(S1) There is some bounded diamond $\bar{O}$ such that

$$\omega_s(A) = \omega_-(A), \text{ for all } A \in \mathcal{O}(\bar{O}'), \text{ for all } \bar{O}' \subset \bar{O}_L \ ;$$

$$\omega_s(A) = \omega_+(A), \text{ for all } A \in \mathcal{O}(\bar{O}''), \text{ for all } \bar{O}'' \subset \bar{O}_R \ .$$

(For $\omega_{\bar{s}}$ exchange + and -).

(S2) The states ω_s and $\omega_{\bar{s}}$ are <u>Poincaré-covariant</u>, in the sense that the automorphisms defined in (9.9) and (9.10) are implemented by a continuous, unitary representation U_s ($U_{\bar{s}}$) of the Poincaré group on the Hilbert spaces $\mathcal{H}_s$, $\mathcal{H}_{\bar{s}}$ obtained from $(\omega_s, \mathcal{O})$, $(\omega_{\bar{s}}, \mathcal{O})$, resp., by means of the G.N.S. construction.

The idea of how to construct states ω_s, $\omega_{\bar{s}}$ satisfying (S1) and (S2) is inspired by algebraic field theory in the form of [DHR]: We propose to construct <u>automorphisms</u> σ_s and $\sigma_{\bar{s}}$ of the algebra $\mathcal{O}$ with the property that

$$\left.\begin{array}{l} \omega_s(A) = \omega_+ \circ \sigma_s \ (A) \equiv \omega_+(\sigma_s(A)) \\[2ex] \omega_{\bar{s}}(A) = \omega_+ \circ \sigma_{\bar{s}} \ (A) \equiv \omega_- \circ \sigma_s(A), \end{array}\right\} \qquad (9.11)$$

for all $A \in \mathcal{O}$

248

In two space-time dimensions existence of [*]automorphisms $\sigma_s, \sigma_{\bar{s}}$ with the property that ω_s and $\omega_{\bar{s}}$ satisfy (S1), (S2) is a very general fact; see $[F7, \S5-6]$.

We attempt to define the automorphisms $\sigma_s, \sigma_{\bar{s}}$ explicitly and then outline the verification of (S1), (S2). Our explicit definition is motivated by the heuristic considerations presented in the Introduction (Section 1); (see Fig. 1,2).

Definition 9.2:

Consider the $(\vec{\phi} \cdot \vec{\phi})_2^2$-theory. Let ϕ_1, ϕ_2, π_1 and π_2 be the canonical variables defined above.

Let $\alpha(\vec{x})$ be a function on space with the property that

$$0 \leq \alpha(\vec{x}) \leq \pi \quad (= 3,14\ldots)$$

and, for some finite L,

$$\alpha(\vec{x}) = \pi, \text{ for } \vec{x} \leq -L, \ \alpha(\vec{x}) = 0, \text{ for } \vec{x} \geq L; \tag{9.12}$$

$$\frac{d}{d\vec{x}} \alpha(\vec{x}) \in L^2(\mathfrak{R}).$$

(9.12) is called the "soliton-condition".
The automorphism σ_s is given by the following Bogoliubov transformation

$$\sigma_s(\phi_1(\vec{x})) = \cos\alpha(\vec{x}) \ \phi_1(\vec{x}) + \sin\alpha(\vec{x}) \ \phi_2(\vec{x})$$

$$\sigma_s(\phi_2(\vec{x})) = -\sin\alpha(\vec{x})\phi_1(\vec{x}) + \cos\alpha(\vec{x}) \ \phi_2(\vec{x}) \tag{9.13}$$

+ identical equations for π_1 and π_2.

Let ϱ_π denote the *automorphism taking $\vec{\phi}$ to $-\vec{\phi}$ and $\vec{\pi}$ to $-\vec{\pi}$. We set

$$\sigma_s^- = \sigma_s \circ \varrho_\pi \qquad\qquad (9.14)$$

(Obviously $\omega_\pm = \omega_\mp \circ \varrho_\pi$).

_____

Next we explain why the transformations defined in (9.13), (9.14) are well defined *automorphisms of $\mathcal{O}$. Let O be a finite union of compact intervals, and let $\Psi_O(\vec{x})$ be a C^∞-function of compact support such that

$$\Psi_O(\vec{x}) = 1, \text{ for all } \vec{x} \in O.$$

We set $\alpha_O = \alpha \Psi_O$ and define the operator

$$L(\alpha_O) = \int d\vec{x} \ \alpha_O(\vec{x}) \ [\phi_1(\vec{x}) \pi_2(\vec{x}) - \pi_1(\vec{x}) \phi_2(\vec{x})] \qquad (9.15)$$

Lemma 9.3:

(1) The operator $L(\alpha_O)$ is selfadjoint on $\mathcal{H}_+$, $\mathcal{H}_-$, and $e^{\pm iL(\alpha_O)} \in \mathcal{O}(\text{supp } \psi_O)$.

(2) For all A in some subalgebra weakly dense in $\mathcal{O}(O)$

$$e^{iL(\alpha_O)} A \ e^{-iL(\alpha_O)} = \sigma_s(A),$$

with σ_s given by (9.13).

The proof of Lemma 9.3 is given in $[F7,\S3]$. It tells us that the formal Definition 9.2 really determines a well defined *automorphism of $\mathcal{A}$.

If we now define the states ω_s, $\omega_{\bar{s}}$ by (9.11) we easily see that condition (S1) is satisfied. Furthermore it is straight forward to show that the representations of $\mathcal{A}$ on the spaces $\mathcal{H}_+$, $\mathcal{H}_-$, $\mathcal{H}_s$ and $\mathcal{H}_{\bar{s}}$ are disjoint irreducible representations. Parity is the conjugation mapping $\mathcal{H}_s$ onto $\mathcal{H}_{\bar{s}}$ and vice versa, (i.e. parity is spontaneously broken on the soliton sectors $\mathcal{H}_s$, $\mathcal{H}_{\bar{s}}$). For proofs see [F7,§3]. A heuristic argument for all that is as follows:
Define the conserved "topological" charge

$$Q = \int d\vec{x} \; (\frac{\partial}{\partial \vec{x}} \; \phi_1) \; (\vec{x},t)$$

(9.16)

$$= \int d\vec{x} \; (\frac{d}{d\vec{x}} \; \phi_1) \; (\vec{x})$$

(associated with the conserved current

$$(j^0(x),j^1(x)) = ((\frac{\partial}{\partial \vec{x}} \phi_1)(x), \; (\frac{\partial}{\partial t} \phi_1)(x))) \; .$$

Then $Q \, \mathcal{H}_\pm = \{\vec{0}\}$,

$$Q\psi \quad = 2\phi_c \; \psi, \; \text{for all } \psi \in \mathcal{H}_s$$

$$Q\psi \quad = -2\phi_c \; \psi, \; \text{for all } \psi \in \mathcal{H}_{\bar{s}}$$

(9.17)

Finally, parity takes Q to $-Q$.

Important remarks:

(1) The spaces $\mathcal{H}_s$, $\mathcal{H}_{\bar{s}}$ and the representations of $\mathcal{A}$ on $\mathcal{H}_s$, $\mathcal{H}_{\bar{s}}$ are independent of the specific choice

of the function α: All functions α satisfying the soliton-condition (9.12) (for some finite L) yield _identical_ soliton-sectors.

(2) As a special case of a _general theorem_ proven in $[F7, \S 6]$ we note that $\wp_e$ (the identity automorphism) $\wp_\pi$, σ_s and $\sigma_{\bar s}$ form an abelian group (the soliton group of $(\vec\phi \cdot \vec\phi)_2^2$):

$$\wp_\pi^{\,2} = \sigma_s^{\,2} = \sigma_{\bar s}^{\,2} = \wp_e, \tag{9.18}$$

$$\sigma_s \circ \wp_\pi = \sigma_{\bar s}, \quad \sigma_{\bar s} \circ \wp_\pi = \sigma_s, \quad \sigma_s \circ \sigma_{\bar s} = \wp_\pi$$

Next we propose to discuss and partly prove, for the the special case of the anisotropic $(\vec\phi \cdot \vec\phi)_2^2$ - theory defined in (9.1)-(9.2) the following _main result_

Theorem 9.4: (see $[F7]$)

Let $\#$ be s or $\bar s$, and suppose that

$$\omega_+ (\phi_1) = -\omega_- (\phi_1) = \phi_c > 0.$$

Then
(1) There exists a continuous unitary representation $U_\#$ of the Poincaré group on $\mathcal{H}_\#$ implementing the automorphism group $\{\tau_{(\Lambda, a)}\}$ of $\mathcal{O}$; i.e. condition (S2) holds.

(2) The spectrum of the energy-momentum operator $(H_\#, P_\#)$, the infinitesimal generator of space-time translations, is
(a) contained in the forward light cone
(b) purely continuous; (in particular $\mathcal{H}_\#$ does _not_

contain any vacuum states).

(3) There exist (Q-) charged <u>soliton fields</u> $s(x), (\bar{s}(x))$ with non-vanishing matrix elements between $\mathcal{H}_+$ and $\mathcal{H}_s$ ($\mathcal{H}_+$ and $\mathcal{H}_{\bar{s}}$). These fields can be constructed to be "almost local" fields and such that they are "almost local" relative to <u>even</u> functions of the fundamental field $\vec{\phi}$; (see $[F7,\S4]$).

(4) If $\mathcal{H}_s$ contains a one-particle-state (<u>the quantum soliton</u>) then $\mathcal{H}_{\bar{s}}$ contains a one-particle-state (the <u>anti-soliton</u>) with equal quantum numbers except opposite Q-charge. For sufficently small $\lambda \in (o, \lambda_c)$, the <u>mass</u> of the soliton is <u>positive</u>.

If $N_s (N_{\bar{s}})$ denotes the number of solitons (anti-solitons) in a <u>scattering state</u> (constructed according to the <u>Haag-Ruelle theory</u>) then

(a) $N_s - N_{\bar{s}}$ is even, for all scattering states in $\mathcal{H}_\pm$;

(b) $N_s - N_{\bar{s}}$ is odd, for all scattering states in $\mathcal{H}_\#$.

<u>Remark:</u>

The proof of Theorem 9.4 is given in $[F7, F7']$. It is technically rather difficult. In the next section we sketch what we feel are the main ideas of the proof.

Section 10:

SOME REMARKS ON THE PROOF OF THEOREM 9.4;
CONCLUSIONS

The main physical as well as technical idea behind
the proof of Theorem 9.4 is the following: Construct
the soliton state ω_s as a weak $*$ limit of states representing a pair of a soliton and an anti-soliton, i.e.
of vector states in $\mathcal{H}_+$ or $\mathcal{H}_-$ with Q-charge 0, by
sending the anti-soliton off "to behind the moon",
(mathematically to spatial ∞). Derive properties of ω_s,
in particular Theorem 9.4, by analyzing in detail these
approximating vector states in $\mathcal{H}_\pm$.

Formulated in terms of $*$ automorphisms of the
algebra $\mathcal{O}\!\ell$ this program is expressed as follows:

Represent σ_s (see (9.13) and Lemma 9.3) as a limit
of inner $*$ automorphisms of $\mathcal{O}\!\ell$; (these are then unitarily
implementable on $\mathcal{H}_\pm$). Next we outline how this idea is
realized technically.

Let α be a function satisfying the soliton-condition (9.12), and let O be the smallest compact inter-
val containing supp $(\frac{d}{d\vec{x}}\,\alpha)$, $\bar{O}$ the diamond with base O.

Let $O_{\vec{x}} \equiv \{\vec{y}:\vec{y}-\vec{x}\in O\}$ and $\bar{O}_{\vec{x}}$ denote the translates
of O, $\bar{O}$, respectively, by the vector $\vec{x}$. We define $\alpha_{\vec{x}} =$
$\alpha(\vec{y}-\vec{x})$ and pick $\vec{x}$ such that O and $O_{\vec{x}}$ are disjoint.

Obviously, the support of $\alpha-\alpha_{\vec{x}}$ is compact,(denote
it by $\Lambda_{\vec{x}}$), so that

$$V_{\alpha,\vec{x}} \equiv e^{iL(\alpha-\alpha_{\vec{x}})} \in \mathcal{U}(\Lambda_{\vec{x}}) ; \tag{10.1}$$

clearly $V_{\alpha,\vec{x}}$ is a well defined unitary operator on $\mathcal{H}_{\pm}$, $(\mathcal{F}, \mathcal{H}_s, \mathcal{H}_{\bar{s}})$.

Let $\Omega_{\pm}$ denote the physical vacuum in $\mathcal{H}_{\pm}$. The vector $V_{\alpha,\vec{x}}\Omega_{\pm}$ can be thought of as representing a pair of a soliton and an anti-soliton, and

$$\omega_s(\cdot) = w^* - \lim_{\vec{x}\to-\infty} (V_{\alpha,\vec{x}}\Omega_+, \cdot\; V_{\alpha,\vec{x}}\Omega_+) \tag{10.2}$$

Moreover

$$\sigma_s(A) = n-\lim_{\vec{x}\to-\infty} V_{\alpha,\vec{x}} A V_{\alpha,\vec{x}}^*, \tag{10.3}$$

for all $A \in \mathcal{U}$.

Equations (10.2) and (10.3) make precise how the announced program is to be understood. Let $\xi \equiv (\Lambda, a)$ denote the elements of the Poincaré group $\mathcal{P}_+^\uparrow$.

We define a "propagator" (a so called 1-cocycle; see e.g. [Ro])

$$\Gamma_{\alpha,\vec{x}}(\xi) = \tau_\xi (V_{\alpha,\vec{x}}) V_{\alpha,\vec{x}}^* \tag{10.4}$$

$$(= U_\pm^*(\xi) V_{\alpha,\vec{x}} U_\pm(\xi) V_{\alpha,\vec{x}}^*,$$

as operators on $\mathcal{H}_\pm$). Here τ_ξ denotes the Poincaré automorphism defined in (9.9), (9.10). A straight-forward calculation shows that

$$\Gamma_{\alpha,\vec{x}}(\xi \cdot \xi') = \tau_{\xi'}(\Gamma_{\alpha,\vec{x}}(\xi)) \cdot \Gamma_{\alpha,\vec{x}}(\xi') \qquad (10.5)$$

This is called the cocycle identity.

10.1 The role of local 1-cocycles in proving Theorem 9.4:

In the following we propose to analyze the 1-co-cycles $\Gamma_{\alpha,\vec{x}}(\xi)$ more closely and we show how they are used in the proof of Theorem 9.4. Let $\bar{\Lambda}_{\vec{x}}$ be the diamond with base $\Lambda_{\vec{x}}$, and let $\bar{\Lambda}_{\vec{x},\xi}$, $(\bar{0}_\xi)$ be the space-time region obtained by applying the Poincaré transformation ξ to $\bar{\Lambda}_{\vec{x}}$, $(\bar{0}$, resp.) Since $\bar{\Lambda}_{\vec{x}}$ is compact, so is $\bar{\Lambda}_{\vec{x},\xi}$, and

$$\Gamma_{\alpha,\vec{x}}(\xi) \in \alpha(\bar{\Lambda}_{\vec{x}} \cup \bar{\Lambda}_{\vec{x},\xi}), \qquad (10.6)$$

i.e. $\Gamma_{\alpha,\vec{x}}(\xi)$ is a strictly local observable. By (F2) and the locally Fock property (F1) we may therefore analyze $\Gamma_{\alpha,\vec{x}}(\xi)$ as an operator acting on the Fock space $\mathcal{J}$ of the free field. This permits us to appeal to the following theorem due to Cannon and Jaffe [CJ] (see also Mc Bryan [McB]):

Given some compact neighborhood N of the identityI of the Poincaré group $p_+^\uparrow$ there exist unitary operators $U_{N,\Lambda_{\vec{x}}}(\xi)$ on $\mathcal{J}$ such that, as an operator equation on.$\mathcal{J}$

$$\Gamma_{\alpha,\vec{x}}(\xi) = U_{n,\Lambda_{\vec{x}}}^*(\xi)\, V_{\alpha,\vec{x}}\, U_{N,\Lambda_{\vec{x}}}(\xi)\, V_{\alpha,\vec{x}}^*, \qquad (10.7)$$

for all $\xi \in N$.

The advantage of this representation of the cocycle $\Gamma_{\alpha,\vec{x}}(\xi)$ is that the operators $U_{N,\Lambda_{\vec{x}}}$ are known __explicitly__, and detailed estimates (in particular Trotter product formulas) are available; see $[CJ, McB]$, and §3, Lemma 3 of $[F7]$ for an application.

These estimates combined with

(1) duality for the free field, first proven by Araki $[A]$, (see Lemma 1, §3 of $[F7]$),

(2) the energy - momentum - and boost densities of the anisotropic $(\vec{\phi}\cdot\vec{\phi})_2^2$-theory are __invariant__ under $\mathcal{S}_\pi$ (i.e. under the substitution $\vec{\phi}\to-\vec{\phi}$) are used in $[F7]$, (see Lemma 3, §3 of $[F7]$ and $[F7']$) to prove

__Theorem 10.1:__ $[F7, F7']$

Given O and N, there are compact diamonds

$$B(O,N) \supseteq \bigcup_{\xi \in N} \bar{O}_\xi \quad \text{and} \quad B(O_{\vec{x}},N) = B(O,N)_{\vec{x}} \supseteq \bigcup_{\xi \in N} \bar{O}_{\vec{x},\xi}$$

such that if $|\vec{x}|$ is so large that $B(O,N)$ and $B(O_{\vec{x}},N)$ are __space-like separated__ then

$$\Gamma_{\alpha,\vec{x}}(\xi) = \Gamma_\alpha(\xi)\Gamma_{-\alpha_{\vec{x}}}(\xi), \tag{10.8}$$

for all $\xi \in N$, and

(1) for all $\xi \in N$, $\Gamma_\alpha(\xi) \in \mathcal{O}(B(O,N))$ is a __local cocycle__ that is strongly continuous in ξ ,

(2) $\Gamma_{-\alpha_{\vec{x}}}(\xi) \in \mathcal{O}(B(O_{\vec{x}},N))$ is a local cocycle that is strongly continuous in ξ .

Remark.

(1) Theorem 10.1 is the <u>technical core</u> in the proof of Theorem 9.4.

(2) In §3, Lemmas 3 and 4 of $[F7]$ Theorem 10.1 has been proven for the special case where the group elements ξ are space-time translations. The proof of the general result is very similar to the one of this special case is however technically a little more complicated. It is too long to be reproduced here.

(3) We emphasize again that the existence of <u>locally correct generators</u> for the Poincaré automorphisms, i.e. the results of $[CJ, McB]$, and duality (for the <u>free</u> field only) are basic ingedients of the proof; technically the Trotter product formula and <u>finite propagation speed</u>, i.e. (9.10), play a basic role.

(4) The main point of Theorem 10.1 (in the realization of the announced program) is that, by approximating the * automorphism σ_s by <u>inner</u> * automorphisms implemented by the <u>local</u> operators $V_{\alpha,\vec{x}}$, (as $\vec{x} \to -\infty$; i.e. by approximating ω_s by vector states in $\mathcal{H}_+$), we achieved to construct strictly local 1-cocycles $\Gamma_\alpha(\xi)$, a direct construction of which appears to be difficult.

For the expert it is now clear that Theorem 10.1 immediately yields a proof of Theorem 9.4. The last step is the following.

Corollary 10.2: $[F7, F7']$

(1) The unitary operators $\{U_\pm(\xi)\Gamma_\alpha(\xi): \xi \in \underset{\sim}{p}_+^\uparrow\}$ form a continuous, unitary representation U_s of the

Poincaré group $p^{\uparrow}_{+}$ on $\mathcal{H}_{\pm}$.

(2) For all $A \in \mathcal{O}$

$$\sigma_s(\tau_\xi(A)) = U_\pm(\xi)\Gamma_\alpha(\xi)\sigma_s(A)(U_\pm(\xi)\Gamma_\alpha(\xi))^*,$$

as an operator equation on $\mathcal{H}_\pm$.   (10.9)

(3) The spectrum of the infinitesimal generators of the space-time translation subgroup

$$\{U_\pm((I,a))\Gamma_\alpha((I,a)):a\in \mathcal{R}^2\}$$

is contained in the forward light cone.

The proof of Corollary 10.2 is exactly as in §3, Lemmas 4 and 7 of [F7]. The proof of (3) relies in an essential manner on (10.2), (10.3) and the fact that $\omega_\pm$ are vacuum states satisfying <u>spatial cluster decomposition pro-</u><u>perties</u>. (This stresses once more the usefulness of our approach: (10.2), (10.3)).

Given Corollary 10.2 the proof of Theorem 9.4 is now simple:

Let T_s be an intertwining operator mapping $\mathcal{H}_+$ 1-1 and isometrically onto $\mathcal{H}_s$. Then a continuous unitary re-presentation U_s of the Poincaré group $p^{\uparrow}_{+}$ on $\mathcal{H}_s$ is defined by the equation

$$U_s^*(\xi)T_s \psi = T_s \tilde{U}_s^*(\xi)\psi$$
                    (10.10)

$$= T_s \Gamma_\alpha^*(\xi)U^*(\xi)\psi.$$

This yields Theorem 9.4, (1).

For the proof of Theorem 9.4, (2) see Lemma 7 and "Proof of Theorem 3" in §3 of ref. [F7], and for the one of (3) & (4) see §4 of [F7].

We feel that it has been shown here what the main
ideas of the proof of Theorem 9.4 are and in which manner
local 1-cocycles enter the analysis. We should emphasize
that the present techniques may not apply to the $P(\phi)_2$-
theory if P is not even. The reason is that, for this
case, it is not known, yet, whether different pure phase
vacua are related to each other by * automorphisms of $\mathcal{O}$
(of the sort of ϱ_π) that commute with the energy-momentum - and boost densities. For a general analysis we
refer the reader to §6 of [F7]; (see also [Ro]).

10.2 Conclusions

(1) Presently we are checking whether one can prove
    directly that the spectrum of the energy-momentum
    operator $(H_\#, P_\#)$ on $\mathcal{H}_\#$ has a positive mass gap
    M_s, (for sufficiently small $\lambda \in (0, \lambda_c)$). Heuristic
    arguments suggest that M_s is bounded below by some
    analogue of what one knows as surface tension τ in
    the statistical mechanics of classical spin systems;
    (see e.g. [GM]). This surface tension can be esti-
    mated directly by means of the Peierls argument;
    (see Section 7). The author has shown that it diver-
    ges to $+\infty$ (i.e. $\tau \nearrow +\infty$) as $\lambda \searrow 0$, (with λ the coupling
    constant of the $(\vec{\phi} \cdot \vec{\phi})^2$ - term, and $\sigma > 1$ fixed).

(2) There seems to be an interesting connection between
    the existence of soliton states and the spectrum

of _bound states_ in such two dimensional quantum
field theories. (Example: For fixed $\sigma > 1$, $\lambda \leq \lambda_c$ and
$0 < |\mu| << 1$ the $\lambda (\vec{\phi} \cdot \vec{\phi})^2 - \sigma \phi_1^2 + \mu \phi_1$ model in two
dimensions should have a rich spectrum of _bound
states_ (of "would-be" solitons and -anti-solitons)
which decay into soliton-anti-soliton pairs, as
$\mu \to 0$). This has been briefly discussed in remark 5)
§6 of ref. [F7] (and, independently, by Glimm, Jaffe
and Spencer in a forthcoming paper).

(3) The next urgent task in the theory of quantum solitons might be the study of _quantum vortices_ in
scalar QED in three space-time dimensions; (abelian
Higgs model). On a formal level this looks rather
promising. In the case of _non-abelian_ Yang-Mills
theories in four space-time dimensions (perhaps the
only candidates for soliton-behaviour in four dimensions) the situation does not seem to be well
understood, yet, even on a purely formal level,
although the existence of solutions of the classical field equations like the t'Hooft monopole is
very intriguing.

(4) For refs. to many original papers on quantum solitons
see [Co4, F7]. Since the number of such papers is
rapidly diverging a proper list of refs. could not
be included here.

References:

[A]   Araki,H.: J. Math. Phys.$\underline{4}$, 1343, (1963),
                    J. Math. Phys.$\underline{5}$, 1   , (1964),
           see also K. Osterwalder, Commun. Math. Phys.
           $\underline{29}$, (1973)

[AHR]   Araki, H. K. Hepp and D. Ruelle,
           Helv. Phys. Acta $\underline{35}$, 164, (1962),

[CJ]   Cannon, J.T. and A. Jaffe, Commun. Math.
           Phys., $\underline{17}$, (1970),

[Co1]   Coleman, S., "Secret Symmetry" 1973 Erice
           Lectures (Int. School of Subnuclear Physics,
           "Ettore Majorana").

[Co2]   Coleman, S. Phys. Rev. $\underline{D11}$, 2088, (1975).

[Co3]   Coleman, S., R. Jackiv and L. Susskind,
           Annals of Physics $\underline{93}$, 267, (1975).

[Co4]   Coleman,S., "Classical Lumps and Their
           Quantum Descendants" 1975 Erice Lectures
           (Int. School of Subnuclear Physics "Ettore
           Majorana").

[CQFT]   Symanzik, K., J. Math. Phys. $\underline{7}$, 510, (1966)
           Symanzik, K., "A Modified Model of Euclidean
           Quantum Field Theory",
           N.Y.U. Preprint, 1964.
           Nelson, E., "Quantum Fields and Markoff Fields"
           in: Proceedings of the Summer Inst. on Partial
           Diff. Equ., D. Spencer (ed), Berkeley 1971,
           A. Math. Soc., Providence 1973.
           Nelson, E., J. Funct. Anal. $\underline{12}$, 97, (1973).
           Guerra, F., Phys. Rev. Letters $\underline{28}$, 1213,(1972).

Glimm, J. and A. Jaffe, refs. [GJ1,3]
Glimm, J., A. Jaffe and T. Spencer, refs.
[GJS1,2]
Guerra, F., L. Rosen and B. Simon, ref.[GRS]
Osterwalder, K. and R. Schrader, ref. [OS].
See also refs. [E] [Si] [Y$_2$].

[DHR]   Doplicher, S., R. Haag and J. Roberts ,
Commun. math. Phys. $\underline{23}$, 199, (1971) and $\underline{35}$
49, (1974).

[DS]   Dunford, N. and J. Schwartz, "Linear Operators",
Interscience, New York, 1963.

[DN]   Dunlop, F. and C. Newman, Commun. math. Phys.
$\underline{44}$, 223, (1975).

[EEF]   Eckmann, J.-P., H. Epstein and J. Fröhlich,
"Asymptotic Perturbation Expansion for the
S-Matrix and ...." to appear in Ann. Inst.
H. Poincaré, 1976.

[E]   "Constructive Quantum Field Theory", G.Velo
and A.S. Wightman (eds.), Springer Lecture
Notes in Physics, vol. $\underline{25}$, (1973).

[ESw]   Ezawa, H. and J. Swieca, Commun. math. Phys.
$\underline{5}$, 330, (1967).

[Fe]   Feldman, J., Commun. math. Phys. $\underline{37}$, 93,
(1974).

[FeO1]   Feldman, J., and K. Osterwalder, "The Wight-
man Axioms and the Mass Gap for Weakly Coupled
$(\phi^4)_3$ Quantum Field Theories", to appear in
Annals of Physics.

[FeO2]   Feldman, J. and K. Osterwalder, "The Con-
          struction of $\lambda(\phi^4)_3$ Quantum Field Models"
          Proceedings of the Int. Colloqu. on Math.
          Methods of QFT, Marseille 1975.

[Fö]   Föllmer, H., "Phase Transition and Martin
          Boundary", in: Seminaire de Probabilités IX,
          Université de Strasbourg, p. 305, Springer
          Lecture Notes in Math..vol. **465**, (1975).

[F1]   Fröhlich, J., "Quantum Sine-Gordon Equation
          and Quantum Solitons in Two Space-Time
          Dimensions", in: "Renormalization Theory"
          Int. School of Math. Physics" Ettore Majorana".
          1975, Reidel, Dordrecht - Boston, 1976.

[F2]   Fröhlich, J., "Classical and Quantum Statis-
          tical Mechanics in One and Two Dimensions:
          ......", to appear in Commun. math. Phys. 1976.

[F3]   Fröhlich, J., "Phase Transitions in Two Dimen-
          sional Quantum Field Models", ZiF, University
          of Bielefeld, Preprint 1976; (extended version
          in preparation).

[F4]   Fröhlich,J., "Poetic Phenomena in Two Dimen-
          sional Quantum Field Theory:.....", Pro-
          ceedings of the Int. Colloqu. on Math.Methods
          of QFT, Marseille 1975.

[F5]   Fröhlich, J., "The Pure Phases, the Irreducible
          Quantum Field.....", to appear in Annals of
          Physics, 1976.

[F6]   Fröhlich, J., "Existence and Analyticity in
          the Bare Parameters of the $[\lambda(\vec{\phi}\cdot\vec{\phi})^2-\sigma\phi_1^2-\mu\phi_1]$-
          Quantum Field Models" to appear.

264

[F7]   Fröhlich, J., "New Super-Selection Sectors
            ("Soliton-States") in Two Dimensional Bose
            Quantum Field Models", to appear in Commun.
            math. Phys., 1976, and paper in preparation
            (referred to as [F7']).

[FSe]   Fröhlich, J. and E. Seiler, "The Massive
            Thirring-Schwinger Model (QED_2): Convergence
            of Perturbation Theory and Particle Structure",
            to appear in Helv. Phys. Acta.

[FS]   Fröhlich, J. and B. Simon, "Pure States for
            General $P(\phi)_2$ Theories: Construction,
            Regularity and Variational Equality", sub-
            mitted to Ann. Math.

[FSS]   Fröhlich, J., B.Simon and T. Spencer, "Infrared
            Bounds, Phase Transitions and Continuous
            Symmetry Breaking", Princeton University,
            Preprint 1976; results announced in Phys. Rev.
            Letters 36, 804, (1976).

[GM]   Gallavotti, G. and A. Martin-Löf, Commun. math.
            Phys. 25, 87, (1972).

[GJ1]   Glimm, J. and A. Jaffe, Fortschritte d. Physik
            21, 327, (1973).

[GJ2]   Glimm, J. and A. Jaffe, Commun. math. Phys. 22
            1, (1971)

[GJ3]   Glimm, J. and A. Jaffe, (ϕ_2^4 II,III, IV),
            Ann. Math. 91, 362, (1970), Acta Math. 125,
            203, (1970), J. Math. Phys. 13, 1568, (1972).

[GJS1]   Glimm, J., A. Jaffe and T. Spencer,
            "The Particle Structure of The Weakly Coupled
            $P(\phi)_2$ Model and Other Applications of High

Temperature Expansions, Part II: The Cluster
Expansion", contribution in ref. [E] and refs.
given there.

[GJS2]   Glimm, J., A. Jaffe and T. Spencer, Commun.
         math. Phys. 45, 203, (1975), and paper in
         preparation.

[Gu]   Guerra, F., "Exponential Bounds in Lattice
         Field Theory", Proceedings of the Int.
         Colloqu. on Math. Methods of QFT, Marseille
         1975.

[GRS]   Guerra, F., L. Rosen and B. Simon, Ann. Math.
         101, 111, (1975).

[HK]   Haag, R. and D. Kastler, J. Math. Phys. 5, 848,
         (1964).

[He]   Herbst, I. "Remarks on Canonical Quantum
         Field Theory", Princeton University, Preprint
         1975.

[H]   "Statistical Mechanics and Quantum Field
         Theory", Les Houches 1970 C. De Witt and R. Stora
         (eds.), Gordon and Breach, New York, 1971.

[Jo]   Jost, R., "The General Theory of Quantized
         Fields", A. Math. Soc., Providence, R.I., 1965.

[K]   Kac, M., "On Applying Mathematics: Reflections
         and Examples", Quarterly of Appl. Math. 30, 17,
         (1972).

[Ku]   Kunz, H., private communication.

[La]   Lamb Jr. G., Rev. Modern Physics 43, 99, (1971).

[LP]   Lebowitz, J., and O. Penrose, Phys. Rev.
         Letters 35, 549, (1975).

266

[MaSé]   Magnen, J. and R. Sénéor, "The Infinite Volume
           Limit of the $(\phi^4)_3$ Model", to appear in Ann.
           Inst. H. Poincaré.

[Ma]   Malishev, V.A. Commun. math. Phys. $\underline{40}$, 75,
           (1975).

[McB]   McBryan, O., Nuovo Cimento, 18A, 654, (1973).

[M]   Mermin, N.D., J. Math. Phys. $\underline{8}$, 1061, (1967).

[N]   Nelson, E., "Probability Theory and Euclidean
           Field Theory", contribution in ref. [E].

[OS]   Osterwalder, K. and R. Schrader, Commun. math.
           Phys. $\underline{31}$, 83, (1973) and $\underline{42}$, 281, (1975).

[OSé]   Osteralder, K. and R. Sénéor, "The S-Matrix
           is Non-Trivial in $(\phi^4)_2$....", to appear in
           Helv. Phys. Acta.

[Pa]   Park, Y.M., "Convergence of Lattice Approxi-
           mations and Infinite Volume Limit in the
           $(\lambda\phi^4-\sigma\phi^2-\mu\phi)_3$ Field Theory", Schladming
           Lectures, 1976, and J. Math. Phys. $\underline{16}$,1065,
           (1975), and to appear in J. Math. Phys.

[Pe]   Peierls, R., Poc. Cambridge Phil. Soc., $\underline{32}$,
           477, (1936).

[PT]   general refs. on phase transitions: See refs.
           [H], [R],[FSS],[GM] and refs. given there.
           Results that are somewhat complementary to
           what is discussed in these lectures may be
           found in L. Onsager, Phys. Rev. $\underline{65}$, 117,
           (1944), (Ising model), H. Araki and E.J. Woods,
           J. Math. Phys. $\underline{4}$, 637, (1963), (free Bose gas)
           T.H. Berlin and M. Kac Phys. Rev. $\underline{86}$, 821, (1952)

E. Lieb and C.J. Thompson, J. Math. Phys. 10,
1403, (1969), and refs given there; (spherical
model).

N.N. Bogolinbov, Soviet Physics JETP, 7, 41,
(1958),

W. Thirring and A. Wehrl, Commun. math. Phys.
4, 301, (1966); (BCS model).

K. Hepp and E. Lieb, Ann. Phys. 76, 360,
(1972), and contribution to ref. [E]; (Laser
models).

R. Griffiths, Phys. Rev. 136A, 437, (1967),

R. Dobrushin, Funct. Anal. Appl. 2, 44, (1968),
(Peierls argument)

R. Israel, Princeton University Series in
Physics, Monograph, to appear; (general results
on phase transitions), and Commun. math. Phys.
43, 59, (1975).

F. Dyson, Commun. math. Phys. 12, 91, (1969).

[Ro]   Roberts, J., "Local Cohomology and Super-
selection Structure", Centre de Physique
Théorique, C.N.R.S. - Marseille, Preprint
1976.
For results concerning the solitons of the
free massless, scalar fields in two dimen-
sions, see R.F. Streater and I.F. Wilde
Nuclear Physics B24, 561, (1970); and
P. Bonnard and R.F. Streater, ZiF,
University of Bielefeld, Preprint 1975.

[R]   Ruelle, P. "Statistical Mechanics - Rigorous
Results", Benjamin, New York, 1969.

[SeSi] Seiler, E. and B. Simon, "Nelson's Symmetry
and All That in the $(Yukawa)_2$ and $(\phi^4)_3$

Field Theories", to appear in Annals of Phys.

[Si]   Simon B., "The $P(\phi)_2$ Euclidean (Quantum) Field Theory, Princeton University Press, Princeton 1974.

[SG]   Simon, B. and R. Griffiths, Commun. math. Phys. $\underline{33}$, (1973).

[St E]   St. Exupéry, A. "Le Petit Prince".

[St W]   Streater, R.F. and A.S. Wightman, "PCT, Spin and Statistics and All That", Benjamin, New York, 1964.

[SvB]   Sylvester G. and H. van Beijeren, "Phase Transitions for Continuous Spin Ising Ferromagnets", Yeshiva University, Preprint 1975.

[T]   See refs. [H] [R]; also K. Huang, "Statistical Mechanics", Wiley and Sons, New York-London 1963.

[Y$_2$]   Refs. on the Euclidean description of the (Yukawa)$_2$ quantum field model are:
E. Seiler, Commun. math. Phys. $\underline{42}$, 163, (1975).
E. Seiler and B. Simon, J. Math. Phys. to appear ("finite mass renormalizations") and ref. [SeSi].
E. Seiler and B. Simon, Commun. math. Phys. $\underline{45}$, 99, (1975).
O. McBryan, Commun. math. Phys. $\underline{44}$, 237, (1975).
O. McBryan, Commun. math. Phys. $\underline{45}$, 279, (1975).

O. McBryan, Contribution to the Proceedings
of the Int. Colloqu. on Math. Methods of
QFT, Marseille, 1975.

J. Magnen and R. Sénéor, "The Wightman Axioms
for the Weakly Coupled Yukawa Model in Two
Dimensions", ZiF, University of Bielefeld,
Preprint, 1975.

A. Cooper and L. Rosen, "The Weakly Coupled
(Yukawa)$_2$ Field Theory: Cluster Expansion
and Wightman Axioms", Princeton University,
Princeton, 1976.

Acta Physica Austriaca, Suppl. XV, 271–321 (1976)
© by Springer-Verlag 1976

CONSTRUCTION OF $(\lambda\phi^4 - \sigma\phi^2 - \mu\phi)_3$ QUANTUM
FIELD MODELS[+]

by

Y. M. PARK[++]
Department of Theoretical Physics
University of Bielefeld
48 Bielefeld 1, F.R. Germany
and
Department of Mathematics[+++]
Yonsei University, Seoul, Korea

TABLE OF CONTENTS

[+]   Lecture given at XV.Internationale Universitätswochen für
      Kernphysik,Schladming,Austria, February 16-27, 1976.
[++] Supported in part by Korean Traders Scholarship Foundation.
[+++] Permanent Adress after September 1, 1976.

1. INTRODUCTION

This lecture is intended to introduce the audience
to some constructions of theories for the boson field
models in three dimensional space-time, which exhibit
ultraviolet divergences. Constructive quantum field
theory has developed rapidly in the past few years. The
polynomial interactions in two dimensional space-time
(the $P(\phi)_2$ models) are the best behaved models and its
detailed structure is now well-known [2, 12, 16, 18, 19,
23, 24, 28, 30, 36, 37, 40]. Most of you may have already
been exposed in the construction of the $P(\phi)_2$ field model
elsewhere (at least, you will have a change again in Prof.
Challifour's lecture at this school) and so I will not
discuss that subject. The $(\lambda\phi^4 - \sigma\phi^2 - \mu\phi)$ interactions in
three dimensional space-time (the $(\lambda\phi^4 - \sigma\phi^2 - \mu\phi)_3$ models),
which we are considering here, are the next well behaved
models. These differ from $P(\phi)_2$ by having ultraviolet

divergences and by requiring ultraviolet mass and wave functions as well as vacuum energy renormalizations [14, 15]. For the $(\lambda\phi^4-\sigma\phi^2-\mu\phi)_3$ models, the existence of theories has been firmly established [9, 10, 13, 26, 34] and the development of the theories are rapidly approaching to the level of the $P(\phi)_2$ models.

$P(\phi)_2$ and $(\lambda\phi^4-\sigma\phi^2-\mu\phi)_3$ are the only superrenormalizable polynomial boson self-interactions. The next singular theory is $(\lambda\phi^4)_4$ which is renormalizable but not super-renormalizable. We hope that deep understanding and good control of the ultraviolet divergences in the $(\lambda\phi^4)_3$ theory may be useful in the construction of the $(\lambda\phi^4)_4$ theory. But, new techniques or methods seem to be necessary to deal with the $(\lambda\phi^4)_4$ model [17, 38]. When preparing this lecture I found out that B. Simon gave a lecture on the construction of $(\lambda\phi^4-\sigma\phi^2-\mu\phi)_2$ at this school in 1973. Three years after that time we have the infinite volume limit theory of the $(\lambda\phi^4-\sigma\phi^2-\mu\phi)_3$ models. I hope someone will be able to give a lecture on the existence of the $(\lambda\phi^4)_4$ theory at this school in the near future, say within 10 or 20 years!

We will only use (discuss) the Euclidean approach to construct the model under consideration. The use of imaginary time in constructive quantum field theory dates to Symanzik's Euclidean field formalism [39] and Nelson [28] and the references therein . This approach was used to prove most of the key technical estimates of the $P(\phi)_2$ and ϕ^4_3 boson models as well as of the Yukawa model in two dimensions. The covariance formalism in terms of Euclidean fields on Green's functions is an important advance both conceptually, yielding Euclidean axioms [29], and technically in proving estimates. Furthermore, the Euclidean

framework makes clear the connection of field theory
and statistical mechanics [23, 28, 36]. The lattice
approximation [23, 37] becomes natural and it relates
boson interactions to (generalized) Ising model. In this
lecture we will survey the results and discuss briefly
ideas of the proofs in the construction of the $(\lambda\phi^4 - \sigma\phi^2 -$
$\mu\phi)_3$ models in the framework of the Euclidean approach.
We will not discuss results and methods related to the
Hamiltonian approach [14, 5, 6, 2, 4]. Because of the
broadness of the subject we will just try to sketch the
overall structure of the construction and so the detailed
proofs of the results are missing, except for one section.
In section 5 we prove in detail that the Schwinger functions (Euclidean Green's functions) are uniformly bounded
in boson field models (in two and three dimensions). Since
the method is independent of the model and also very
simple, I feel that it is useful to give a complete proof.

The success of a purely Euclidean approach in constructing the $P(\phi)_2$ models suggested that it might be
easier to construct the $(\lambda\phi^4 - \sigma\phi^2 - \mu\phi)_3$ models through
their Euclidean Green's function (the Schwinger function)
as well. The first successful use of the Euclidean method
in $(\lambda\phi^4)_3$ is due to Glimm and Jaffe [15] who showed the
semi-boundedness of the spatially cut-off Hamiltonian
H(g). Their method provided a major tool in controlling
the ultraviolet divergences in the $(\lambda\phi^4 - \sigma\phi^2 - \mu\phi)_3$ models.
Modifying Glimm and Jaffe's method Feldman [7] has shown
the existence of the ultraviolet limit of the (momentum
and space) cut-off Schwinger functions. Park [31, 34]
has shown the convergence of the lattice approximation
of the model with various classical boundary conditions
in a finite box as the lattice spacing tends to zero.

Consequently, correlation inequalities and the Lee-Yang theorem have been established for the theories in a finite box. The existence of the infinite volume limit for the weakly coupled $(\lambda\phi^4)_3$ models has been established by Feldman and Osterwalder [9] and Magnen and Seneor [26] independently. Recently Park [32, 33] and Seiler and Simon [35] have established uniform bounds of the pressures and the Schwinger functions with free and periodic boundary conditions. Using correlation inequalities and uniform bounds of the Schwinger functions, Fröhlich [13] and Park [34] have constructed the infinite volume limit of the strongly coupled $(\lambda\phi^4 - \sigma\phi^2 - \mu\phi)_3$ models. The work on the models is now largely aimed at determing physical properties and simplifying earlier works.

We now discuss the construction program briefly. Let $S_{\Lambda,\kappa}^{(X)}(f_1,\ldots,f_n)$ and $S_{\Lambda,\delta}^{(X)}(f_1,\ldots,f_n)$ be the momentum cut-off Schwinger functions and the lattice cut-off Schwinger functions respectively for the $(\lambda\phi^4 - \sigma\phi^2 - \mu\phi)_3$ models in a finite box, where κ,δ and X refer to the momentum cut-off parameter, the lattice spacing parameter and the boundary condition on the boundary of Λ. We will give the detailed definitions later. Throughout this lecture we assume that $\lambda > 0$, $\sigma, \mu \in R$ and $X = F$, D, N, P (free, Dirichlet, Neumann and periodic boundary condition respectively). The main goal is to show that the limit

$$S^{(X)}(f_1,\ldots,f_n) = \lim_{\Lambda \to R^3} \lim_{\kappa \to \infty} S_{\Lambda,\kappa}^{(X)}(f_1,\ldots,f_n)$$

$$= \lim_{\Lambda \to R^3} \lim_{\delta \to \infty} S_{\Lambda,\delta}^{(X)}(f_1,\ldots,f_n)$$

exists for some boundary conditions, where $f_i \in S(R^3)$, and the set $\{S^{(X)}(f_1,..,f_n)\}$ satisfy the axioms of Osterwalder-Schrader [29]:

(E0)   A distribution property

(E1)   Euclidean invariance

(E2)   Positivity

(E3)   Symmetry

(E4)   Clustering

The Osterwalder-Schrader reconstruction theorem [29] gives the existence of a (unique) Wightman theory having the S^X as its Euclidean Green's functions. The basic construction program can be summarized by the following diagram:

$$S^{(X)}_{\Lambda,\kappa}(f_1,...,f_n)$$

$$\downarrow \kappa \to \infty$$

$$S^{(X)}_{\Lambda}(f_1,...,f_n) \xleftarrow{\ \delta \to 0\ } S^{(X)}_{\Lambda,\delta}(f_1,...,f_n)$$

$$\downarrow \Lambda \to R^3$$

(correlation inequalities and Lee-Yang theorem)

$$S^{(X)}(f_1,...,f_n)$$

(O - S axioms, physical properties, etc.)

The organization of this lecture is as follows: In Section 2, we introduce notation, definitions and the models we are interested in. We also introduce the momentum cut-off Schwinger functions and the lattice cut-off Schwinger functions in a finite box Λ. In Section 3,

we survey the removal of the momentum cut-off as well
as the lattice cut-off and so establish the finite
volume theory (and correlation inequalities). In Section 4, we discuss the construction of the infinite
volume theories. Finally, we give a simple proof of
the uniform bounds of the Schwinger functions in boson
field models (in two and three dimensions). In Section
6, we take a brief outlook at unsolved problems.

2. NOTATION, DEFINITION AND MODELS

In this section we first introduce notation, definitions and models. We then define the momentum cut-off
Schwinger functions and the lattice cut-off Schwinger
functions in a finite box. The Schwinger functions for
the models are given formally by

$$S(x_1,..,x_n) = \text{normalized constant} \times$$

$$\times \int \phi(x_1)..\phi(x_n) e^{-\int dx : P(\phi(x)):} : e^{-\frac{1}{2}\int dx [\phi(x)(-\Delta+m_o^2)\phi(x)]} \prod_{x \in R^d} d\phi(x)$$

where Δ be the Laplacian and $:$ $:$ denotes Wick ordering.
For the $(\lambda\phi^4-\sigma\phi^2-\mu\phi)_3$ models we have

$$\int dx : P(\phi(x)): = \int dx :(\lambda\phi(x)^4-\sigma\phi(x)^2-\mu\phi(x)): + \text{counter terms.}$$

The counter terms (in addition to Wick ordering) required

to define the Schwinger functions properly is the following:

1) a linearly divergent, second order vacuum counter term (corresponding partly to a vacuum energy renormalization and partly to a wave function renormalization)

2) a logarithmically divergent third order vacuum counter term (corresponding to a vacuum energy renormalization)

3) a logarithmically divergent (second order) mass counter term.

The definition of the Schwinger functions given above is, of course, formal and we must give it a precise meaning.

The Euclidean free theory is given on the path space $L^2(S'(R^3),d\mu)$ where $d\mu$ is the Gaussian measure with mean zero and covariance

$$C(x,y) \equiv (-\Delta+m_o^2)^{-1}(x,y) = (2\pi)^{-3}\int d^3k \ e^{ik\cdot(x-y)} (k^2+m_o^2)^{-1}$$

$$= \frac{1}{|x-y|} e^{-m_o|x-y|} \ . \tag{2.1}$$

The Euclidean free fields are the linear functions on $S_R'(R^3)$:

$$\phi(f)(q) = <q,f>$$

for all $q \ \epsilon \ S_R'(R^3)$ and $f \ \epsilon \ S(R^3)$. We write $\phi(f) = \int \phi(x) f(x) dx$. The momentum cut-off free fields are defined by

$$\phi_\kappa(x) = \int \phi(y) \delta_\kappa(y-x) d^3y \tag{2.2}$$

where $\delta_\kappa(x)$ be an approximation of $\delta(x)$ function, i.e., $\lim_{\kappa \to \infty} \delta_\kappa(x) = \delta(x)$. Let $\Lambda \subset R^3$ be a finite box centered at the origin and let $\partial\Lambda$ be the boundary of Λ. We write

$$:\phi_\kappa^n:(\Lambda) = \int_\Lambda d^3x : \phi_\kappa(x)^n : \qquad \qquad (2.3)$$

The doubly (momentum and volume) cut-off Euclidean inter-acting action for the $(\lambda\phi^4 - \sigma\phi^2 - \mu\phi)_3$ models, $\lambda > 0$, μ, $\sigma \in R$, is defined as

$$V(\Lambda,\kappa) = V_I + V_C$$

$$V_I = :(\lambda\phi_\kappa^4 - \sigma\phi_\kappa^2 - \mu\phi_\kappa):(\Lambda)$$

$$V_C = E_2 + E_3 + \frac{\lambda^2}{2} \delta m_\kappa^2 : \phi_\kappa^2 : (\Lambda)$$

$$E_2 = \frac{\lambda^2}{2} \int (:\phi_\kappa^4:(\Lambda))^2 \, d\mu \qquad \qquad (2.4)$$

$$E_3 = -\frac{\lambda^3}{6} \int (:\phi_\kappa^4:(\Lambda))^3 d\mu$$

$$\delta m_\kappa^2 = 4^2 x 6 \int d^3x \, [C_\kappa(x,y)]^3$$

$$C_\kappa(x,y) = (2\pi)^{-3} \int d^3k \, e^{ik\cdot(x-y)} (k^2 + m_0^2)^{-1} \kappa(k)^2$$

where $\kappa(k)$ is a cut-off function (its Fourier transformat-ion is $\delta_\kappa(x)$). The cut-off partition function and the cut-off Schwinger functions are defined by

$$Z_{\Lambda,\kappa} = \int e^{-V(\Lambda,\kappa)} \, d\mu$$

$$S_{\Lambda,\kappa}(f_1,\ldots,f_n) = (Z_{\Lambda,\kappa})^{-1} \int \phi(f_1)..\phi(f_n) e^{-V(\Lambda,\kappa)} d\mu \qquad (2.5)$$

for $f_i \in S(R^3)$. The above objects are well defined because of momentum cut-off [15, 36].

We have defined above the interacting action, the partition function and the Schwinger function with <u>free</u> (F) boundary conditions on $\partial\Lambda$ [36]. Frequently, we will write these by $V^{(F)}$, $Z^{(F)}$ and $S^{(F)}$ to indicate the free boundary condition. In the theory of statistical mechanical systems, such as the Ising model, the clever use of one or more kinds of boundary conditions plays a major role. A similar situation occures in the $P(\phi)_2$ models [28, 23, 24, 36]. As for the $P(\phi)_2$ models we introduce boundary conditions in $\{S_{\Lambda,\kappa}\}$ by imposing boundary conditions on the Laplacian . Placing zero Dirichlet data or periodic boundary conditions on $\partial\Lambda$, the surface of the region of interaction, has proved particularly useful. Let $d\mu_{\Lambda}^X$ be the Gaussian measure on $S'(\Lambda)$ with mean zero and covariance given by $C^X(x,y) = (-\Delta_X + m_0^2)^{-1}(x,y)$, where Δ_X be the Laplacian with X - boundary condition on $\partial\Lambda$. We will be interested in free (F), Dirichlet (D) periodic (P) and Neumann (N) boundary condition, i.e., X = F, D, P, N. For the proper renormalization theory with boundary condition, it is important that Wick ordering and vacuum counter terms remain matched to the covariance (for example, $:\phi(x)^2:_X = \phi(x)^2 - C^X(x,x)$). But we keep the coefficient δm^2 of the mass counter term fixed (the differences of δm^2 between the different boundary conditions are finite [34]):

$$\delta m^2 = \delta m^{(X)2}, \quad X = F, D, P, N . \qquad (2.6)$$

Let $V^{(X)}(\Lambda,\kappa)$ be the interacting action corresponding to X-boundary condition. We then define the partition function and the Schwinger function with X-boundary condition by

$$Z_{\Lambda,\kappa}^{(X)} = \int e^{-V^{(X)}(\Lambda,\kappa)} \, d\mu_\Lambda^X$$

$$S_{\Lambda,\kappa}^{(X)}(f_1,\ldots,f_n) = (Z_{\Lambda,\kappa}^{(X)})^{-1} \int \phi(f_1)\cdots\phi(f_n) e^{-V^{(X)}(\Lambda,\kappa)} d\mu_\Lambda^X \qquad (2.7)$$

For more detailed definitions we refer to [34].

Finally, we introduce the lattice approximation of the $(\lambda\phi^4 - \sigma\phi^2 - \mu\phi)_3$ field theories in a box Λ. The main purpose of the lattice approximation in boson field theory is to apply statistical mechanical ideas, namely, the method of correlation inequalities [23, 28, 36]. Let δ be the lattice spacing parameter for the lattice $L_\delta = \{n\delta : n\epsilon Z^3\}$. We denote $\Lambda_\delta = \Lambda \cap L_\delta$. We introduce the lattice fields $\phi_\delta(n)$ as the real Gaussian process indexed by the lattice in Λ_δ with mean zero and covariance given by $C_\delta^X(n,m) = (-\Delta_{X,\delta} + m_0^2)^{-1}(n,m)$, where $\Delta_{X,\delta}$ is the lattice approximation of the Laplacian Δ_X with X-boundary condition on $\partial\Lambda$ [28, 29, 34, 36]. For example,

$$(-\Delta_\delta + m_0^2)^{-1}(n,m) \equiv (-\Delta_{F,\delta} + m_0^2)^{-1}(n,m)$$

$$= (2\pi)^{-3} \int_{-\pi/\delta}^{\pi/\delta} e^{ik\cdot(n-m)} \delta_{\mu_\delta}(k)^{-2} d^3k \qquad (2.8)$$

where

$$\mu_\delta(k)^2 = \delta^{-2}[6 - 2 \sum_{i=1}^{3} \cos(\delta k^{(i)})] + m_o^2$$

$$\sim \mu(k)^2 \qquad \text{as} \qquad \delta \to 0$$

$$\mu(k)^2 = k^2 + m_o^2 . \tag{2.9}$$

Let $d\mu_{\Lambda,\delta}^X$ be the corresponding Gaussian measure. The lattice cut-off interacting actions are given by

$$V^{(X)}(\Lambda,\delta) = \delta^3 \sum_{n\delta \in \Lambda_\delta} : (\lambda\phi_\delta(n)^4 - \sigma\phi_\delta(n)^2 - \mu\phi_\delta(n)):_X$$

$$+ E_2^{(X)}(\Lambda,\delta) + E_3^{(X)}(\Lambda,\delta) + \frac{\lambda}{2}\delta m_\delta^2 \sum_{n\delta \in \Lambda_\delta} :\phi_\delta(n)^2:_X \tag{2.10}$$

where $E_2^{(X)}(\Lambda,\delta)$, $E_3^{(X)}(\Lambda,\delta)$ and δm_δ^2 are defined analogously to the corresponding quantities in (2.4). See the references [31, 34] for the details. The lattice cut-off part-ition function and Schwinger functions are defined by

$$Z_{\Lambda,\delta}^{(X)} = \int e^{-V^{(X)}(\Lambda,\delta)} d\mu_{\Lambda,\delta}^X$$

$$S_{\Lambda,\delta}^{(X)}(f_1,\ldots,f_n) = (Z_{\Lambda,\delta}^{(X)})^{-1}\int \phi_\delta(f_1)\ldots\phi_\delta(f_n) e^{-V^{(X)}(\Lambda,\delta)} d\mu_{\Lambda,\delta}^X \tag{2.11}$$

For the details on the lattice approximation of boson field models we refer to [23, 24, 25, 28, 36, 37] . In Section 3.2, we will discuss the connection of lattice field theories with ferromagnetic systems.

3. THEORIES IN A FINITE VOLUME

3.1 Removal of the Momentum Cut-off

We will discuss the control of the ultraviolet divergences for the models in this section. In contrast to $P(\phi)_2$ in which the only problem was the existence of thermodynamic limit theories, the control of the ultraviolet divergences in the $(\lambda\phi^4 - \sigma\phi^2 - \mu\phi)_3$ models was one of the most challenging problems in construction of the theory. The first successful control of the ultraviolet divergences in the framework of the Euclidean method was due to Glimm and Jaffe [15] who developed the method of the "phase space cell expansion (PSCE)" to prove an upper bound of the momentum cut-off partition function $Z_{\Lambda,\kappa}^{(F)}$ uniformly in κ (this implied the positivity of the space cut-off Hamiltonian). Their model has been extended by Feldman [7] who showed the existence of the (unnormalized) Schwinger function in a finite volume. Employing the PSCE Park [31, 34] has established the convergence of the lattice approximation of the models with various boundary conditions in a finite box and so established correlation inequalities (see the following section). Furthermore, Feldman and Osterwalder [9] and Magnen and Seneor [26] have applied the method in proving the existence of the infinite volume limit of the weakly coupled $(\lambda\phi^4)_3$ model (see the discussion in Section 4.2). We devote this section to a discussion of the PSCE.

For the sake of simplification we only discuss in detail the $(\lambda\phi^4)_3$ model with free (F) boundary conditions and suppress F in the notation. We will only give the extended results to general $(\lambda\phi^4 - \sigma\phi^2 - \mu\phi)_3$ models and also to various boundary conditions. We first give the results of Glimm and Jaffe [15] and Feldman [7]:

__Theorem 3.1.1.__ (a)(Glimm and Jaffe [15])

$$Z_{\Lambda,\kappa} \leq e^{O(|\Lambda|)}$$

uniformly in κ, where $|\Lambda|$ is the volume of Λ.
(b) (Feldman[7]) The partition function $Z_{\Lambda,\kappa}$ and the (unnormalized) Schwinger functions

$$Z_{\Lambda,\kappa} \, S_{\Lambda,\kappa} \, (f_1,\ldots,f_n) = \int \phi\,(f_1)\ldots\phi\,(f_n)\,e^{-V(\Lambda,\kappa)}\,d\mu$$

converge as $\kappa \to \infty$. The limits are continuous in λ and satisfy

$$|\,Z_\Lambda S_\Lambda\,(f_1,\ldots,f_n)\,| \leq n!\,(\prod_{i=1}^{n}\,\|f_i\|\,e^{O(|\Lambda|)})$$

for a suitable Schwartz space norm $\|\cdot\|$. Furthermore, the (unnormalized) Schwinger functions $\{\,Z_\Lambda S_\Lambda\,(f_1,\ldots,f_n)\}$ are the moments of an unique (unnormalized) physical measure on $S'(R^3)$.

From the continuity in λ and $Z_\Lambda = 1$ for $\lambda = 0$ [7], it follows that for Λ fixed and λ sufficiently small (depending on Λ)

$$Z_\Lambda \geq 1/2 \quad.$$

Thus, in the above case the normalized Schwinger functions exist and satisfy

$$|S_\Lambda(f_1,\ldots,f_n)| \leq \text{const}(\Lambda) \, n! \prod_{i=1}^{n} \|f_i\| \, .$$

Since we are interested in the $\Lambda \rightarrow R^3$ limit of these Schwinger functions, we need to know whether

$$Z_\Lambda > 0$$

for all $\lambda > 0$ and $|\Lambda| < \infty$ to define the volume cut-off (normalized) Schwinger functions. Indeed, we have

Proposition 3.1.2. (Park [32, 34] and Seiler and Simon [35]) For all $\lambda > 0$

$$Z_\Lambda^{(X)} \geq e^{-O(|\Lambda|)} \quad , \quad X = F, P \tag{3.1.1}$$

$$Z_\Lambda^{(X)} > 0 \qquad , \quad X = D, N \, . \tag{3.1.2}$$

This proposition has been established in [32, 34] . Seiler and Simon [35] have also shown (3.1.1) for $X = F$ by a method similar to that in [32]. We do not discuss the proof in this lecture since it involves considerable notational complications. Using the above result and following the method used in proving Theorem 3.1.1., we have [34]

Theorem 3.1.3. For the $(\lambda\phi^4-\sigma\phi^2-\mu\phi)_3$ models and for $X = F$, D, N, P, the limit

$$S_\Lambda^{(X)}(f_1,\ldots,f_n) = \lim_{\kappa \to \infty} S_{\Lambda,\kappa}^{(X)}(f_1,\ldots,f_n)$$

exist and satisfy

$$|S_\Lambda^{(X)} (f_1, \ldots, f_n)| \leq (K(\Lambda))^n n! \prod_{i=1}^{n} \|f_i\|$$

for a Schwartz space norm $\|\cdot\|$. Furthermore, the limit Schwinger functions are the moments of an unique probability measure dq_Λ^X on $S_R'(\Lambda)$ (or $S_R'(R^3)$), namely,

$$S_\Lambda^{(X.)} (f_1, \ldots, f_n) = \int \phi(f_1) \ldots \phi(f_n) \, dq_\Lambda^X$$

$$dq_\Lambda^X = \lim_{\kappa \to \infty} (Z_{\Lambda,\kappa}^{(X)})^{-1} e^{-V^{(X)}(\Lambda,\kappa)} d\mu_\Lambda^X \; . \qquad (3.1.3)$$

Remark: In fact, the constant $K(\Lambda)$ is independent of Λ for $X = F, P$ [33, 36]. See Section 4.1.

In the rest of this section we sketch the basic idea and the method employed in the proof of Theorem 3.1.1 [15, 7]. Since each reasonably detailed description of the "phase space cell expansion" developed by Glimm and Jaffe [15] and modified by Feldman [7] involves highly technical complications, we will skip the details and give only a bare description. Anyone who is interested in details is referred to [15, 7]. A summary version of the PSCE can be found in [18]. We follow partly the discussion in [18].

For simplicity we only discuss the proof of Theorem 3.1.1 (a), i.e., we consider the partition function $Z_{\Lambda,\kappa}$. In fact, the proof of Theorem 3.1.1 is reduced to uniform bounds of the integrals of the form (3.1.4) below in [7]. Hence, our discussion can be applied in the proof of Theorem 3.1.1 (b). We fix the volume Λ (and suppress Λ

in the notation) and investigate how Z depends on the
ultraviolet cut-off κ. The basic aims of PSCE are to obtain a convergent expansion in a given space time (phase
space) volume $\Delta \subset \Lambda$, and polynomial decoupling of different localization regions. The semi-boundedness of the interacting action $V(\kappa)$ is used to truncate the perturbation expansion to yield a convergent expansion and smooth
cut-offs in momentum space are used to obtain polynomial
decouplings. In order to truncate the perturbation expansion we introduce a lower momentum cutoff ρ into the
action V. Each momentum component in $V(\kappa,\rho)$ lies in the
interval $[\rho,\kappa]$. We consider an integral of the form

$$\int R(\phi)\ e^{-V(\kappa,\rho)}\ d\mu \tag{3.1.4}$$

where $R(\phi)$ is a polynomial function in ϕ. We perform our
expansion on the integrals of the form (3.1.4) inductively.
At the start of the expansion, $R(\phi) = 1$, $\rho = 0$ and $\kappa = \kappa_0$
(Notice that we have Z_{Λ,κ_0} for some κ_0 to prove Theorem
3.1.1 (a)). Each expansion step replaces (3.1.4) by a sum
of similar terms. Each step consists of three main steps:

(a) a high momentum (P-C) expansion
(b) a lower momentum (Wick) expansion
(c) combinatoric estimates.

The high momentum expansion is used to lower κ and the
lower momentum expansion is used to raise ρ. Combinatoric
estimates are used to bound the number of terms generated
during these expansions. The expansion terminates when
$\rho = \kappa$, i.e., $V(\kappa,\rho) = 0$ and (3.1.4) is reduced to a sum
of Gaussian integral $\int R(\phi)\,d\mu$. The partition function is

then expressed as a sum of these Gaussian integrals.
One estimates this sum uniformly in κ_o. The rule for
alternating the expansion step is complicated. The main
idea is to obtain a small contribution from each high
energy vertex in R, by performing explicit renormalization cancellations. We avoid a divergent perturbation expansion by truncating it by the lower momentum expansion.

(a) The high momentum expansion: We use this expans-
ion to lower the upper momentum cut-off κ in the exponent
$V(\kappa,\rho)$. Consider $V_1(\bar{\kappa},\rho)$ and $V_o(\underline{\kappa},\rho)$ for $\bar{\kappa} \geq \underline{\kappa}$. We write
$V_\alpha = \alpha V_1 + (1 - \alpha)V_o$. Then

$$e^{-V_1} = e^{-V_o} - \int_o^1 (V_1 - V_o) \, e^{-V_\alpha} \, d\alpha \quad . \tag{3.1.5}$$

The first term in (3.1.5) has the desired form. The second
term has the same upper cut-off and a new vertex $\delta V = V_1 - V_o$
in R. Since δV has a lower momentum cut-off $\underline{\kappa}$, we desire
that δV contribute to a convergence factor $\underline{\kappa}^{-\varepsilon}$ to final
estimates. The proof of this factor can be obtained only
after performing renormalization cancellations of the
divergence counter terms δV_C in δV. We use the following
integration by part formula [15]:

$$\int \phi(x) F(\phi) \, d\mu = \int dy [\int C(x,y) \, \frac{\partial F(\phi)}{\partial \phi(y)} \, d\mu] \quad . \tag{3.1.6}$$

The above formula can be easily obtained by studying finite
dimensional approximations to the functional space integration. For details, see [15]. For Wick ordering we have [15]

$$\int :\phi(x_1) \, .. \, \phi(x_n) : F(\phi) \, d\mu =$$

$$= \int :\phi(x_2)\ldots\phi(x_n): [\int dx_1' \ C(x_1,x_1') \ \frac{\partial F(\phi)}{\partial \phi(x_1')} \ ] \, d\mu \qquad . \qquad (3.1.7)$$

The above formula yields

$$\lambda \int :\phi(x)^4: e^{-\lambda \int_\Lambda :\phi^4:} \, d\mu = [\lambda^2 \int_{\Lambda \times \Lambda} dx dy C(x,y)^4] \int e^{-\lambda \int_\Lambda :\phi^4:} \, d\mu \ (3.1.8)$$

$$+ \ \text{other terms.}$$

Using (3.1.7) we integrate by parts δV_I, namely, $\lambda \phi^4$ part
of δV. We also integrate any new V_I produced in R as a result of differentiating the exponent. We thus obtain in
closed form (and we cancel) the ultraviolet divergence
part of δV. For instance, in (3.1.8) we displayed the
second order vacuum contribution. The third order vacuum
and mass contributions occur among other terms in (3.1.8).
The vacuum contributions cancel exactly with the corresponding counterterms in δV_C, and the mass renormalization
diagram after cancellation leave a convergence factor $\underline{\kappa}^{-\epsilon}$.
The remaining other terms from this procedure are conver-
gent and so contribute $\underline{\kappa}^{-\epsilon}$ to the final estimates. This
step gives a renormalized perturbation expansion. Because
of the large number of terms, it is necessary to truncate
this expansion after introducing $\underline{\kappa}^{-\delta}$ vertices
δV in R.

     (b) <u>The low momentum expansion</u>: We truncate the
perturbation expansion series by raising the lower cut-

off in the exponent. We note that

$$e^{-V(\Lambda,\kappa)} \leq e^{O(\kappa^2)|\Lambda|} \ . \tag{3.1.9}$$

This "Wick bound" follows by integrating

$$:\phi_\kappa(x)^4: \ = \ (\phi_\kappa(x)^2 - 3C_\kappa)^2 - 6C_\kappa^2 \geq \ - \ 6C_\kappa^2$$

over the space time volume, where $C_\kappa = \int \phi_\kappa(x)^2 d\mu =$ $C_\kappa(x,x) \leq O(\kappa)$. We use the Wick bound to raise ρ from $\underline{\rho}$ to $\bar{\rho}$. We apply it in a space time cube $\Delta \subset \Lambda$ on which (3.1.9) remains bounded, i.e., cubes for which

$$\bar{\rho}^2|\Delta| \leq O(1) \ .$$

This means that the localization length $L = |\Delta|^{1/3}$ satisfies

$$L \leq O(\bar{\rho}^{-2/3})$$

and defines our phase space localization. On the other hand, the uncertainty principle requires that

$$O(\bar{\rho}^{-1}) \leq L$$

for the localization to be proper. The above conditions are compatible and this is the formal reason why the Wick construction is useful. For the interaction the corresponding compatibility is marginal. The analysis has shown that we must treat separately cubes Δ belonging to a space time cover of the interaction region in the exponent. Thus,

one actually deals with upper and lower cut-off funct-
ions $\kappa(\Delta), \rho(\Delta)$ which are functions of Δ (also some Δ's
tend to zero as $\kappa_0 \to \infty$). To control distance factors in
sum over phase cells, one needs (at least) a polynomial
decay property of the approximating covariance $C_{\kappa,\rho}(x,y)$.
We use smooth momentum cut-off to obtain polynomial decay.

(c) <u>Combinatoric estimates:</u> The expansion of (3.1.4)
is complete when inductive use of the high and lower
momentum expansions has removed the exponent of (3.1.4)
completely and so when (3.1.4) has been reduced to a sum
of Gaussian integrals $\int R(\phi) d\mu$. We perform Gaussian integration to obtain [15]

$$\left| \int R(\phi) e^{-V} d\mu \right| \leq \sum_\alpha \left| \int R_\alpha d\mu \right| \leq \sum_G |I(G)| \qquad (3.1.10)$$

where $I(G)$ is the elementary integration labeled by Feynman
graph G. The final estimate of (3.1.10) is completed by
first using the combinatoric factors to bound the number
of terms in the sum of graph G and then bounding the size
of each term [15]. We pose the combinatoric estimates in
terms of combinatoric factors $C(G)$. For example,

$$\sum_G |I(G)| \leq \left(\sum_G C(G)^{-1} \right) \sup_G C(G) |I(G)|$$

$$\leq \sup_G C(G) |I(G)| \qquad (3.1.11)$$

where the combinatoric factors $C(G)$ are positive numbers
satisfying $\sum C(G)^{-1} \leq 1$. We assign appropriate combinatoric
factors $C(G)$ to certain vertices, lines and leges as the
final part of each inductive step (or substep) [15, 7].

The total combinatoric factor C(G) is the product of
these elementary factors. The final estimate is to bound
(3.1.11) by the method of graph estimates [15]. For
further detailed discussion of the PSCE, we refer to
[15].

Finally, we remark that the above expansion method
can be applied (and extended) to periodic, Dirichlet and
Neumann boundary conditions to yield Theorem 3.1.3 [34].

3.2 Convergence of the Lattice Approximation and Correlation Inequalities

One of the main tools available in constructive
field theory is the approximation of boson field theories
by "generalized Ising models" and the resulting correlation inequalities [23, 28, 36, 37]. The earlier application of the approximation in the $P(\phi)_2$ field theory was
due to Nelson [28], Guerra, Rosen, Simon [23] and Simon
and Griffiths [37], etc. The basis of the Ising method
is two approximation: the lattice approximation [23, 28]
and the classical Ising approximation [37]. The classical
Ising approximation proceeds in two steps. One first passes
to the lattice approximation and then approximates the
lattice approximation (spin approximation) by classical
Ising spins. The latter part causes no troubles in three
and four dimensions [37] and so the convergence of the
lattice approximation is the only problem in higher dimensions. In two dimensional models [1, 23, 36] the proof of
the convergence is relatively easy because of no ultra-
violet divergences. For the $(\lambda\phi^4 - \sigma\phi^2 - \mu\phi)_3$ models the con-

vergence of the lattice approximation has been proven
by Park [31, 34]. This allows one to develop the models
parallel to $P(\phi)_2$. In this section we discuss the results and basic idea of the proof.

We now consider the lattice approximation of boson
field models in more detail. By a "generalized Ising
model" we mean a finite collection of random variables
whose joint probability distribution has the form

$$e^{\sum\limits_{i \neq j} a_{ij} q_i q_j} \, dv_1(q_1) \ldots dv_N(q_N) \tag{3.2.1}$$

for measure v_i on R. If each $a_{ij} \geq 0$, the model is called
ferromagnetic. The purpose of the lattice approximation
in the Euclidean boson field theory is to exploit the
ferromagnetic nature of the Euclidean lattice field. To
see this we consider the lattice cut-off interaction
measure defined by

$$dq_{\Lambda, \delta}^X = (Z_{\Lambda, \delta}^{(X)})^{-1} \, e^{-V^{(X)}(\Lambda, \delta)} \, d\mu_{\Lambda, \delta}^X \tag{3.2.2}$$

where $V^{(X)}(\Lambda, \delta)$ and $Z_{\Lambda, \delta}^{(X)}$ are the interacting action and
the partition function corresponding to the model. In
general, one has (for example, see (2.10))

$$V^{(X)}(\Lambda, \delta) = \delta^d \sum_{n\delta \in \Lambda_\delta} :P(\phi_\delta(n)):_X + \text{scalar counterterms} .$$

Identifying $\phi_\delta(n) = q_n$, (3.2.2) can be written as [23, 24]

$$dq^X_{\Lambda,\delta} = \text{const } e^{-\sum \delta^d : P(q_n): } d\mu^X_{\Lambda,\delta} \text{ ,}$$

where the Gaussian measure is given by [23, 24]

$$d\mu^X_{\Lambda,\delta} = \text{const } e^{-\frac{1}{2}q \cdot (C^X_{\Lambda,\delta})^{-1} \cdot q} d^N q \text{ .}$$

Here N = the number of lattice points in Λ_δ, and $C^X_{\Lambda,\delta}$ is NXN matrix which approximates $(-\Delta_X + m_o^2)^{-1\delta}$ 23, 24 . The matrix $(C^X_{\Lambda,\delta})^{-1}$ links only the nearest neighbours and $(C^X_{\Lambda,\delta})^{-1}_{ij} \geq 0$, $i \neq j$. Thus, the measure (3.2.2) is ferromagnetic. For precise connection between lattice field models and ferromagnetic models, we refer to [23, 24, 36].

For ferromagnetic systems, the first and second Griffiths inequalities (G-I and G-II) [20], the Fortuin-Kastelyn-Ginibre inequalities (FKG) [11], the Griffith-Hurst-Sherman inequalities (GHS) [21], the Lebowitz inequality [25] and generalizations of these inequalities are known. The first three types (G-I, G-II, FKG) hold for quite general Ising models (in particular, continuous spin) while the GHS and Lebowitz inequalities seem to apply only to classical Ising models. We write

$$< \cdot >^{(X)}_\Lambda = \int \cdot \, dq^X_\Lambda$$

$$dq^X_\Lambda = \lim_{\delta \to 0} dq^X_{\Lambda,\delta} \text{ , } X = F,D,P,N, \tag{3.2.3}$$

if the limit exists (in weak sense) for a given boson field model. As a consequence of the convergence of the lattice approximation and classical Ising approximation, G-I, G-II, FKS, GHS and Lebowitz inequalities hold for the $(\lambda\phi^4-\sigma\phi^2-\mu\phi)_2$ models [23, 24, 36, 37]: For X = F, D, P, N and for $f_i \geq 0$

(i)   First Griffiths inequalities (G-I)

$$<\phi(f_1)..\phi(f_n)>_{\Lambda}^{(X)} \geq 0$$

(ii)   Second Griffiths inequalities (G-II)

$$<\phi(f_1)..\phi(f_{n+m})>_{\Lambda}^{(X)} \geq <\phi(f_1)..\phi(f_n)>_{\Lambda}^{(X)} .$$

$$\cdot <\phi(f_{n+1})..\phi(f_{n+m})>_{\Lambda}^{(X)}$$

(iii) Fortuin-Kastelyn-Ginibre inequalities (FKG):
Let F and G be increasing functions of the fields (see [36]). Then

$$<FG>_{\Lambda}^{(X)} \geq <F>_{\Lambda}^{(X)} <G>_{\Lambda}^{(X)}$$

(iv) Griffiths-Hurst-Sherman (GHS) and Lebowitz inequalities: If $\mu > 0$, then the truncated three point function satisfies

$$<\phi(f_1)\phi(f_2)\phi(f_3)>_{\Lambda}^{(X)T} \leq 0$$

If $\mu = 0$, then

$$< \phi (f_1) \phi (f_2) \phi (f_3) \phi (f_4) >_{\Lambda}^{(X)T} \leq 0 .$$

Finally, we introduce Lee-Yang theorem:

(v)   Lee-Yang theorem [36, 37]: For $(\lambda \phi^4 - \sigma \phi^2)_2$ models we write

$$F(\mu) = < e^{\mu \phi (f)} >_{\Lambda}^{(X)}$$

for $f \geq 0$. Then $F(\mu) \neq 0$ if Re $\mu > 0$.

Remark: The first three types (G-I , G-II, FKG) hold for the $(P_e (\phi) - \mu \phi)_2$ model, where $P_e (x)$ is an even semibounded polynomial. The GHS and Lebowitz inequalities and Lee-Yang theorem seem to apply only to the $(\lambda \phi^4 - \sigma \phi^2 - \mu \phi)_2$ models [36, 37].

We now turn to the convergence of the lattice approximation for the $(\lambda \phi^4 - \sigma \phi^2 - \mu \phi)_3$ models. We first list the result:

Theorem 3.2.1: (Park [31, 34]). For the $(\lambda \phi^4 - \sigma \phi^2 - \mu \phi)_3$ models, the limit

$$Z_{\Lambda}^{(X)} = \lim_{\delta \to 0} Z_{\Lambda, \delta}^{(X)}$$

$$S_{\Lambda}^{(X)} (f_1, \ldots, f_n) = \lim_{\delta \to 0} S_{\Lambda, \delta}^{(X)} (f_1, \ldots, f_n)$$

exist for X = F, D, P, N and coincide with the corresponding quantities of Theorem 3.1.3.

As a consequence of the theorem (and non-vanishness of the partition functions, Theorem 3.1.2, for the Lee-Yang theorem) we obtain

Theorem 3.2.2: Correlation inequalities of the types G-I, G-II, FKG, GHS and Lebowitz and Lee-Yang theorem hold for the $(\lambda\phi^4 - \sigma\phi^2 - \mu\phi)_3$ field theory with X = F, D, N, P, in a finite box Λ.

We give the main idea of the proof of Theorem 3.2.1. The removal of the lattice cut-off in three dimensions is somewhat complicated because of the ultraviolet divergences of the models. A direct proof of the convergence

$$S_\Lambda^{(X)} = \lim_{\delta\to 0} S_{\Lambda,\delta}^{(X)}$$

seems to be very difficult. Therefore, we introduce momentum cut-off to the lattice cut-off Schwinger functions by replacing $\phi_\delta(n)$ with $\phi_{\delta,\kappa}(n)$ in $V(\Lambda,\delta)$. It is easy to show that

$$S_{\Lambda,\delta,\kappa}^{(X)} \;\to\; S_{\Lambda,\delta}^{(X)} \qquad \text{as} \quad \kappa \;\to\; \infty$$

$$S_{\Lambda,\delta,\kappa}^{(X)} \;\to\; S_{\Lambda,\kappa}^{(X)} \qquad \text{as} \quad \delta \;\to\; 0$$

because of the virtue of the momentum cut-offs δ and κ. Hence, if we prove that

$$S_{\Lambda,\delta,\kappa}^{(X)}(f_1,\ldots,f_n) \xrightarrow[\kappa\to\infty]{} S_{\Lambda,\delta}^{(X)}(f_1,\ldots,f_n) \quad \text{uniformly in } \delta \qquad (3.2.4)$$

for $\delta \geq 0$, the convergence of the lattice approximation follows from the standard 3ε argument:

$$|S_{\Lambda,\delta}^{(X)}-S_{\Lambda}^{(X)}| \leq |S_{\Lambda,\delta}^{(X)}-S_{\Lambda,\delta,\kappa}^{(X)}| + |S_{\Lambda,\delta,\kappa}^{(X)}-S_{\Lambda,\kappa}^{(X)}| + |S_{\Lambda,\kappa}^{(X)}-S_{\Lambda}^{(X)}|$$

$$\leq 3\varepsilon \qquad \text{for sufficiently small } \delta.$$

The problem is thus reduced to show (3.2.4). An advantage of this reduction is that, since the lattice spacing parameter $\delta \geq 0$ and the region of interaction Λ are fixed, we might be able to apply the method of the phase space cell expansion which we have discussed already. For simplicity, we consider the partition function $Z_{\Lambda,\delta}$ for $X = F$. We write

$$Z_{\Lambda,\delta,\kappa} = \int e^{-V(\Lambda,\delta,\kappa)} d\mu_\delta$$

where $V(\Lambda,\delta,\kappa)$ is the triple cut-off interacting action. Replacing $V(\Lambda,\kappa)$ by $V(\Lambda,\delta,\kappa)$ (see the previous section) we follow an expansion procedure similar to that of Glimm and Jaffe [15] (and Feldman [7]). In the final estimates for bounding the elementary integrations labeled by the Feynman graphs G (see (3.1.10) and (3.1.11)), one needs to carry out the estimates of Glimm and Jaffe [15], and Feldman [7] uniformly in δ. This can be done [31, 34]. For example, we consider the (Feynman) propagator for the lattice cut-off two point function in the momentum space:

$$\mu_\delta(k)^{-2} = [\delta^{-2}(6-2\sum_{i=1}^{3}\cos(\delta k^{(i)})) + m_o^2]^{-1}$$

$$\to \mu(k)^{-2} = (k^2 + m_o^2)^{-1} \qquad \text{as} \qquad \delta \to 0 \; .$$

From the inequality [23]

$$2\pi^{-2} y^2 \leq 1 - \cos y \leq 2y^2 \text{ if } y \in [-\pi, \pi]$$

it follows that

$$[\frac{\mu(k)}{\mu_\delta(k)}]^{\pm 1} \leq O(1) \text{ if } k_i \in [-\frac{\pi}{\delta}, \frac{\pi}{\delta}] .$$

Since the kernel of a given Feynman graph is a multiple of propagators and factors coming from conservation of momentum, one expects to get uniform estimates in δ.

The complete proof of Theorem 3.2.1 is rather complicated. For the details we refer to [31, 34].

4. THE INFINITE VOLUME LIMIT THEORIES

4.1 Uniform Bounds of the Pressures and the Schwinger Functions

We have considered the construction of the $(\lambda \phi^4 - \sigma \phi^2 - \mu \phi)_3$ models in a finite box in the previous sections. Our next concern is the construction of the thermodynamic limit theory of the models. Two main methods have been employed for this construction. The first is the method of the cluster expansion for weakly coupled models, and the second is the method of correlation inequalities for the arbitrary (strongly) coupled models. In Section 4.2 we will discuss the method of cluster expansion and then in Section 4.3 we will give a brief discussion of the method of correlation inequalities. In this section we summarize some results on uniform bounds of the pressures and the Schwinger functions (uniform with respect to Λ).

The first step to remove the volume cut-off Λ for the models (at least, for strong coupling) involves uniform bounds of the object we are interested in. We define the pressure by

$$\alpha_\Lambda^{(X)} = \frac{1}{|\Lambda|} \log Z_\Lambda^{(X)} \quad . \tag{4.1.1}$$

We now give the uniform bounds of the pressures and the Schwinger functions for the $(\lambda\phi^4 - \sigma\phi^2 - \mu\phi)_3$ models.

Theorem 4.1.1: (Glimm und Jaffe [15], Park [32], and Seiler and Simon [35]). For X = F, P,

$$|\alpha_\Lambda^{(X)}| \leq \text{const}$$

where the constant is independent of Λ.

Theorem 4.1.2: (Seiler and Simon [35], Park [33]). There is a constant $K(\lambda,\sigma,\mu)$ independent of Λ such that

$$\left| \overline{\lim_{\Lambda \to R^3}} S_\Lambda^{(F)} (f_1,\ldots,f_n) \right| \leq K^n n! \prod_{i=1}^{n} \| f_i \| \tag{4.1.2}$$

$$\left| S_\Lambda^{(P)} (f_1,\ldots,f_n) \right| \leq K^n n! \prod_{i=1}^{n} \| f_i \| \tag{4.1.3}$$

for a suitable Schwartz space norm $\| \cdot \|$.

The uniform upper bound of the pressure (for X=F)
follows from Glimm and Jaffe's result in Theorem 3.1.1.
Park [32] (for X = F, P) and Seiler and Simon [35] have
extended this theorem by showing the uniform lower bounds
(see Section 3.1, proposition 3.1.2). The uniform bound
of the Schwinger functions for X = F was first established
by Seiler and Simon [35]. The result for X = P is due to
Park [33]. Since the proofs of the results involve detailed understanding of the models and also some mathematical
prerequirements, we do not discuss the proofs here. Fortunately the proof of (4.1.3) turns out to be very easy.
In Section 5 we will produce a simple proof of (4.1.3)
by using the FKG inequalities established in Theorem
3.2.2 and the uniform bound of the pressures for X = P.
The use of FKG inequalities has been suggested by Guerra
[22].

Seiler and Simon [35] have also established the
existence of the infinite volume limit of $\alpha_\Lambda^{(F)}$: The
limit

$$\alpha^{(F)} = \lim_{\Lambda \to R^3} \alpha_\Lambda^{(F)}$$

exists. Their method can be easily extended to $\alpha_\Lambda^{(P)}$. It
should be interesting to know if the limits

$$\alpha^{(X)} = \lim_{\Lambda \to R^3} \alpha_\Lambda^{(X)}, \qquad X = F, D, P, N,$$

exist and are independent of boundary conditions. In
$P(\phi)_2$ this has been proved in [24].

4.2 Infinite Volume Limit: Weakly Coupled Models

The existence of the infinite volume limit of the weakly coupled $(\lambda\phi^4)_3$ field models have been established by Feldman and Osterwalder [9] and Magnen and Seneor [26] independently. Their results can be easily extended to the $(\lambda\phi^4 - \sigma\phi^2 - \mu\phi)_3$ models provided that λ, σ and μ are sufficiently small and m_0 is sufficiently large [13]. For simplicity we only consider the $(\lambda\phi^4)_3$ models. The proofs were based on the cluster expansion of Glimm, Jaffe and Spencer [18] together with the PSCE [15, 7]. We first list the results and then discuss the proofs.

__Theorem 4.2.1:__ [9, 26]. For the $(\lambda\phi^4)_3$ field models we assume that λ is sufficiently small and m_0 is sufficiently large. Then the no-cut-off limit

$$S^{(F)}(f_1,\ldots,f_n) = \lim_{\Lambda \to R^3} \lim_{\kappa \to \infty} S^{(F)}_{\Lambda,\kappa}(f_1,\ldots,f_n)$$

exist. The $\{S^{(F)}(f_1,\ldots,f_n)\}$ satisfy all of the Oster-walder-Schrader axioms [29]. Hence, they determine a unique Wightman theory. The Wightman theory has a non-zero mass gap m.

__Theorem 4.2.2:__ [9, 26]. The $\{S^{(F)}(f_1,\ldots,f_n)\}$ are the moments of a unique probability measure on $S'(R^3)$. They are C^∞-functions in $\lambda \in [0, \lambda_0]$. The perturbation theory provides an asymptotic expansion for $\{S^{(F)}\}$.

__Remark:__ (a) Presumably, under the assumptions of Theorem 4.2.1 the limit

$$S^{(X)}(f_1,\ldots,f_n) = \lim_{\Lambda \to R^3} S_\Lambda^{(X)}(f_1,\ldots,f_n), \quad X = P, D, N$$

exist and coincide with $S^{(F)}(f_1,\ldots,f_n)$.

(b) Magnen and Seneor [27] have extended their method to construct the infinite volume limit theory for the weakly coupled Yukawa$_2$ field model.

The main tools are two types of expansion - the cluster expansion to prove exponential clustering and convergence of the infinite volume limit, and the phase space cell expansion (PSCE) to control the ultraviolet divergences in each cell in phase space. The cluster expansion is analogous with a high temperature expansion in statistical mechanics and was first used to establish exponential clustering for the weakly coupled $P(\phi)_2$ models by Glimm, Jaffe and Spencer [18]. It is based on the observation that, if there is no coupling between different space-time regions, the infinite volume limit would be trivial. In particular, supposed we partition space-time into disjoint unit cubes $\{\Delta_i \mid i \in Z^3\}$ and place zero data on the surface of these cubes. Then our covariance no longer couples different cubes, i.e., $(-\Delta_D + m_o^2)^{-1}(x,y)=0$ if $x \in \Delta_i$, $y \in \Delta_j$, $i \neq j$. Hence, if support $\{f_1,\ldots,f_n\} \subset \Lambda_1$ where Λ_1 is a union of unit cubes,

$$S(f_1,\ldots,f_n) = \lim_{\Lambda \to R^3} S_\Lambda(f_1,\ldots,f_n)$$

$$= \lim_{\Lambda \to R^3} Z_\Lambda^{-1} \int \phi(f_1)\ldots\phi(f_n) e^{-V(\Lambda)} d\mu$$

$$= \lim_{\Lambda \to R^3} (Z_\Lambda Z_{\Lambda \sim \Lambda_1})^{-1} (\int \phi(f_1)..\phi(f_n) e^{-V(\Lambda_1)} d\mu)$$

$$\cdot (\int e^{-V(\Lambda \sim \Lambda_1)} d\mu)$$

$$= S_{\Lambda_1} (f_1, \ldots, f_n).$$

Of course, we do not have this complete decoupling, but the covariance decouples exponentially (see (2.1)). The cluster expansion gives the original coupled theory as a perturbation of the complete decoupled theory. We replace the covariance $(-\Delta + m_o^2)^{-1}$ in the Gaussian measure by $(-\Delta_\Gamma + m_o^2)^{-1}$, where Γ is a set of unit cubes and Δ_Γ is the Laplacian with zero data on Γ. We thus obtain a new Gaussian measure $d\mu_\Gamma$ which decouples across the surface of Γ. If Γ contains all unit cubes in R^3, then $d\mu_\Gamma$ would factor into a product measure over unit cubes (the case of the complete decoupled theory we discussed above). The idea is to introduce the boundary condition on one unit cube at a time and to control the error terms arising from change of covariances. The basic formula is the following change of covariances: Let C_1 and C_o be given covariances and let $C_\alpha = \alpha C_1 + (1-\alpha) C_o$. Then [18]

$$\int F(\phi, C_1) d\mu_{C_1} = \int F(\phi, C_o) d\mu_{C_o} + \int_o^1 d\alpha \frac{d}{d\alpha} \int F(\phi, C_\alpha) d\mu_{C_\alpha}$$

$$= \int F(\phi, C_o) d\mu_{C_o}$$

$$+ \int_o^1 d\alpha \int \{\frac{1}{2} (C_1(x,y) - C_o(x,y)) \frac{\partial^2 F(\phi, C_\alpha)}{\partial \phi(x) \partial \phi(y)} \} d\mu_{C_\alpha}$$

$$+ \int_{0}^{1} d\alpha \int \{ \frac{d}{d\alpha} F(\phi, C_\alpha) \} d\mu_{C_\alpha} \quad .$$

For the method to work the following properties are required [18, 9, 26]:

(a) The covariance $(-\Delta + m_o^2)^{-1}$ decouples exponentially.

(b) The Euclidean action $V(\phi)$ is local.

In $P(\phi)_2$ the above requirements are satisfied because it has no ultraviolet cut-off. In $(\lambda\phi^4)_3$ one would like to do all the estimates with the doubly cut-off Schwinger functions $S_{\Lambda, \kappa}$. However, the ordinary methods of cutting off the high momenta make the theory violate the requirements (a) and (b). The problem was how one could approximate the covariance $(-\Delta + m_o^2)^{-1}$ by approximating covariances satisfying (a) and (b). Feldman and Osterwalder [9, 8] and Magnen and Seneor [26] introduced cut-offs which simultaneously suppress high momenta, preserve locality of the interacting action and exponential decay of the propagator and which allow for the introduction of boundary conditions. For notational convenience we only give the rough idea of Feldman and Osterwalder [9, 8]. The propagator can be represented by

$$(-\Delta + m_o^2)^{-1}(x,y) = \frac{1}{|x-y|} e^{-m_o|x-y|}$$

$$= \int_{0}^{\infty} dt \ e^{-m_o^2 t} \int P_{xy}^t (d\omega) B(\omega), (B(\omega) = 1),$$

where

$$P_{xy}^t (d\omega) = \text{conditional Wiener measure}$$

$B(\omega)$ determines boundary conditions.

We define an auxiliary Gaussian field

$\psi(t,x)$, $t \, \varepsilon \, (0,\infty)$, $x \, \varepsilon \, R^3$

whose two point function is given by

$$<\psi(t,x)\psi(s,y)> = \delta(t-s)e^{-m_o^2 t} \int P_{xy}^t (d\omega) B(\omega) \; .$$

Then the familiar field $\phi(x)$ is given by

$$\phi(x) = \int_0^\infty \psi(t,x) dt$$

and an ultraviolet cut-off field $\phi_\kappa(x)$ is given by

$$\phi_\kappa(x) = \int_{\kappa^{-2}}^\infty \psi(t,x) \; dt \; .$$

We note that

$$<\phi_\kappa(x)\phi_\kappa(y)> = \int d^3k \; e^{ik \cdot (x-y)} \; \frac{e^{-\kappa^{-2}(k^2+m_o^2)}}{k^2 + m_o^2}$$

and so high momenta obeying $k^2 > \kappa^2$ are effectively suppressed. One may easily check that the above cut-off satisfies the requirement (a) and (b).

Feldman and Osterwalder [9] used these cut-offs in PSCE to derive ultraviolet uniform estimates which in

turn were plugged into the cluster expansion to yield
estimates that are independent of both ultraviolet and
volume cut-offs. For detailed proof we refer to [9, 26].

4.3 Infinite Volume Limit: Strongly Coupled Models

The construction of the infinite volume limit theory
without restriction of the magnitude of the coupling constant is based on an application of correlation inequalities
together with uniform bounds of the Schwinger functions.
One of Griffiths's original applications of his inequalities was to prove the existence of infinite volume limit
for the correlation functions of a spin system [20].
Nelson extended this idea to construct an infinite volume
$P(\phi)_2$ theory for P = even $-\mu x$ and (half) Dirichlet states.
Glimm and Jaffe [16] have combined correlation inequalities with the cluster expansion to construct an infinite
volume theory - the so called "Weak Coupling Boundary
Condition Theory"- in these P.

Since we have established correlation inequalities
for the $(\lambda\phi^4 - \sigma\phi^2 - \mu\phi)_3$ field models, we may apply the
ideas of Nelson and Glimm and Jaffe to construct in-
finite volume theories for arbitrary coupling constants.
The first result was due to Fröhlich [13] who followed
the idea of Glimm and Jaffe [16] and constructed the infinite volume (weak coupling boundary) state. The basic
results he used are the convergence of the cluster expansion [9, 26], the uniform bounds of the Schwinger functions
(4.1.2) [35] and the convergence of lattice approximation
[31] (when he studied the construction, the convergence
of the lattice approximation with Dirichlet boundary con-

dition was not established yet). In the meantime, Park
[34] completed the proof of the convergence of the
lattice approximation with various (including Dirichlet)
boundary conditions. This allowed him to construct the
infinite volume Dirichlet state. We summarize these results:

Theorem 4.3.1: For the $(\lambda\phi^4 - \sigma\phi^2 - \mu\phi)_3$ theory with Dirichlet
(D) and weak coupling (W) boundary conditions, the infinite
volume limit Schwinger functions $\{S^{(X)}(f_1,...,f_n)\}$ exist for
$f_i \in S(R^3)$. These Schwinger functions satisfy the axioms
of Osterwalder and Schrader with the possible exception of
clustering. They are moments of a unique probability measure
dq^X on $S'(R^3)$. The theory (for $X = D,W$) satisfies all the
Wightman axioms with the possible exceptions of uniqueness
of the vacuum for $\mu = 0$ and the mass gap.

Remark: (a) the correlation inequalities, and the Lee-Yang
theorem listed in Section 3.2 carry over to the infinite
volume theories for $X = D, W$.
(b) We conjecture that

$$S^{(D)}(f_1,...,f_n) = S^{(W)}(f_1,...,f_n) \qquad \text{if} \quad \mu \neq 0 .$$

The conjecture is based on the fact that the vacuum is
unique for $\mu \neq 0$.
(c) Fröhlich [13] has analyzed further properties of
the infinite volume theory.

   For convenience, we only discuss the construction
for $X = D$. For the Weak Coupling state we refer to [13].
Since we have fixed the coefficient δm^2 of the mass re-

normalization counterterm, the G-II inequalities give us (see [36])

$$S_{\Lambda,\delta}^{(D)}(f_1,\ldots,f_n) \leq S_{\Lambda,\delta}^{(P)}(f_1,\ldots,f_n)$$

$$S_{\Lambda,\delta}^{(D)}(f_1,\ldots,f_n) \leq S_{\Lambda',\delta}^{(D)}(f_1,\ldots,f_n) \qquad \text{if} \quad \Lambda \subset \Lambda'$$

for $f_i \geq 0$. As a consequence of the convergence of the lattice approximation (Theorem 3.2.1) and the uniform bounds of the Schwinger functions (Theorem 4.1.2) it follows that

$$S_\Lambda^{(D)}(f_1,\ldots,f_n) \leq S_\Lambda^{(P)}(f_1,\ldots,f_n) \leq K^n n! \prod_i \|f_i\|$$

$$S_\Lambda^{(D)}(f_1,\ldots,f_n) \leq S_{\Lambda'}^{(D)}(f_1,\ldots,f_n) \qquad \text{if} \quad \Lambda \subset \Lambda' \; .$$

Hence, for $f_i \geq 0$ the infinite volume limit exists and satisfies

$$S^{(D)}(f_1,\ldots,f_n) = \lim_{\Lambda \uparrow R^3} S_\Lambda^{(D)}(f_1,\ldots,f_n)$$

$$\leq K^n \, n! \prod_i \|f_i\| \; .$$

For general functions f_i's we decompose each function into its positive and negative parts to yield the convergence and the bounds. The other Euclidean axioms except the Clustering follow as in $P(\phi)_2$. The rest of the theorem

then follows from the Osterwalder and Schrader reconstruction theorem and Lee-Yang theorem (for uniqueness of vacuum
for $\mu \neq 0$ [36]).

5. UNIFORM BOUNDS OF THE SCHWINGER FUNCTIONS IN
BOSON FIELD MODELS: A SIMPLE PROOF

In the Euclidean strategy in Constructive Quantum
Field Theory the proof of uniform bounds of the Schwinger
functions for a given model with (lattice and) volume cut-
off is the first step to control the infinite volume limit
and so to complete the program of constructing relativistic
quantum field theory. See [1, 17, 23, 28, 33, 35, 36] and
the references therein. In this section we give a simple
proof of uniform bounds of the Schwinger functions (with
periodic boundary conditions) for various boson field
models.

It has been suggested that control of the pressures
gives control of the Schwinger functions [15, 18]. In [33]
we have shown that uniform bounds of the pressures give
uniform bounds for the corresponding Schwinger functions
with periodic boundary conditions for boson field models.
The basic methods we used were the transfer matrix formula
for periodic states, the Griffiths inequalities and the
translation invariance of periodic states. See [33] for
details. The transfer matrix formula was used to get exponential bounds of the fields. Recently Guerra [22] observed that the exponential bounds can be obtained by
FGK inequalities. This makes the proof considerable
shorter. We will adapt Guerra's idea and the method used

in [33].

The boson field models we are dealing with are the following:

(a) The $P(\phi)_2$ field models, where P = even poly - μx.

(b) The $(\lambda\phi^4 - \sigma\phi^2 - \mu\phi)_3$ field models.

(c) The lattice cut-off exponential type interactions in d-dimensional space-time: The lattice $(\int dv(\alpha) e^{\alpha\phi})_d$ field models, where $dv(\alpha)$ is an even measure on R.

In principle, our method extends also to the ϕ^4_4 theory, provided the relevant problems of renormalization and uniform bounds of the corresponding pressures can be solved. We remark that the models in (c) have been extensively studied by Albeverio and Höegh-Krohn [1] for supp $v(\alpha) \subset (-\sqrt{4\pi}, \sqrt{4\pi})$ and d = 2. They have shown the uniform bounds for the lattice cut-off Schwinger functions by Griffiths inequalities [1]. Let $<\cdot>_\Lambda^{(P)}$ be the expectation with respect to the interacting measure dq_Λ^P with periodic boundary conditions for a given model and let $z_\Lambda^{(P)}(s)$ be the corresponding pressure with a linear external field $s\phi$. For the $(\Lambda\phi^4 - \sigma\phi^2 - \mu\phi)_3$ models,

$$<\cdot>_\Lambda^{(P)} = \int \cdot \, dq_\Lambda^P , \qquad dq_\Lambda^P = \lim_{\delta\to 0} dq_{\Lambda,\delta}^P$$

$$z_\Lambda^{(P)}(s) = \lim_{\delta\to 0} z_{\Lambda,\delta}^{(P)}(s), \quad z_{\Lambda,\delta}^{(P)}(s) = \int e^{s\phi_\delta(\Lambda) - V^{(P)}(\Lambda,\delta)} d\mu_{\Lambda,\delta}^P \quad (5.1)$$

where $dq_{\Lambda,\delta}^P$ has been defined in (3.2.2). We define the pressure by

$$\alpha_\Lambda^{(P)}(s) = \frac{1}{|\Lambda|} \log z_\Lambda^{(P)}(s) . \qquad (5.2)$$

We also write $\phi(\Delta) = \phi(\chi_\Delta)$ where χ_Δ is the characteristic function of $\Delta \subset \Lambda$.

In any case (including the cases (a), (b) and
(c)) for a state $<\cdot>$ we make the following definitions
[22]:

Definition 5.1. An interaction is stable if the corresponding pressures satisfy

$$|\alpha_\Lambda^{(P)}(1) - \alpha_\Lambda^{(P)}(0)| \leq const, \qquad (5.3)$$

where the constant is independent of Λ (and δ for
lattice field models).

Definition 5.2. A state $<\cdot>$ is called FKG state if for
any two increasing functions F, G of the fields the expectation values of F, G and FG are correlated by

$$<FG> \geq <F> <G>$$

whenever the averages have a meaning.

We now prove exponential bounds for the models in
which the corresponding states are FKG states.

Proposition 5.3. Let $<\cdot>_\Lambda^{(P)}$ be a periodic FKG state.
Assume $|\Lambda| = n_1 x..xn_d$ with n_i integers. Let $\Delta \subset \Lambda$ be a
unit cube with sides parallel to the sides of Λ and let
$s \geq 0$. Then

$$<e^{s\phi(\Delta)}>_\Lambda^{(P)} \leq e^{[\alpha_\Lambda^{(P)}(s) - \alpha_\Lambda^{(P)}(0)]} .$$

Proof. Since $e^{s\phi(\Delta)}$ is a monotonic increasing function for $s \geq 0$, we use translation invariance and the FKG property of the state to obtain

$$e^{|\Lambda|[\alpha_\Lambda^{(P)}(s) - \alpha_\Lambda^{(P)}(0)]} = \langle e^{s\phi(\Lambda)} \rangle_\Lambda^{(P)}$$

$$\geq (\langle e^{s\phi(\Delta)} \rangle_\Lambda^{(P)})^{|\Lambda|} .$$

This proves the proposition.

We next show the uniform bounds of the Schwinger functions for a boson field model.

Theorem 5.4. Let the interaction be stable and let the corresponding periodic state $\langle \cdot \rangle_\Lambda^{(P)}$ be a FKG state. Furthermore, assume that the Griffiths inequalities (G-I) hold for the state $\langle \cdot \rangle_\Lambda^{(P)}$, and that $\langle e^{\tau\phi(\Delta)} \rangle_\Lambda^{(P)}$ is analytic in τ for $|\tau| \leq 1$. Then

$$|\langle \phi(f_1)\cdots\phi(f_n) \rangle_\Lambda^{(P)}| \leq K^n \, n! \prod_{i=1}^{n} \|f_i\|$$

for a suitable Schwartz space norm $\|\cdot\|$.

Proof. We only need to prove the theorem for $f_i > 0$. For general functions $\{f_i\}$ one only needs to decompose each function into its positive and negative parts to prove the theorem (the constant K increases to 2K). Hence, we assume that each function f_i in the following proof is positive.

Let $\Delta^{(i)} \subset \Lambda$ be a unit cube centered at $j \in \mathbb{Z}^d$.

Proposition 5.3, the Cauchy integral formula and Griffiths inequalities yield

$$<\phi (\Delta^{(j)})^m>_\Lambda^{(P)} \le (\text{const})^m m! \; e^{[\alpha_\Lambda^{(P)}(1) - \alpha_\Lambda^{(P)}(0)]}$$

$$\le K^m m! \quad . \tag{5.4}$$

Here we have used G-I inequalities to get $|<e^{\tau \phi (\Delta)}>| \le <e^{\phi (\Delta)}>$ for $|\tau| \le 1$. Let f_i have support in some unit cube $\Delta^{(i)} \subset \Lambda$. From G-I and (5.4) it follows that

$$<\phi (f_i)^m>_\Lambda^{(P)} \le \| f_i \|_\infty^m \; <\phi (\Delta^{(i)})^m>_\Lambda^{(P)}$$

$$\le K^m m! \, \| f_i \|_\infty^m \quad . \tag{5.5}$$

Let each f_i have support in some unit cube $\Delta^{(i)}$. The Hölder's inequality and (5.5) yield

$$< \prod_{i=1}^n \phi (f_i)>_\Lambda^{(P)} \le \prod_{i=1}^n (<\phi (f_i)^n>_\Lambda^{(P)})^{1/n}$$

$$\le K^n n! \prod_{i=1}^n \| f_i \|_\infty \quad . \tag{5.6}$$

For $f_i \, \epsilon \, S(\Lambda)$ we write $f_i = \sum_{j_i} f_i^{(j_i)}$, where $f^{(j_i)}$ has support in the unit cube centered at $j_i \, \epsilon \, Z^d$. Then, from (5.6) it follows that

$$< \prod_{i=1}^{k} \phi(f_i) >_{\Lambda}^{(P)} = \sum_{j_i \in Z^d_{\cap \Lambda}} < \prod_{i=1}^{k} \phi(f_i^{(j_i)}) >_{\Lambda}^{(P)}$$

$$\leq \sum_{j_i} K^k \, k! \, \prod_{i=1}^{k} \| f_i^{(j_i)} \|_{\infty}$$

$$\leq K^k \, k! \, \prod_{i=1}^{k} \left(\sum_{j_i} \| f_i^{(j_i)} \|_{\infty} \right)$$

$$\leq K^k \, k! \, \prod_{i=1}^{k} \| f_i \|$$

for some Schwartz space norm $\| \cdot \|$. This completes the proof.

The boson field models (a) - (c) we are interested in, satisfy all the assumptions in Theorem 5.4 (for example, see Theorem 3.2.2 and Theorem 4.1.1) [32, 36]. As a corollary of Theorem 5.4 we obtain

Corollary 5.5. The periodic Schwinger functions for the $(\lambda\phi^4 - \sigma\phi^2 - \mu\phi)_3$ - $P(\phi)_2$ and - lattice $(\int dv(\alpha) e^{\alpha\phi})_d$ models in a finite volume Λ are bounded uniformly in Λ (and δ) by

$$|S_{\Lambda}^{(P)}(f_1, \ldots, f_n)| \leq K^n \, n! \, \prod_{i=1}^{n} \| f_i \|$$

for a suitable Schwartz space norm $\| \cdot \|$.

6. OUTLOOK

So far, most of the studies in the $(\lambda\phi^4 - \sigma\phi^2 - \mu\phi)_3$ field theory has been devoted to the construction of the infinite volume limit. The next step in the work on the models is to determine detailed physical properties. Our study has shown that many of the techniques and methods developed in $P(\phi)_2$ can be extended to the $(\lambda\phi^4 - \sigma\phi^2 - \mu\phi)_3$. We believe that virtually all the results known for the $(\lambda\phi^4 - \sigma\phi^2 - \mu\phi)_2$ models can be established for the $(\lambda\phi^4 - \sigma\phi^2 - \mu\phi)_3$ models. In particular, we are interested in the following problems:

(1) Existence of an upper mass gap for weak coupling.
(2) Non-triviality of the S-matrix[+].
(3) Existence of phase transitions[++].
(4) Borel summability for the Schwinger functions.

The above problems have been solved in two dimensions. For example, see [3, 19, 30]. The next problems which are open for the d = 2,3 are

(5) Asymptotic completeness.
(6) Duality of local algebras.

The next problem is the question of four space-time dimensions or in other words how to deal with renormalizable interactions. Clearly, this is our most challenging problem, for example:

(7) Existence of ϕ_4^4.

Our present methods in dealing with ultraviolet divergences

+, ++ See Addenda to the Note.

have been tied to superrenormalizability and for renormalizable models new ideas seem to be required. We ask, can an understanding of the renormalization group or of multiplicative renormalization [38] be an aid in constructing renormalizable models?

ADDENDA TO THE NOTE

+) According to private communication with Professor Fröhlich, the non-triviality of the S-matrix for weakly coupled models can be established by a method similar to that used in [3].

++) Recently Fröhlich, Simon and Spencer proved the existence of phase transitions. See J. Fröhlich, Phase Transitions, Goldstone Bosons and Dynamical Superselection Rules, Lecture Note for XI. Internationale Universitätswochen für Kernphysik, Schladming, Feb. 16-27, 1976.

REFERENCES

1. S. Albeverio and R. Høegh-Krohn, The Wightman Axioms and the Mass Gap for Strong Interaction of Exponential Type in Two-Dimensional Space-Time, J. Funct. Anal. 16, 39 (1974).

2. J.P. Eckman, Representations of the C.C.R. in the $(\phi^4)_3$ Model; Independent of Space Cutoff, Comm. Math. Phys. 25, 1 (1972).

318

3. J.P. Eckman, H. Epstein and J. Fröhlich, Asymptotic Perturbation Expansion for the S-Matrix and Definition of Time-Ordered Functions in Relativistic Field Models, University of Geneve, Preprint (1975).

4. J.P. Eckman and K. Osterwalder, On the Uniqueness of the Hamiltonian and of the Representation of the C.C.R. for the Quartic Boson Interaction in Three Dimensions, Helv. Phys. Acta $\underline{44}$, 884 (1971).

5. J. Fabrey, Exponential Representations of Canonical Commutation Relations, Comm. Math. Phys. 1 (1970).

6. J. Fabrey, Weyl Systems for the $(\phi^4)_3$ Model, J. Math. Phys. $\underline{37}$, 93 (1974).

7. J. Feldman, The $\lambda\phi_3^4$ Field Theory in a Finite Volume, Comm. Math. Phys. $\underline{37}$, 93 (1974).

8. J. Feldman, The Nonperturbative Renormalization of $(\lambda\phi^4)_3$, M.I.T. preprint (1975).

9. J. Feldman and K. Osterwalder, The Wightman Axioms and the Mass Gap for Weakly Coupled $(\phi^4)_3$ Quantum Field Theories, Harvard University, Preprint, February (1975).

10. J. Feldman and K. Osterwalder, The Construction of $\lambda\phi_3^4$ Quantum Field Models, Proceedings of the International Colloquium on Mathematical Methods of Quantum Field Theory, Marseille (1975).

11. C. Fortuin, P. Kastelyn and J. Ginibre, Correlation Inequalities on Some partially Ordered Sets, Comm. Math. Phys. $\underline{22}$, 89 (1971).

12. J. Fröhlich, New Super-Selection Sections ('Soliton-States') in Two Dimensional Bose Quantum Field Models, Princeton University, Preprint (1975).

13. J. Fröhlich, Existence and Analyticity in the Bare

parameters of the $(\lambda \, (\vec{\phi} \cdot \vec{\phi})^2 - \sigma \phi_1^2 - \mu \phi_1)$ - Quantum Field Models, I, Princeton University, Preprint (1975).

14. J. Glimm, Boson Fields with the $:\phi^4:$ Interaction in the Three Dimensions, Comm. Math. Phys. $\underline{10}$, 1 (1968).

15. J. Glimm and A. Jaffe, Positivity of the ϕ_3^4 Hamiltonian, Fortschritte der Physik, $\underline{21}$, 327 (1973).

16. J. Glimm and A. Jaffe, On the Approach to the Critical Point, Harvard University, Preprint (1974).

17. J. Glimm, A. Jaffe, A Remark on the Existence of ϕ_4^4, Phys. Rev. Letter, $\underline{33}$, 440 (1974).

18. J. Glimm, A. Jaffe and T. Spencer, The Particle Structure of the Weakly Coupled $P(\phi)_2$ Model and Other Applications of High Temperature Expansions, Contribution to [40].

19. J. Glimm, A. Jaffe and T. Spencer, Phase Transitions for the ϕ_2^4 Quantum Field Theory, Comm. Math. Phys. $\underline{45}$, 203 (1975).

20. R. Griffiths, Correlation in Ising Ferromagnetics, I, II, III, J. Math. Phys. $\underline{8}$, 478-483, 484-489 (1967).

21. R. Griffiths, C. Hurst and S. Sherman, Concavity of Magnatization of an Ising Ferromagnet in a positive External Field, J. Math. Phys. $\underline{11}$, 790 (1970).

22. F. Guerra, External Field Dependence of Magnetization and Long Range Order in Quantum Field Theory, I.A.S., Preprint, December (1975).

23. F. Guerra, L. Rosen and B. Simon, The $P(\phi)_2$ Euclidean Quantum Field Theory as Classical Statistical Mechanics, Ann. of Math. $\underline{101}$, 111 (1975).

320

24. F. Guerra, L. Rosen and B. Simon, Boundary Condit-
    ions in the $P(\phi)_2$ Euclidean Quantum Field Theory,
    Princeton University, Preprint (1975).

25. J. Lebowitz, GHS and Other Inequalities, Comm. Math.
    Phys. $\underline{35}$, 87 (1974).

26. J. Magnen and R. Seneor, Infinite Volume Limit of
    the ϕ_3^4 Model, Ecole Polytechnique, Preprint, Febru-
    ary (1975).

27. J. Magnen and R. Seneor, The Wightman Axioms for the
    Weakly Coupled Yukawa Model in Two Dimensions, Ecole
    Polytechnique, Preprint (1975).

28. E. Nelson, Probability Theory and Euclidean Field
    Theory, Contribution to [40].

29. K. Osterwalder and R. Schrader, Axioms for Euclidean
    Green's Functions, I and II, Comm. Math. Phys. $\underline{31}$,
    83 (1973), Comm. Math. Phys. $\underline{42}$, 281 (1975).

30. K. Osterwalder and R. Seneor, The Scattering Matrix
    is Nontrivial for Weakly Coupled $P(\phi)_2$ Models,
    Preprint (1975).

31. Y. Park, Lattice Approximation of the $(\lambda\phi^4 - \mu\phi)_3$
    Field Theory, J. Math. Phys. $\underline{16}$, 1065 (1975); The
    $\lambda\phi_3^4$ Euclidean Quantum Field Theory in a periodic
    Box, J. Math. Phys. $\underline{16}$, 2183 (1975).

32. Y. Park, Uniform Bounds of the Pressures of the $\lambda\phi_3^4$
    Field Model, to appear in J. Math. Phys.

33. Y. Park, Uniform bounds of the Schwinger Functions
    in Boson Field Models, to appear in J. Math. Phys.

34. Y. Park, Convergence of Lattice Approximations and
    Infinite Volume Limit in the $(\lambda\phi^4 - \sigma\phi^2 - \mu\phi)_3$ Field
    Theory, University of Bielefeld (ZiF), preprint,
    December (1975).

35. E. Seiler and B. Simon, Nelson's Symmetry and All
    That in the Yukawa$_2$ and $(\phi^4)_3$ Field Theories,
    Princeton University, Preprint (1975).

36. B. Simon, The $P(\phi)_2$ Euclidean (Quantum) Field
    Theory, Princeton University Press (1974).

37. B. Simon and R. Griffiths, The $(\phi^4)_2$ Field Theory
    as a classical Ising Model, Comm. Math. Phys. <u>33</u>,
    145 (1973).

38. R. Schrader, A Possible Constructive Approach to ϕ_4^4,
    Free University, Preprint (1975).

39. K. Symanzik, Euclidean Quantum Field Theory, in:
    Local Quantum Theory, ed. R. Jost, Academic press,
    New York (1969).

40. G. Velo and A. Wightman, Constructive Quantum Field
    Theory, Lecture Notes in Physics, Vol. 25, Springer-
    Verlag, Berlin (1973).

Acta Physica Austriaca, Suppl. XV, 323–335 (1976)
© by Springer-Verlag 1976

THE CANONICAL STRUCTURE OF A CLASSICAL THEORY,
QUANTIZATION PROCEDURES AND NON-EQUILIBRIUM
QUANTUM STATISTICAL MECHANICS[+]

by

K. JEZUITA[++]

Institute of Nuclear Research
00-681 Warsaw, Hoza 69, Poland

1. INTRODUCTION

The aim of this lecture is to show that the harmonic oscillator model of the non-equilibrium quantum statistical mechanics which was presented by Emch [1] we can obtain by the natural quantization of the two interesting transformations in the phase-space of the one-dimensional classical harmonic oscillator:

$$\Pi_{\beta/2} : \begin{bmatrix} z \\ z^* \end{bmatrix} \longrightarrow \begin{bmatrix} e^{-\frac{\beta\omega}{2}} z \\ e^{\frac{\beta\omega}{2}} z^* \end{bmatrix} \tag{1}$$

[+] Seminar given at XV. Internationale Universitätswochen für Kernphysik, Schladming, Austria, February 16–27, 1976.

[++] Supported in part by N.S.F. grant No. GF-41958.

$$\gamma(s) : \begin{bmatrix} z \\ z^* \end{bmatrix} \longrightarrow \begin{bmatrix} e^{-\lambda s} z \\ e^{-\lambda s} z^* \end{bmatrix} , \quad s \geq 0 \qquad (2)$$

where $z = \omega^{1/2} q + i\omega^{-1/2} p$, the physical constants $\hbar = 1$, $k = 1$, and β is the inverse temperature.

In Section 2, we present the natural quantization scheme and illustrate it in case of the time evolution of the quantum harmonic oscillator. In Section 3 and 4 we quantize the transformations $\Pi_{\beta/2}$ and $\gamma(s)$.

In case of the transformation $\Pi_{\beta/2}$ we get the representation of the Weyl group which describes the quantum harmonic oscillator interacting with light in equilibrium at the temperature β^{-1}. Our construction of the equilibrium state and derivation of the Planck's formula is exactly the mathematical formulation of the Einstein's laws of radiation [2]. In case of the transformation $\gamma(s)$ we get the model of an infinite chain of coupled quantum oscillators and the dissipative evolution of the chosen oscillator with the Markov property.

2. NATURAL QUANTIZATION SCHEME FOR HARMONIC OSCILLATOR

For simplicity we work in one dimension. We would like to quantize linear transformations in the phase-space of the classical harmonic oscillator, in particular the time evolution

$$
u(t) \ : \quad \begin{bmatrix} z \\ z^* \end{bmatrix} \longrightarrow \begin{bmatrix} e^{-i\omega t} \, z \\ e^{i\omega t} \, z^* \end{bmatrix} \tag{3}
$$

and the translations in the position q and the momentum p

$$
W(z) \ : \quad \begin{bmatrix} z_o \\ z_o^* \end{bmatrix} \longrightarrow \begin{bmatrix} z_o + z \\ z_o^* + z^* \end{bmatrix} \tag{4}
$$

which are distinguished in the quantization scheme. Let us denote by $\hat{W}(z)$ the quantum version of $W(z)$. We assume that the operators $\hat{W}(z)$ form a unitary irreducible representation of the Weyl group:

$$
\hat{W}(z) = \exp\{-i(q\hat{P} + p\hat{Q})\}, \qquad [\hat{Q},\hat{P}] = iI \tag{5}
$$

$$
\hat{W}(z_1)\hat{W}(z_2) = \hat{W}(z_1+z_2) \, \exp\{\tfrac{i}{2} \, \text{Im}(z_1^* z_2)\} \quad . \tag{6}
$$

The natural quantization scheme is the following simple modification of the Segal method, the GNS construction [3], [4] of a representation of the Weyl algebra or the Nagy's theorem [5] on minimal unitary dilatations for the Weyl group. For a given transformation $\alpha(s)$ in the classical phase-space

$$
\alpha(s) \ : \quad \begin{bmatrix} z \\ z^* \end{bmatrix} \longrightarrow \begin{bmatrix} \alpha(s)[z] \\ \alpha(s)[z^*] \end{bmatrix} \tag{7}
$$

we look first for an algebra of operators $\hat{W}(z)$ [5], [6] in some Hilbert space H and a transformation $\hat{\alpha}(s)$ in this algebra

$$\hat{\alpha}(s) : \hat{W}(z) \longrightarrow \hat{\alpha}(s) [\hat{W}(z)] \tag{8}$$

such that the algebraic relation

$$\hat{\alpha}(s) [\hat{W}(z)] = \hat{W} (\alpha(s)[z]) \tag{9}$$

is satisfied. Next we look for an invariant state $\phi \in H$ under the transformation $\hat{\alpha}(s)$. The invariance condition we write in the form

$$<\phi, \hat{W}(\alpha(s)[z])\phi> = <\phi, \hat{W}(z)\phi> \quad . \tag{10}$$

The space of quantum states is the subspace of H spanned by the vectors of the form $\hat{W}(z)\phi$ which we call the generalized coherent states [6]. We do not assume that the transformation $\hat{\alpha}(s)$ is unitarily implemented and therefore we work in the Heisenberg picture. If it is possible we define the Schrödinger picture, the transformation $\hat{\alpha}_*(s)$ in the Hilbert space H

$$\hat{\alpha}_*(s) : \psi \rightarrow \psi(s) \tag{11}$$

in such a way that

$$<\psi(s), \hat{W}(z)\psi(s)> = <\psi, \hat{\alpha}(s)[\hat{W}(z)]\psi> \quad . \tag{12}$$

The simplest realization of the above scheme is the quantization of the time evolution u(t), [3], of the harmonic oscillator. The quantum harmonic oscillator is described by the unitary irreducible representation of the Weyl group $\hat{W}(z)$, (5), (6), in the Hilbert space H_o.

The quantum time evolution in the Schrödinger picture is done by the unitary group

$$\hat{U}_*(t) = \exp(-it\hat{H}) \, , \quad \hat{H} = \frac{1}{2}(\hat{P}^2 + \omega^2 \hat{Q}^2) \, . \tag{13}$$

The eigenstates Φ_N of the energy operator $\hat{H}$

$$\hat{H}\Phi_N = (N + \frac{1}{2})\omega\Phi_N \tag{14}$$

form the orthonormal base in the space H_o. In the Heisenberg picture we have

$$\hat{U}(t) \, [\hat{W}(z)] = e^{it\hat{H}} \, \hat{W}(z) \, e^{-it\hat{H}} \, . \tag{15}$$

Using (3), (5), and the known formula

$$e^Y \, e^X \, e^{-Y} = e^{X+[Y,X] + \frac{1}{2}[Y,[Y,X]] + \ldots} \tag{16}$$

we see that the relation (9) is satisfied

$$e^{it\hat{H}} \, \hat{W}(z) \, e^{-it\hat{H}} = \hat{W}(e^{-i\omega t}z) \, . \tag{17}$$

Every energy eigenstate Φ_N is the invariant state under $\hat{U}(t)$

$$<\Phi_N, \hat{W}(e^{-i\omega t}z)\Phi_N> = <\Phi_N, \hat{W}(z)\Phi_N> = f_N(|z|^2) \tag{18}$$

in particular the ground-state ϕ_o for which

$$\langle \phi_o, \hat{W}(z)\phi_o \rangle = \exp\left(-\tfrac{1}{4}|z|^2\right) . \qquad (19)$$

The states of the form $\hat{W}(z)\phi_N$ have similar properties as the coherent states [7]

$$\hat{W}(z)\phi_o = \exp\left(-\tfrac{1}{2}|z|^2\right) \sum_{N=0}^{\infty} \frac{z^N}{(N!)^{1/2}} \phi_N . \qquad (20)$$

The time evolution of such states looks like classical

$$e^{-it\hat{H}} (\hat{W}(z)\phi_N) = e^{-it\omega(N+\tfrac{1}{2})} \hat{W}(e^{-i\omega t}z)\phi_N \qquad (21)$$

and they span the Hilbert space H_o for a given ϕ_N.

3. QUANTUM HARMONIC OSCILLATOR AT
THERMAL EQUILIBRIUM

Let us quantize the transformation $\Pi_{\beta/2}$, (1), which we call the imaginary time evolution since Π_{it} is just the time evolution of the classical harmonic oscillator. The transformation $\Pi_{\beta/2}$ preserves the time invariant scalar product

$$(z_1, z_2) = z_1^* z_2 \qquad (22)$$

but does not commute with the complex conjugation.

It is easy to see from (1), (3), (14) and (17) that for any $\zeta_+, \zeta_- \in R^1$

$$e^{\frac{\beta\hat{H}}{2}} \hat{W}(\zeta_+ z) e^{-\frac{\beta\hat{H}}{2}} = \hat{W}(\zeta_+ e^{-\frac{\beta\omega}{2}} z)$$

$$e^{-\frac{\beta\hat{H}}{2}} \hat{W}(\zeta_- z^*) e^{\frac{\beta\hat{H}}{2}} = \hat{W}(\zeta_- e^{\frac{\beta\omega}{2}} z^*) \quad . \tag{23}$$

In the space $H_o \otimes H_o$ of the tensor product of two irreducible representations $\hat{W}(z)$ and $\hat{W}(z^*)$ we define the operators

$$\hat{V}(z) = \hat{W}(\zeta_+ z) \otimes \hat{W}(\zeta_- z^*) \tag{24}$$

which form the representation of the Weyl group (6) if

$$\zeta_+^2 - \zeta_-^2 = 1 \tag{25}$$

and the quantum imaginary time evolution $\hat{\Pi}_{\beta/2}$ in the Heisenberg picture

$$\hat{\Pi}_{\beta/2} [\hat{V}(z)] = \tilde{\Pi}_{\beta/2}^{-1} \hat{V}(z) \tilde{\Pi}_{\beta/2} \tag{26}$$

where

$$\tilde{\Pi}_{\beta/2} = e^{-\frac{\beta\hat{H}}{2}} \otimes e^{\frac{\beta\hat{H}}{2}} \quad . \tag{27}$$

As a result of (23) the relation (9) is satisfied

$$\hat{\Pi}_{\beta/2}[\hat{W}(\zeta_+ z) \otimes \hat{W}(\zeta_- z^*)] = \hat{W}(\zeta_+ e^{-\frac{\beta\omega}{2}} z) \otimes \hat{W}(\zeta_- e^{\frac{\beta\omega}{2}} z^*). (28)$$

Using (18) we see that for every N the states $\psi_N = \phi_N \otimes \phi_N$ are invariant under $\hat{\Pi}_{\beta/2}$

$$<\psi_N, \ \hat{W}(\zeta_+ e^{-\frac{\beta\omega}{2}} z) \otimes \hat{W}(\zeta_- e^{\frac{\beta\omega}{2}} z^*) \ \psi_N> =$$

$$<\psi_N, \ \hat{W}(\zeta_+ z) \otimes \hat{W}(\zeta_- z^*) \psi_N > \tag{29}$$

but only if

$$\zeta_+^2 e^{-\beta\omega} = \zeta_-^2 . \tag{30}$$

For a given β the parameters ζ_+, ζ_- are fixed

$$\zeta_+^2 = \frac{e^{\beta\omega}}{e^{\beta\omega}-1} , \quad \zeta_-^2 = \frac{1}{e^{\beta\omega}-1} , \quad \zeta_+^2 + \zeta_-^2 = \coth\frac{\beta\omega}{2} . \tag{31}$$

The cyclic representation $\hat{V}_\beta^N(z)$ which is characterized by ψ_N and (24), (25), (30) describes the quantum harmonic oscillator in the state ϕ_N interacting with light in equilibrium at temperature β^{-1}. The first component of the tensor product corresponds to the oscillator and the second to light. The energy operator of this system

$$\hat{H}_\beta = \zeta_+^2 \hat{H} \otimes I + \zeta_-^2 I \otimes \hat{H} \tag{32}$$

consists of three parts

$$\hat{H}_\beta = \hat{H} \otimes I + \zeta_-^2 \hat{H} \otimes I + \zeta_-^2 I \otimes \hat{H} \tag{33}$$

corresponding to spontaneous emission, induced emission and absorption. The average energy of light is

$$<E> = \frac{<\psi_N, \zeta_-^2 I \otimes \hat{H}\psi_N>}{N + \frac{1}{2}} = \frac{\omega}{e^{\beta\omega}-1} . \tag{34}$$

The above derivation of the Planck's formula (34) is equivalent to the Einstein's derivation and his law of radiation [2]. In place of the physical assumption that the induced emission probability and the absorption probability are equal we have the condition (25) which means that $\hat{V}_\beta^N(z)$ is the representation of the canonical commutation relations.

4. DIFFUSION FOR COUPLED QUANTUM OSCILLATORS

Let us quantize the transformation $\gamma(s)$, (2), which does not preserve the scalar product (22). In such case it is impossible to realize the conditions (9) and (10) for an irreducible representation $\hat{W}(z)$. If we would like to have an irreducible representation we must change the relations (8), (9), (10). We define the transformation $\hat{a}'(s)$ in the algebra of operators $\hat{W}(z)$

$$\hat{\alpha}'(s) : \frac{\hat{W}(z)}{<\Phi,\hat{W}(z)\Phi>} \quad \longrightarrow \quad \frac{\hat{W}(\alpha(s)[z])}{<\Phi,\hat{W}(\alpha(s)[z])\Phi>} \quad . \tag{35}$$

This definition is the union of the relations (8), (9), (1o) and depends of the state Φ. In case of the transformation $\gamma(s)$ we use the irreducible representation $\hat{V}_\beta(z)$, (32), and the ground state $\Phi' = \Phi_o \otimes \Phi_o$. As result of (2), (19), (24), (31) and (35) we have

$$\hat{\gamma}'(s):\hat{W}(z) \rightarrow \hat{W}(e^{-\lambda s}z)\exp\{-\tfrac{1}{4}|z|^2(1-e^{-2\lambda s})\coth\tfrac{\beta\omega}{2}\} . \tag{36}$$

This is the forward $s \geq 0$, time evolution in the Heisenberg picture which describes the diffusion process and has the Markovian property [1], [8]

$$\hat{\gamma}'(s_1) \, \hat{\gamma}'(s_2) = \hat{\gamma}'(s_1 + s_2) \, , \quad s_1,s_2 \geq 0 . \tag{37}$$

A physical interpretation of this system can be given after the construction of a representation $\hat{W}(z)$ such that the conditions (9), (1O) are satisfied [1]. First, using the Nagy's theorem [5] on minimal unitary dilatations, we extend the phase-space of the classical harmonic oscillator to the minimal space T in which the extension $\Gamma(s)$ of $\gamma(s)$ is unitary. The space T is the Hilbert space $L^2(R^1,dx)$ which we can interpret as the phase-space of an infinite chain of classical harmonic oscillators and

$$\Gamma(s) : f(x) \rightarrow e^{-ixs}f(x), \quad f \in L^2(R^1,dx), \quad s \in R^1. \tag{38}$$

Our chosen oscillator corresponds to the vector $f_o \in L^2(R^1, dx)$

$$f_o(x) = \frac{1}{\pi} \left(\frac{1}{\lambda^2 + x^2} \right)^{1/2} \tag{39}$$

and the transformation $\gamma(s)$ is the orthogonal projection of $\Gamma(s)$ onto f_o

$$(f_o \ \Gamma(s) \ f_o) = e^{-\lambda s} , \quad s \geq 0 . \tag{40}$$

Next we quantize the transformation $\Gamma(s)$ together with two transformations in the phase-space T

$$U(t)[f](x) = e^{-i\omega t} f(x), \quad U(t)[f^*](x) = e^{i\omega t} f^*(x) \tag{41}$$

$$\Pi_{\beta/2}[f](x) = e^{-\frac{\beta\omega}{2}} f(x), \quad \Pi_{\beta/2}[f^*](x) = e^{\frac{\beta\omega}{2}} f^*(x) \tag{42}$$

induced by the transformations (1), (3). The special solution of our scheme is the cyclic representation $\hat{W}_\beta(f)$ of the CCR on T, presented by Emch [1], in the tensor product of two usual Fock space representations $\hat{W}_F(f)$, $\hat{W}_F(f^*)$ of the CCR on

$$\hat{W}_\beta(f) = \hat{W}_F(\zeta_+ f) \otimes \hat{W}_F(\zeta_- f^*) \tag{43}$$

associated to the vacuum state $\Phi = \Phi_o \otimes \Phi_o$. The quantum transformations in the Hilbert space H of the states of the infinite chain of quantum harmonic oscillators

$$\hat{U}(t) \; : \; \hat{W}_\beta(f) \; \rightarrow \; \hat{W}_\beta(e^{-i\omega t}f)$$

$$\hat{\Pi}_{\beta/2} \; : \; \hat{W}_F(\zeta_+f) \otimes \hat{W}_F(\zeta_-f^*) \; \rightarrow \; \hat{W}_F(\zeta_+e^{-\frac{\beta\omega}{2}}f) \otimes \hat{W}_\beta(\zeta_-e^{\frac{\beta\omega}{2}}f^*)$$

$$\hat{\Gamma}(s) \; : \; \hat{W}_\beta(f) \; \rightarrow \; \hat{W}_\beta(\Gamma(s)[f]) \qquad\qquad\qquad (44)$$

satisfy the conditions (9), (10) for the state $\Phi = \Phi_o \otimes \Phi_o$. We can recover $\hat{\gamma}'(s)$ from $\hat{\Gamma}(s)$ by the orthogonal projection onto the subspace generated by the vector f_o [1]. The diffusion process of the single oscillator characterized by $\hat{\gamma}'(s)$ is the reduced description of the complete system of infinite oscillators with the deterministic evolution $\hat{\Gamma}(s)$.

5. SUMMARY

We have presented the derivation, from the classical theory, of two very simple interacting quantum systems. There are two interesting points in this quantization scheme. A given interaction is induced by some transformation in the classical phase-space and we can quantize even transformations which do not preserve energy without the assumption that they must be unitarily implemented. Therefore we hope that this approach will be useful for the understanding of such problems as symmetry breaking and many body systems.

ACKNOWLEDGEMENTS

I am very grateful to the Organizing Committee of the School and, in particular to Professor P. Urban, for the hospitality during my stay in Schladming, and to Professor G. G. Emch for his lecture which induced this work.

REFERENCES

1. G.G. Emch, Non-Equilibrium Quantum Statistical Mechanics, Lecture Notes, Schladming 1976.

2. R.P. Feynman, R.B. Leighton, M. Sands, The Feynman Lectures on Physics, Vol. I. Addison-Wesley, 1963.

3. G.G. Emch, Algebraic Methods in Statistical Mechanics and Quantum Field Theory, Wiley-Interscience, New York, 1972.

4. I.E. Segal, Foundations of the Theory of Dynamical Systems of Infinitely Many Degrees of Freedom: I, Mat. Fys. Medd. Dan. Vid. Selsk. $\underline{31}$, No. 2 (1959); II, Can.J. Math. $\underline{13}$, 1-18, (1961); III, Ill. J. Math. $\underline{6}$, 500-523 (1962).

5. B.Sz.-Nagy, C.Foias, Harmonic Analysis of Operators on Hilbert Space, Budapest, 1970.

6. A.M. Perelomov, Coherent States for Arbitrary Lie Groups, Comm. Math. Phys. $\underline{26}$, 222-236 (1972).

7. J.R. Klauder, E.C.G. Sudarshan, Fundamentals of Quantum Optics, Benjamin, New York, 1968.

8. E.B. Davies, Diffusion for Weakly Coupled Quantum Oscillators, Comm.Math.Phys.$\underline{27}$, 309-325 (1972).

Acta Physica Austriaca, Suppl. XV, 337–354 (1976)
© by Springer-Verlag 1976

STABILITY OF MATTER[+]

by

W. THIRRING
Institut für Theor. Physik der
Universität Graz

1. INTRODUCTION

One of the fundamental properties of real matter
is that its energy is an extensive quantity because the
chemical forces are saturating. This ought to be a consequence of nonrelativistic quantum mechanics where a
system composed of electrons and nuclei is described by
a Hamiltonian

$$H_N = \sum_{i=1}^{N} \frac{p_i^2}{2m_i} + \sum_{i>j} (e_i e_j - \kappa m_i m_j)|x_i - x_j|^{-1} \qquad (1)$$

(Notation: $(x_i, p_i, m_i; e_i)$ are position, momentum, mass
and charge of the ith particle, N is their total number,

[+] Lecture given at XV.Internationale Universitätswochen für
Kernphysik,Schladming,Austria,February 16 - 27, 1976.

κ the gravitational constant.

Thus one wants energy $A > 0$ independent of N which gives the lower bound

?
$$H_N \geq - AN. \tag{2}$$

Operator inequalities mean that the inequality holds for all expectation values and thus (2) is equivalent to the statement that the spectrum of R_N is above $- AN$. A result of the type (2) has first been derived by Dyson and Lenard (1). I shall present a simplified proof due to E. Lieb and myself (2). It also gives a considerable improvement in A although we have not yet reached what we believe to be the best possible constant.

The difficulty of proving (2) comes from the fact that (2) is not generally true, but only under very special circumstances.

REMARKS

1.) Since H_N contains a double sum with $\sim N^2$ terms there must be a remarkable cancellation if we want something $\sim N$. In particular if the system has a total charge

$$\sum_{i=1}^{N} e_i = Q \neq 0$$

and is in a volume with radius R there will be an electrostatic energy $\sim Q^2/R$ which may not be $\sim N$. (Since this is >0, (2) would still be satisfied.)

2.) Since gravity does not neutralize there is no hope of proving (2) if $\kappa > 0$. In fact it can be shown that in this case (if $Q = 0$) $H_N \sim - N^3$ for bosons and $\sim - N^{7/3}$ for fermions. This behaviour can be understood as follows. The ground state energy is the best possible compromise between the quantum mechanical zero point energy and the potential energy. For sufficiently high N gravity compresses the system to a high density plasma so that the electric potential is neutralized and the gravitational energy becomes $\sim -N^2/R$. (We shall consider the equal mass case and use units $m_i = \hbar = 1$). The zero point energy is $N/(\Delta x)^2$ where Δx is the linear dimension of the space available for a particle to move around. For bosons we may identify Δx with R, the radius of the system. Then the energy becomes $\sim \frac{N}{R^2} - \kappa \frac{N^2}{R}$ which assumes its minimum $\sim - N^3$ for $R \sim N^{-1}$. For fermions the exclusion principle requires $(\Delta x)^3 = R^3/N$. In that case the energy $\sim \frac{N^{5/3}}{R^2} - \kappa \frac{N^2}{R}$ has its minimum $\sim - N^{7/3}$ at $R \sim N^{-1/3}$.

3.) Even if $Q = \kappa = 0$ H_N is not $\sim N$ if positive and negative bosons are present. This can be made plausible by a similar argument. For a neutral system the potential energy per particle will be of the order $-1/r$ if now $e_i = 1$ and r is the distance to the next neighbour: $r = R/N^{1/3}$. Thus for fermions the energy will be $\frac{N^{5/3}}{R^2} - \frac{N^{4/3}}{R}$. This has its minimum $\sim - N$ at $R \sim N^{1/3}$ which is fine. For bosons we have $H_N \sim \frac{N}{R^2} - \frac{N^{4/3}}{R}$ which goes like $- N^{5/3}$ for $R \sim N^{-1/3}$. Of course, these heuristic considerations are not conclusive. In particular, up to date it is undecided whether for bosons the ground state energy goes $\sim - N^{5/3}$. To get an upper bound for it by using trial functions one has found only $- N^{7/5}$. This comes about because the required correlations between the particles cost extra kinetic energy.

4.) One should be able to prove stability by using Fermi statistics for the electrons only: Helium or deuterium appear to be perfectly stable. The relevant energy being Ry with the electron mass, (2) should remain valid in the limit nuclear mass $\to\infty$.

5.) If one uses the relativistic kinetic energy the situation becomes even more catastrophic. In this case for $Q = 0$, $\kappa > 0$ and N sufficiently big H_N is no longer bounded from below. The reason is, that then the kinetic energy is $\sim |\vec{p}| \sim 1/R$ and thus for $\kappa N^2 \geq N$ (or $N^{4/3}$ for fermions) the infimum $-\infty$ is approached for $R\to 0$.

6.) The situation of relativistic electrodynamics but N not too large, so that we may put $\kappa = 0$, has not yet been formulated precisely enough, let alone be resolved. I guess all formal Hamiltonians proposed for this case are singular like the magnetic spin-spin interaction, and not essentially selfadjoint. They may have stable and unstable extensions and it will not be clear, which extension is the one realized in nature. If one just replaces the kinetic energy by $|\vec{p}|$, as in 5.) one would conclude that the energy for fermions goes like $\frac{N^{4/3}}{R} - Ze^2 \frac{N^{4/3}}{R}$, which would be stable if $Ze^2 \leq 1$. A proof that this is acutally so is missing. The empirical argument that matter appears to be stable is not convincing because we may be separated by a sufficiently thick potential barrier from the abyss.

7.) The danger for stability does not come from the long range of the Coulomb potential, but from its $1/r$ singularity. Smoothing it at $r = 0$ by $1/r \to (1-e^{-\mu r})/r = v(r)$ renders it stable (even for bosons). To see this one has to realize that $v(r)$ is a function of positive type (i.e.

its Fourier transform $\tilde{v}(k) > 0$). Now

$$\sum_{i>j} e_i e_j v(x_i - x_j) = -\frac{N}{2} v(0) + \int \frac{d^3k}{(2\pi)^3} \tilde{v}(k) \left| \sum_j e_j e^{i\vec{k}\vec{x}_j} \right|^2$$

and the last term is > 0. Thus, if $\kappa = 0$, the potential energy and hence H_N is bounded from below by $-\frac{N}{2} v(0)$. In case the energy is not extensive the system must collapse to take advantage of the $1/r$ singularity. As a consequence the stability problem for a Yukawa potential is as serious as for a Coulomb potential.

One might argue that in reality nuclei have a form factor and if the real potential looks more like the smoothed one stability is trivial. However, this cannot be the true reason for it because if $\mu = (1/$ nuclear radius) then $v(0)$ is $\sim$ MeV and we want a bound $- N \cdot$ Rydberg.

2. THOMAS-FERMI THEORY

The real clue to our problem is Thomas-Fermi theory. It does not lead to chemical binding and thus immediately to stability. We shall see that it gives, with slightly changed constants, a lower bound to the true ground state energy of H_N, which settles the question.

To commence the demonstration of these statements we rewrite (1) with $\kappa = 0$, M nuclei with masses M_α, charges Z_α, coordinates X_α, P_α and N electrons with x, p as coordinates, and $e = 2m = 1$.

$$H_N = \sum_{j=1}^{N} p_j^2 - \sum_{\alpha,j} \frac{Z_\alpha}{|x_j - X_\alpha|} + \sum_{i>j} |x_i - x_j|^{-1} +$$

$$+ \sum_{\alpha > \beta} Z_\alpha Z_\beta |X_\alpha - X_\beta|^{-1} + \sum_{\alpha=1}^{M} \frac{p_\alpha^2}{2m_\alpha} . \tag{3}$$

In Thomas-Fermi theory this operator is replaced by a functional $E_N(\sigma)$ of the electron density $\sigma(x)$ depending also on the nuclear coordinates and charges

$$E_N(\sigma) = \frac{3}{5}(3\pi^2)^{2/3} \int d^3x \ \sigma^{5/3}(x) - \sum_\alpha Z_\alpha \int \frac{d^3x\,\sigma(x)}{|x-X_\alpha|} +$$

$$+ \frac{1}{2} \int d^3x\,d^3y \frac{\sigma(x)\,\sigma(y)}{|x-y|} + \sum_{\alpha > \beta} Z_\alpha Z_\beta |X_\alpha - X_\beta|^{-1} . \tag{4}$$

Thus we see that the electron kinetic energy is replaced by $\sim \int \sigma^{5/3}$, the potential energy by the obvious electro-static expressions and the nuclear kinetic energy is dropped. For a lower bound this is allright since the kinetic energy is positive. Furthermore we want to allow the nuclei to be bosons, so not much is gained by keeping it. In this section, we will show that $E(\sigma)$ as funct-ion of the X_i approaches its infimum for $|X_i - X_j| \to \infty$. Thus it is bounded by the sum of the energies of isolated atoms. $E(\sigma)$ reaches its minimum, (X_i fixed) if σ obeys the Thomas-Fermi equation.

$$3\pi^2 \ \sigma(x) = [\Phi(x)]^{3/2}$$

$$\Phi(x) = \sum_{k=1}^{M} \frac{Z_k}{|x-X_k|} - \int \frac{d^3y\sigma(y)}{|x-y|} > 0 . \qquad (5)$$

For a demonstration of this claim as well as the existence
and uniqueness of solutions see (3). As it is intuitively
expected this minimizing density σ and hence the potential
Φ increases if the Z_i increase. To see this consider two
sets of nuclear charges $Z_i^{(1)} \geq Z_i^{(2)}$, $i = 1,\ldots M$ (the X_i
the same) and the corresponding solutions $\sigma^{(1,2)}$, $\Phi^{(1,2)}$.
The set

$$S = \{x|\sigma^{(1)} < \sigma^{(2)}\} = \{x|\Phi: = \Phi^{(1)} - \Phi^{(2)} < 0\} \qquad (6)$$

is open (Φ is continuous in S) and obviously does not contain the X_i. Thus, on S Φ obeys the equation $\Delta\Phi = \sigma^{(1)} -$
$- \sigma^{(2)} < 0$. Therefore Φ reaches its minimum over S on the
boundary of S or at infinity. There it is zero and we conclude that S must be empty.

Next we shall show that the energy is decreased if
the system is broken into two parts. For this purpose call
$E(Z_1\ldots, Z_M)$ the minimum of $E(\sigma)$ over the appropriate
function-space for σ (we shall vary the Z_k and keep the
X_k fixed). Then we assert

Theorem (1)

$$E(Z_1,\ldots,Z_M) > E(Z_1,\ldots,Z_k,0,\ldots,0) + E(0,\ldots,0,Z_{k+1},\ldots,Z_M)$$

$$\forall k = 1,\ldots, M - 1.$$

Proof:
Consider the Z_i as variable but their ratios fixed:

$z_i = Z z_i$, and denote

$$E(Z) = E (Z z_1, Z z_2, \ldots \quad Z z_M)$$

$$E_1(Z) = E (Z z_1, \ldots Z z_k, 0 \ldots 0)$$

$$E_2(Z) = E (0, \ldots 0, Z z_{k+1}, \ldots Z z_M).$$

We have shown $\sigma \geq \sigma^{(1,2)}$, $\phi \geq \phi^{(1,2)}$ for the corresponding solutions of the Thomas-Fermi equation. The derivatives obey a kind of Hellmann-Feynman theorem

$$\frac{\partial E}{\partial Z} = \sum_{i=1}^{M} z_i \{ Z \sum_{j \neq i} \frac{z_j}{|X_i - X_j|} - \int \frac{d^3 x \sigma(x)}{|x - X_i|} \} =$$

$$= \sum_{i=1}^{M} z_i \lim_{x \to X_i} [\phi(x) - \frac{Z z_i}{|x - X_i|}] \tag{7}$$

and similarly for $E_{1,2}$: Thus we have altogether

$$\frac{\partial E}{\partial Z} - \frac{\partial E_1}{\partial Z} - \frac{\partial E_2}{\partial Z} = \sum_{i=1}^{k} z_i [\phi(X_i) - \phi^{(1)}(X_i)] +$$

$$+ \sum_{i=K+1}^{M} z_i [\phi(X_i) - \phi^{(2)}(X_i)] \geq 0; \tag{8}$$

since $E(0) = E_1(0) = E_2(0) = 0$ we have $E(Z) \geq E_1(Z) + E_2(Z)$ and hence the theorem.

Continuing the fragmentation we see that $E(Z_1 \ldots Z_M)$

is above the sum of the energies $E(Z_k)$ of the individual atoms. The latter is known to be (in our units) $-3.68 \cdot Z_k^{7/3}$ which gives for all X_k, Z_k a bound of $E(\sigma)$:

$$\frac{3}{5\gamma} \int d^3x\sigma^{5/3} - \sum_{k=1}^{M} Z_k \int \frac{d^3x\sigma(x)}{|x-X_k|} + \frac{1}{2} \int \frac{d^3xd^3y}{|x-y|}\sigma(x)\sigma(y) +$$

$$+ \sum_{n>m} \frac{Z_n Z_m}{|X_n-X_m|} \geq -3.68 \sum_{k=1}^{M} Z_k^{7/3} \cdot \gamma \quad \forall \gamma > 0 . \tag{9}$$

REMARKS

1.) Without the repulsion of the nuclei the in-equality (8) goes in the other direction. Thus there are chemical binding forces in Thomas-Fermi theory but they are never strong enough to overcome the Coulomb repulsion.

2.) (9) can be expressed by saying that the observed chemical binding energies can never exceed the error made by the Thomas-Fermi theory.

3. THE EXTENSIVE NATURE OF THE ENERGY

Trying to relate the Hamiltonian (3) to the Thomas-Fermi functional (4) it appears obvious that σ should be

$$\sigma(x_1) = \int d^3x_2 \ldots d^3x_N |\psi(x_1, x_2, \ldots, x_N)|^2 \qquad (10)$$

ψ being the ground state wave function. If we want to show
something like $\langle\psi|H_N\psi\rangle \geq E(\sigma)$ two problems arise:

1.) what has $\int\sigma^{5/3}$ got to do with $\int d^3x_1 \ldots d^3x_N \sum_i |\nabla_i\psi|^2$.

2.) what is the error in replacing

$$\langle\psi| \sum_{i>j} |x_i - x_j|^{-1} \psi\rangle \qquad \text{by} \qquad \frac{1}{2}\int \frac{d^3x\, d^3y}{|x-y|}\sigma(x)\sigma(y),$$

that is neglecting correlations.
Curiously enough we may use (9) to solve the second problem.
Its special case $Z_i = 1$, $M = N$ can be written

$$\sum_{i>j} |x_i - x_j|^{-1} \geq \sum_i \int \frac{d^3x\,\sigma(x)}{|x_i - x|} - \frac{1}{2}\int \frac{d^3x\, d^3y}{|x-y|}\sigma(x)\sigma(y) -$$

$$- 3.68\gamma N - \frac{3}{5\gamma}\int d^3x\,\sigma^{5/3}(x) . \qquad (11)$$

This is an electrostatic inequality and tells you that the
Coulomb repulsion between equal charges is always more
than their energy in a potential produced by a charge
distribution minus three correction terms, two depending
on σ. It allows us to minorize (3) by a σ-dependent
Hamiltonian $H(\sigma)$ where the electrons do no longer inter-
act but all move in a common potential V_σ.

$$H \geq H_\sigma := \sum_{i=1}^N h_i - \frac{1}{2}\int \frac{d^3x\, d^3y}{|x-y|}\sigma(x)\sigma(y) - \frac{3}{5\gamma}\int d^3x\,\sigma^{5/3} -$$

$$- 3.68\gamma N + \qquad (12)$$

$$+ \sum_{n>m} \frac{Z_n Z_m}{|X_n - X_m|} \quad , \qquad h_i = p_i^2 - \sum_{k=1}^{M} \frac{Z_k}{|x_i - X_k|} + \int \frac{d^3x \sigma(x)}{|x - x_1|} \quad .$$

Thus we have reduced the problem to solving the 3-dimensional Schrödinger equation with V_σ and then choose σ judiciously. At this point the Fermi-statistics of the electrons enter because it allows us to use the sum of the lowest N eigenvalues of $h(\sigma)$ to get a bound for H_N.

A rather crude bound for this sum is obtained by using a result of Birman and Schwinger for the number of bound states.

Theorem (2):

Let e_j be the j-th negative eigenvalue of the Hamiltonian $p^2 + V(x)$. Then

$$\sum_j |e_j| \leq \frac{4}{15\pi} \int d^3x |V|_-^{5/2} \tag{13}$$

where $|V_-| = \begin{cases} |V(x)| & \text{if} \quad V(x) < 0 \\ 0 & \text{if} \quad V(x) \geq 0 \end{cases}$.

Proof:

First we notice that all eigenvalues of $p^2 + V$ are lowered if V is replaced by its negative part $- |V|_-$. Furthermore the number of negative eigenvalues is not decreased if V is multiplied by $\lambda \geq 1$ because

$$p^2 + \lambda V = \lambda(p^2 + V) + (1 - \lambda) p^2 < \lambda(p^2 + V) \quad . \tag{14}$$

Hence the number of negative eigenvalues of $p^2 + V$ is majorized by the one of $p^2 - \lambda |V|_-$, $\lambda \geq 1$. Rewriting the corresponding Schrödinger equation as $(p^2 - E)^{-1} |V|_- \psi = \frac{1}{\lambda}\psi$ we realize that the number $N(E)$ of eigenvalues of $p^2 + V$ below E is $\leq$ the number of eigenvalues ≥ 1 of $(p^2 - E)^{-1}|V|_-$.

We could also use the operator $|V|_-^{\frac{1}{2}} \frac{1}{p^2-E} |V|_-^{\frac{1}{2}}$, which is > 0 for $E \leq 0$, or even better

$$|V-E/2|_-^{1/2} \frac{1}{p^2-E/2} \cdot |V-E/2|_-^{1/2} .$$

In particular $N(E)$ is less than the trace of the square of this later operator. In an x-representation we have

$$N(-\alpha) \leq \mathrm{Tr} \frac{1}{p^2+\alpha/2} |V + \alpha/2|_- \cdot \frac{1}{p^2+\alpha/2} |V + \frac{\alpha}{2}| =$$

$$= (4\pi)^{-2} \int d^3x\, d^3y\, |V(x) + \frac{\alpha}{2}|_- |x-y|^{-2} e^{-\sqrt{2\alpha}|x-y|} |V(y) + \frac{\alpha}{2}|_- \leq$$

$$\leq \frac{1}{4\pi\sqrt{2}\,\alpha} \int d^3x\, |V(x) + \frac{\alpha}{2}|_-^2 . \tag{15}$$

In the last step Youngs inequality

$$\|f * g\|_r \leq \|f\|_p \|g\|_q , \quad 1 + \frac{1}{r} = \frac{1}{p} + \frac{1}{q}$$

has been used. To complete the proof one only has to realize that $\sum_i |e_i| = \int_{-\infty}^{0} dE\, N(E)$ and carry out the integration in (15).

REMARKS

1.) The constant in (13) is not the best possible one and has in fact been improved by almost a factor 2.

2.) The classical expression

$$\int_{h<0} d^3x \, d^3p \, |h(x,p)| = \frac{1}{15\pi^2} \int d^3x |v|^{5/3} \qquad (16)$$

is smaller than (13) by 4π. Experience gained from computer studies lead us (4) to conjecture that (16) is actually a bound.

3.) Using the present strategy we will not have to answer question (1). However it is not hard to see that (13) is equivalent to

$$\int \sum_i |\nabla_i \psi|^2 \, dx_1 \ldots dx_N \geq \frac{1}{4\pi} \frac{3}{5} (3\pi^2)^{2/3} \int d^3x [\rho(x)]^{5/3} \, . \qquad (17)$$

With (13) we get the following numerical bound for H_N (with a factor 2 for spin)

$$H(\sigma) \geq - \frac{4.2}{15\pi} \int d^3x | \sum_n \frac{-Z_n}{|x-X_n|} + \int \frac{d^3y\sigma(y)}{|x-y|} |^{5/2} -$$

$$- \frac{1}{2} \int \frac{d^3x d^3y}{|x-y|} \sigma(x)\sigma(y) - 3.68\gamma N -$$

$$- \frac{3}{5\gamma} \int d^3x \sigma^{5/3}(x) + \frac{1}{2} \sum_{n>m} \frac{Z_n Z_m}{|X_n - X_m|} \, . \qquad (18)$$

If we now choose $\sigma(x)$ to obey the Thomas Fermi equation (5) we may express $\int v^{5/2}$ as $\int \sigma^{5/3}$ or $\int \sigma V$ and rewrite (18) in a form similar to (4)

$$H_N \geq \frac{3}{5}[(\frac{3\pi}{4})^{\frac{2}{3}} - \frac{1}{\gamma}] \int d^3x \sigma^{5/3}(x) - 3.68\gamma N -$$

$$- \int d^3x \sum_n \frac{Z_n \sigma(x)}{|x-X_n|} + \frac{1}{2} \int \frac{d^3x d^3y}{|x-y|} \sigma(x)\sigma(y) +$$

$$+ \frac{1}{2} \sum_{n>m} \frac{Z_n Z_m}{|X_n-X_m|} \ . \tag{19}$$

Using again the inequality (9) and optimizing with respect to γ finally gives

$$H_N \geq - 3.68[\gamma N + ((\frac{3\pi}{4})^{\frac{2}{3}} - \frac{1}{\gamma})^{-1} \sum_n z_n^{7/3}] \quad \rightarrow$$

$$H_N \geq - 2.08 \ N \ [1 + (\frac{1}{N} \sum_{n=1}^{M} z_n^{7/3})^{1/2}]^2 \ . \tag{20}$$

REMARKS

1.) If we have a neutral system with equal nuclear charges $z_i = N/M$ we get, for large Z, a bound $\sim M \ z^{7/3}$. This corresponds to the Z-dependence of the energies of atoms. For $N \ll ZM$ our bound is not optimal, one should expect something $\sim NZ^2$.

2.) If we have q species of fermions the same argument leads, in the case $z_i = 1$, M = N, to a bound $\sim q^{2/3}$ N. If q = N this means that without symmetry requirement on ψ we have a bound $\sim N^{5/3}$. A fortiori this proves the lower bound for bosons of our intuitive argument. The same conclusions can also be reached from the trivial fact $|e_i| \leq \sum_j |e_j|$.

3.) If (16) were a bound the constant in front of (20) would be the one for single atoms in Thomas-Fermi theory. This is the best possible constant in (20) since for Z→∞ this value is actually approached.

4. THE EXTENSIVE NATURE OF THE VOLUME

As discussed in 1. instability is always accompanied by a collaps of the system. Conversely one would like to see that the volume of a stable system goes $\sim N$. Since we do not dispose of any wave function to calculate expectation values with, it may seem a difficult task to prove an inequality of the type $<|\vec{x}|> \geq cN^{1/3}$. Operator inequalities are the only thing we have got. In a sense, they are actually better than wave functions since they contain information about all possible expectation values. Thus we can indeed prove $<\vec{x}^2> \geq cN^{2/3}$.

We start with a little auxiliary exercise. Consider N Fermions in the potential of a harmonic force with frequency ω (in 3 dimensions). Filling the oscillator levels for the ground state energy of the system we arrive at the inequality

$$\sum_{i=1}^{N} (\vec{p}_i^2 + \omega^2 \vec{x}_i^2) \geq c\omega \, N^{4/3} \; , \qquad\qquad c = a \text{ constant} . \qquad (21)$$

The $N^{4/3}$ behaviour comes about because we calculate

$$\sum_{n_i \geq 0} (n_1 + n_2 + n_3)$$

where the upper limit for the summation is determined by $\sum_{n_i > 0} 1 = N$. The inequality means

$$\sum_i \langle \vec{x}_i^2 \rangle \geq \frac{cN^{4/3}}{\omega} - \frac{\sum_i \langle \vec{p}_i^2 \rangle}{\omega^2} \qquad (22)$$

for all expectation values and all ω's. In particular we may choose

$$\omega = 2 \, N^{-4/3} \; \langle \sum_i \vec{p}_i^2 \rangle / c \qquad \text{to obtain}$$

$$\sum_i \langle \vec{x}_i^2 \rangle \geq \frac{N^{8/3}}{\sum_i \langle \vec{p}_i^2 \rangle} \frac{c^2}{4} \; . \qquad (23)$$

This is a kind of uncertainty relation for N fermions and contains no further reference to the oscillator. For the ground state wave function of the Coulomb-system we know that the virial theorem holds, namely $\langle \psi |$ kinetic energy of the electrons $| \psi \rangle + \langle \psi |$ kinetic energy of the nuclei $| \psi \rangle = -$ ground state energy $\leq \tilde{c} \cdot N$. Thus $\sum_i \langle p_i^2 \rangle \leq \tilde{c} N$ and we may continue (23) as

$$\sum_i <x_i^2> \geq c'N^{5/3} \tag{24}$$

or

$$[\frac{1}{N} \sum_i <x_i^2>]^{\frac{1}{2}} \geq c''N^{1/3} \tag{25}$$

which is the kind of relation we havebeen looking for.

REMARKS

1.) For convenience we calculated $<x^2>$ but the same argument would work for other moments.

2.) If the system is in a box there is an external virial and this reasoning does not apply. Clearly in this case the seize of the box will enter into a lower bound for $<x^2>$.

3.) With our assumptions no upper bound $\sim N^{1/3}$ for $<x^2>$ can be obtained. Since neutrality is not used it is possible that some electrons are blown to infinity by an excess electronic charge. Thus (25) shows stability in the sense that the system does not collaps and not in the sense that it does not explode.

4.) The unit of length in (25) is the Bohr radius. (25) gives the correct order of magnitude but is rather crude.

REFERENCES

1. F.J. Dyson, A. Lenard, J. Math. Phys. $\underline{8}$, 423 (1967).

2. E.H. Lieb, W.E. Thirring, Phys. Rev. Lett. $\underline{35}$, 687 (1975). See ibid. 1116 for errata.

3. E. Lieb, B. Simon, Phys. Rev. Lett. $\underline{31}$, 681 (1973) and Princeton preprint.
    W. Thirring, Vorlesungen über Mathematische Physik, T8.

4. E. H. Lieb, W. E. Thirring, Inequalities for the Moments of the Eigenvalues of the Schrödinger Hamiltonian and their relation to Sobolev inequalities. In the volume dedicated to V. Bargmann, Princeton University Press 1976.

Acta Physica Austriaca, Suppl. XV, 355–358 (1976)
© by Springer-Verlag 1976

SUMMARY - FIRST WEEK[+]

by

H. REEH
Institut für Theoretische Physik
Universität Göttingen

The lectures of the first week of this years
Schladming Winter School have come to an end, and I am
supposed to give a short summary. I start with the
lectures.

They were devoted to three fundamental questions
of theoretical physics: Non-trivial models of relativistic quantum field theory. Stability of matter. Non-equilibrium statistical mechanics. Common to all talks of
this week was the use of a precise mathematical language.

Most time was spent on the first subject: We had
the lectures by Prof.s Streit, John Challifour, Fröhlich
and Park, starting with an introduction to the tools of
probability theory, probability measures, Schwinger functionals, lattice approximations - somewhat continuously

---

[+] Summary given at XV. Internationale Universitätswochen
für Kernphysik,Schladming,Austria,February 16-27, 1976.

speeding up - and ending with the latest results concerning theories with coupled bosons in a space-time of dimension 3 exhibiting ultraviolet problems, and just a few minutes ago about the quantum sine-Gordon-equation (space time dimension 2).

We were told how the Euclidean approach works in particular in combination with lattice approximations and how inequalities of statistical mechanics can be used for the benefit of field theory. In addition we saw that results concerning symmetry breaking for Ising like lattice models can be used to show that for sufficiently large coupling one gets field theories with a degenerate vacuum, and in case of space dimension 2 even the heuristic Goldstone picture can be established and massless particles occur.

The present status of constructive field theory seems to be as follows:For $P(\phi)_2$ almost everything is worked out by now including nontriviality of scattering. Not yet solved is the full unitarity of the S-matrix (asymptotic completeness), and in the strong coupling limit: the details of the mass spectrum (bound states?) and scattering theory. As we learned from the talks by Fröhlich and Park, there seems to be continuous progress to reach for $(\phi^4)_3$ the same state of affairs rather soon.

Why does one struggle so much to formulate a model of relativistic quantum field theory? To my opinion the reason is not alone that one wants to have a consistent mathematical theory leading to an S-matrix which describes the processes observed in nature: One would like to know in addition whether there really exists an interpolating local Heisenberg field as a basic physical notion for our understanding of particle physics.

There were two remarkable predictions made by
two of the experts concerning the date when we will
have a $(\phi^4)_4$-theory: Prediction of Park: within the
next 10 - 20 years (no proof indicated). Predition
of Challifour: by 1982 (proof by the method of in-
complete induction).

This seems to be a remarkable progress compared
to an estimate made by Streater half a year ago of a
lower limit at around the year 2000. The progress certainly must be due to the stimulating atmosphere at the
ZIF, Bielefeld.

There had been a few other propositions during
lectures in this field without a proof or a reference
to a proof, concerning money, blackboards and drying of
pencils. I should like to add a similar proposition to
my summary:

Do not (allow to) bring prepared transparencies to
the lectures (unless in case of a very experienced teacher).

Now to the other subjects: We had two impressive
lectures by Prof. Thirring on the stability of ordinary
matter, that is matter composed of nuclei (Bosons or
Fermions) and electrons. If the number N of particles
is not of astronomical order so that gravitation does
not cause a collapse, then the Hamiltonian of the non-
relativistic model containing only Coulomb interactions
of the particles, is bounded below by $-2,08$ N, and the
volume per particle is bounded below. The astute method
of proof of Lieb and Thirring contains a come-back of
Thomas-Fermi-theory. The progress compared to the classical

paper by Dyson and Lenard is measurable: Reduction of
the length of proof at least by a factor 15, and an
improvement of the estimate by a factor of order 2^{40}
(Moral of the story: Have physical intuition to find
the suitable mathematical method). It is a remarkable
fact that non-relativistic quantum mechanics provides
us with a rather good picture of ordinary matter which
has no intrinsic need for corrections at small distances.

Prof. Emch presented an algebraic scheme for non-
equilibrium quantum statistical mechanics employing
different time evolutions. This then was illustrated
by an interpretation in terms of a van Hove limit of
an infinite chain of harmonic oszillators. I have to
confess that I still have difficulties in understanding
intuitively how the weak coupling enters the game (unfortunately I missed the first lecture). In the following lectures Prof. Emch presented and discussed a quantum mechanical notion of a regular K-flow and dynamical
entropy.

So much for the lectures. I believe, we all had a
fine week, we had marvellous weather and excellent well
preserved powder snow besides of the slopes in the woods.
To my knowledge nobody made use of the excellent local
hospital.

I should like to conclude by expressing our thanks
to the speakers of this first week for the trouble they
took in preparing and delivering the lectures and lecture
notes (I hope we will get those of Prof. Fröhlich too)
and to Prof. Urban and his perfect organizers of this
meeting.

Acta Physica Austriaca, Suppl. XV, 359–363 (1976)
© by Springer-Verlag 1976

PHENOMENOLOGY OF NEUTRAL CURRENTS[+]

by

J.J.SAKURAI[++]

CERN,Geneva

In September 1975 I gave a series of lectures at
the DESY Summer Institute for Theoretical Physics en-
titled "Neutral Currents without Gauge Theory Prejudices",
the written version of which is now available as a CERN
preprint (TH 2099) and will eventually be published in
the Proceedings of the Summer Institute. When I accepted
to give a talk at Schladming, I was hoping that there
would be a great deal of progress in the field of neutral
current phenomenology between September 1975 and February
1976. Unfortunately my expectation was not realized; as a
result, the lectures given at Schladming were substant-
ially the same as my DESY lectures. In this written version,
I make a few supplementary remarks with attention to some

[+] Lecture given at XV.Internationale Universitätswochen
     für Kernphysik,Schladming,Austria,February 16-27,1976.
[++] John Simon Guggenheim Memorial Foundation Fellow, on
     leave from the University of California, Los Angeles.

recent developements.

(i) Both the Gargamelle Collaboration and the Caltech group have continued to study the space-time structure of the hadronic part of neutral currents by measuring the neutral-to-charged current ratios and the E_{had} distributions for both neutrino and antineutrino induced reactions. Both groups find that the best fit is somewhere between pure V/pure A and V-A. The Gargamelle Collaboration is closer to V-A than to pure V/pure A while the Caltech group is closer to pure V/pure A than to V-A. It is probably fair to say that neither pure V/pure A nor pure V-A is conclusively ruled out.

(ii) A new astrophysical argument has been advanced against a sizable tensor interaction in neutral currents (or "densities"!) by Ruderman and collaborators [1]. Their reasoning goes as follows. Suppose we have a tensor-type interaction. We then expect a finite magnetic moment for a neutrino due to Fig. 1, which gives

$$\mu_\nu = \frac{2}{\pi^2} \frac{2G_T}{\sqrt{2}} m_\ell \ln \left(\frac{\Lambda}{m_\ell}\right)$$

where G_T is the tensor coupling constant defined by

$$\frac{2G_T}{\sqrt{2}} (\bar{\nu} \, \sigma_{\lambda\tau} \nu) (\bar{\ell} \, \sigma_{\lambda\tau} \, \ell) \quad .$$

Note that ℓ can also be a "quark". Now a photon in astrophysical plasma can be visualized as having a small effective mass

$$m_{eff} \ c^2 = \bar{h} \ \omega_p$$

where ω_p is the plasma frequency. With a finite magnetic moment for ν, a plasma photon can decay into a $\nu\bar{\nu}$ pair. This mechanism may be relevant in stellar evolution where a transition from the high luminosity stage to the white dwarf stage takes place. Too high a ν production rate due to

$$\gamma_{plasma\ photon} \ \rightarrow \ \nu \ + \ \bar{\nu}$$

implies that the cooling of a white dwarf would be too rapid, resulting in a marked deficiency in the distribution of white dwarfs. For quantitative estimates everything depends on what one uses for $\ell^{\pm}$ and the cut off Λ, but knowing that there are "quarks" at ~ 300 MeV, heavy leptons at 1.8 GeV etc., we may conservatively set

$$(m_\ell/m_\mu) \ \ln \ (\Lambda/m_\mu) \ \geq \ 6 \ .$$

Then the astrophysical cooling time inferred from the distribution of white dwarfs implies

$$G_T/G \ \lesssim \ 1/15 \ ,$$

which is too small to account for the observed neutral current interactions. Since pure SP was already ruled out by the E_{had} distributions of the Caltech group, the situation does not look too promising for SPT

(iii) There is now further evidence against the
hypothesis that the hadronic part of neutral currents
is pure isoscalar. The Gargamelle Collaboration [2]
has studied the π° - to - π^{-} ratio in single pion
production

$$\stackrel{(-)}{\nu} + N \rightarrow \stackrel{(-)}{\nu} + \pi^{\circ,-} + N .$$

If we have a pure isoscalar current on an isoscalar
target, the π° - to - π^{-} ratio has to be 1:1; this
is clear because we have no preferred direction in
isospin space to start with. Furthermore, this conclusion holds even in the presence of final state
interactions among nucleons and produced pions since
the nuclear charge exchange corrections are expected
to obey charge independence. Experimentally the ratio
seems to be 1.4 ± 0.2 for the ν reaction and 2.1 ± 0.4
for the $\bar{\nu}$ reaction; these numbers should be compared
to 0.9 when we correct for the fact that freon is not
pure isoscalar. It thus appears that there is a significant amount of isovector component. Needless to say,
for a final check it is desirable to see a clean Δ
signal in single pion production.

(iv) We hear that the prospects for constructing high
energy (s $\sim$ 1000 GeV^2) e^+e^- machines by 1980 are
extremely good in both Europe (PETRA at DESY) and
the U.S. (PEP at SLAC). Let us hope that we will
be able to hear about the first experiment to detect
neutral current effects in electron-positron anni-
hilations into muon pairs in the 1981 Schladming
Meeting.

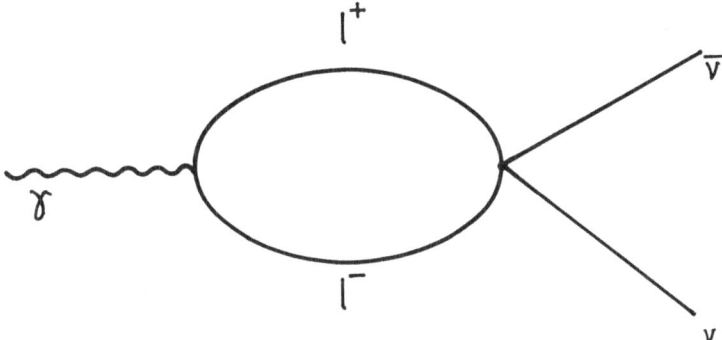

Fig. 1

Diagram giving rise to a finite magnetic moment for a
neutrino when the $(\bar{\ell}\ell)$ $(\bar{\nu}\nu)$ vertex is assumed to contain
the tensor type.

REFERENCES

1. P. Sutherland et al., UPR-0057 T.

2. C.H. Bertrand Coremans et al., CERN/D.Ph.II/Phys.75-5.

Acta Physica Austriaca, Suppl. XV, 365–422 (1976)
© by Springer-Verlag 1976

NEUTRINO INTERACTIONS:

HOW DOES CHARM FARE?[+]

by

M.K. GAILLARD

CERN and Laboratoire de Physique

Theorique et Particules

Elementaires, Orsay

1. CONVENTIONAL CHARM AND NEUTRINO INTERACTIONS

In this lecture we shall confine ourselves to the
conventional charm model of weak interactions as proposed [1] by Glashow, Iliopoulos and Maiani (GIM):

a) Hadronic states are assumed to be built from four
quarks (u,d,s,c) where (u,d,s) are the Gell-Mann-Zweig
quarks and c carries the charm quantum number C = + 1.
All hadronic states currently in the Particle Data
tables have C = O.

[+] Lecture given at XV. Internationale Universitätswochen
für Kernphysik,Schladming,Austria,February 16-27,1976.

b) the hadronic component of the charge rising weak
current operator has the structure

$$J_\mu = \bar{u}\gamma_\mu (1 - \gamma_5)\, d_c + \bar{c}\,\gamma_\mu (1 - \gamma_5)\, s \qquad (1.1)$$

where d_c and s_c are the Cabibbo-rotated "down"
$(I_3 = -1/2)$ and strang quarks

$$d_c = d \cos \theta_c + s \sin \theta_c$$

$$s_c = s \cos \theta_c - d \sin \theta_c \qquad (1.2)$$

where θ_c is the Cabbibo angle; $\sin \theta_c \simeq 0.2$.

Aside from the direct observation of new heavy
hadron-like states which are most popularly interpreted
as $\bar{c}c$ bound states, the strongest evidence that new particles are being produced comes from neutrino interactions. In the following sections we shall discuss the
effects expected in the framework of the GIM model. The
success of the parton model in describing much of the
data in leptoproduction below the assumed charm threshold suggests that it may be applicable to charmed particle production sufficiently above threshold. Therefore we shall briefly recall the rules of the parton
game, and the evidence on the parton content of the
nucleon as extracted from SLAC electroproduction [2]
and Gargamelle neutrino data [3], and comment on its
domain of validity. A further ingredient necessary
for predicting di-lepton yield is an estimate of the

semi-leptonic branching ratio which we shall also discuss. In the following lecture we shall make a detailed
comparison with available data and comment on alternative models.

1.1 THE PARTON MODEL

Consider the process $\nu + p \rightarrow \mu +$ anything. We
define the usual kinematic variables appropriate to
deep inelastic lepto-production (Fig. 1):

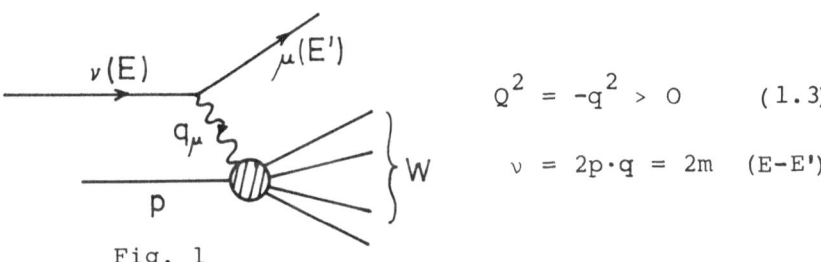

$$Q^2 = -q^2 > 0 \qquad (1.3)$$

$$\nu = 2p\cdot q = 2m \ (E-E')$$

Fig. 1

where q_μ is the four-momentum transfer, m the nuclear
mass and E and E' are the neutrino and muon energies,
respectively, in the laboratory frame. The "scaling"
variables are defined by

$$y = \nu/2mE$$
$$\qquad (1.4)$$
$$x = Q^2/\nu \ .$$

In a domain of Q^2, ν where masses are negligible x and
y vary independently over the range $0 \leq x, y \leq 1$. The
invariant mass W of the final state hadronic system is

(neglecting the proton mass):

$$W^2 = \nu - Q^2 = 2m \, E y \, (1 - x) \, . \qquad (1.5)$$

The parton model is the assumption that for large Q^2 and ν the neutrino is scattered from a single, free, effectively massless constituent of the nucleon, i.e. quark (Fig. 2) and that scattering from individual quarks is incoherent. In this picture the variable x is determined by the fraction of the proton momentum carried by the strack quark (parton)

$$x = P_q/p \qquad (1.6)$$

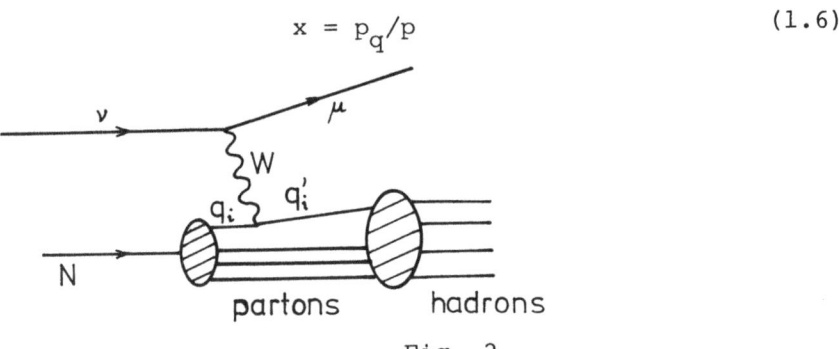

Fig. 2

in a frame where masses and the parton transverse momentum are negligible:

$$E_p^2 \gg m_p^2, \; E_{q_i}^2 \gg m_{q_i}^2 + p_{i\perp}^2 \qquad (1.7)$$

and y is related to the lepton scattering angle in the ν-parton center of mass frame:

$$y = \frac{1}{2} \, (1 - \cos \theta_i) \quad .$$

The ν-p cross section is then an incoherent sum over ν-parton cross sections:

$$\frac{d\sigma}{dxdy} = \sum_i \frac{d\sigma_i(x)}{dxdy} \; P_i(x)$$

(1.8)

where $P_i(x)$ is the probability of finding the ith parton with momentum fraction x. Recalling that for V-A couplings fermion fermion scattering is isotropic and fermion-anti-fermion scattering has a $(1-\cos\theta)^2$ dependence, we find for the ν-parton cross sections

$$\frac{d\sigma_i}{dxdy} \sim x \; \{\frac{1}{(1-y)^2}\} \quad \text{for} \quad \{ \begin{matrix} \nu + q, & \bar{\nu} + \bar{q} \\ \nu + \bar{q}, & \bar{\nu} + q \end{matrix} \} \quad .$$

Neglecting Cabibbo-suppressed contributions the allowed elementary processes below charm threshold are

$$\left. \begin{matrix} \nu + d \rightarrow \mu^- + u \\ \nu + \bar{u} \rightarrow \mu^- + \bar{d} \\ \bar{\nu} + u \rightarrow \mu^+ + d \\ \bar{\nu} + \bar{d} \rightarrow \mu^+ + \bar{u} \end{matrix} \right\} \qquad \frac{d\sigma}{dy} \sim \left\{ \begin{matrix} 1 \\ (1-y)^2 \\ (1-y)^2 \\ 1 \end{matrix} \right. \qquad (1.9)$$

For an I = 0 target $P_u(x) = P_d(x)$, $P_{\bar{u}}(x) = P_{\bar{d}}(x)$ and the cross sections are proportional to:

$$\frac{d\sigma^\nu}{dxdy} = [u(x) + d(x)] + [\bar{u}(x) + \bar{d}(x)] (1-y)^2 + 0(\sin^2\theta_c) \qquad (1.10)$$

$$\frac{d\sigma^{\bar{\nu}}}{dxdy} = [\bar{u}(x)+\bar{d}(x)]+[u(x)+d(x)](1-y)^2+0(\sin^2\theta_c)$$

$$\sigma^{\bar{\nu}}/\sigma^{\nu} = \frac{1}{3}(\frac{1+3r}{1+r/3})$$
(1.11)

where we define

$$u(x) = x P_u(x), \qquad \int_0^1 dx\ u(x) \equiv u \text{ etc.}$$
(1.12)

and

$$r = (\bar{u} + \bar{d})/(u + d)$$
(1.13)

is the relative anti-parton content of the nucleon. The analysis of data from Gargamelle [3] suggests that

$$r \approx 0.1$$
(1.14)

with anti-partons confined to the small x region:

$$\bar{u}(x), \bar{d}(x) \to 0, \qquad x \sim (0.1 - 0.2),$$
(1.15)

a result which is in accord with theoretical prejudice [4]. A similar analysis is made for electroproduction, where the e-parton cross section is proportional to the parton squared charge, and the y distribution is characteristic of the pure V-interaction.

$$\frac{d\sigma^{e\pm q}}{dxdy} = \frac{d\sigma^{e\pm\bar{q}}}{dxdy} \sim Q_q^2 \text{ x } [1 + (1-y)^2]$$

$$\sigma^{eZ(I=0)} \sim \frac{5}{9} [u + \bar{u} + d + \bar{d}] + \frac{2}{9} (s + \bar{s}) . \qquad (1.16)$$

By comparing the experimental values for (1.16) with (1.10) one can in principle extract the relative $s + \bar{s}$ content of the nucleon. The data [2], [3] are compatible with $s + \bar{s} \approx 0$, but a 20 % strange parton content is not ruled out. Intuitively one expects that

$$s, \bar{s} \underset{\sim}{<} \bar{u}, \bar{d} \qquad (1.17)$$

and that the s, $\bar{s}$ distribution will also be confined to small x.

Now let us naively extend these ideas to charmed particle production. The important elementary scattering processes are:

$$\nu_\mu + d \rightarrow \mu^- + c, \quad \sigma \sim \sin^2\theta_c [d(x) + c(x)] \qquad (1.18a)$$

$$\nu_\mu + s \rightarrow \mu^- + c, \quad \sigma \sim 2 \cos^2\theta_c \, s(x) \Bigg\} \rightarrow 0, x \underset{\sim}{>} 0.1 \qquad (1.18b)$$

$$\bar{\nu}_\mu + \bar{s} \rightarrow \mu^- + \bar{c}, \quad \sigma \sim 2 \cos^2\theta_c \, \bar{s}(x) \Bigg\} \qquad (1.18c)$$

where we have assumed an $I = 0$ nuclear target. From the above discussion we see that these two contributions may be comparable:

$$\frac{2s}{u+d} = \frac{2\bar{s}}{\bar{u}+\bar{d}} \underset{\sim}{<} 10 \% , \qquad \sin \theta_c \approx 5 \% \qquad . \qquad (1.19)$$

The $\bar{\nu}\bar{d}$ cross section is suppressed by both the Cabibbo

angle and the sea factor and is negligible. If the
nucleon has any charm content it must be considerably less that the s-content (to be compatible with
the strong Zweig suppression of the ψ/J relative to
the ϕ), and the processes

$$\nu_\mu + \bar{c} \rightarrow \mu^- + s,d$$

$$\bar{\nu}_\mu + c \rightarrow \mu^+ + \bar{s},\bar{d}$$

are further suppressed by a $(1 - y)^2$ factor relative
to (1.18b,c); we shall neglect this contribution.
Then we expect that (sufficiently above threshold),
charm production will be characterized by [5]

a) a flat y-distribution

b) a component with a valence x-distribution for ν's,
contributing roughly 5 % to the total ν cross section.

c) a possibly comparable component confined to small x
and which contributes equally to ν and $\bar{\nu}$ cross sections,
and is thus relatively enhanced by the factor ~ 3 in $\bar{\nu}$
reactions.

These expectations are illustrated [6] in Fig. 3
where the y-distribution has been cut-off at a threshold value (see next section). However, before con-
fronting these predictions with the data, we should
comment on their domain of validity.

1.2 THRESHOLD EFFECTS

In order to produce a system of invariant mass
m_0, we must have $W \geq m_0$, and we see from (1.5) that
this implies:

$$E \geq m_0^2/2m, \quad y \geq m_0^2/2mE, \quad x \leq 1 - m_0^2/2mEy \quad . \quad (1.20)$$

As a new mass threshold is passed, its presence will
first appear in the high y, low x region, thus distort-
ing the distributions described above. Furthermore, we
hardly expect the parton picture to be a reasonable description in the baryon resonance region for W; the results discussed above are relevant for $W \gtrsim 2$ GeV. For
charmed particle production, the lowest baryon state
may have a mass of about 2.5 GeV. Then the threshold
energy is

$$E_c \gtrsim 3 \text{ GeV} . \quad (1.21)$$

However, the charm "resonance region" may extent to 3
or 4 GeV. Thus if we want to cover a reasonable region
of the x, y plane in the scaling region, say $1 > y > 0.5$,
$0 < x < 0.6$ for $W^2 > (10-15)$ GeV we require

$$E_c^{(sc)} \gtrsim (20 - 30) \text{ GeV} \quad (1.22)$$

i.e. a considerably higher threshold is required for
effective scaling. One then has to guess as to how the
scaling threshold might be approached. What has been

done by most authors is simply to multiply the parton
distributions by a θ-function:

$$d\sigma \rightarrow d\sigma \; \theta \; (W - W_c) \qquad\qquad (1.23)$$

where $W_c \simeq$ a few GeV is the anticipated effective
scaling threshold in W.

Another related problem is that in the parton
model described above, the quarks were treated as
massless. It is generally believed that ordinary quarks
are (effectively) light on the scale of ordinary hadron
masses; $m_{u,d,s} \lesssim$ a few hundred MeV. However, the charmed quark apparently has an effective mass of 1.5-2 GeV
(c.f. charmonium spectrum at SLAC). There has recently
emerged some field theoretical justification [7] for an
approach [8] which simply takes the parton model literally for massive quarks. This alters the kinematics in
such a way that the scaling variable x is redefined by:

$$x \rightarrow x' = x + m_c^2/2m \; Ey \qquad\qquad (1.24)$$

if all but the charmed quark mass are neglected. It
happens moreover that the charmed quark mass is on a
scale similar to the expected masses of the charmed
hadrons. For example from the observed rise in

$$R = \frac{\sigma(e^+e^- \rightarrow \text{hadrons})}{\sigma(e^+e^- \rightarrow \mu^+\mu^-)} \qquad\qquad (1.25)$$

at a total energy of about 4 GeV - as well as from

SU(4) mass formulae - one expects for the lowest lying mesons

$$D = (c \, \bar{u} \text{ or } c \, \bar{d}, 0^-)$$

$$m_D \simeq m_F \simeq 2 \text{ GeV} ,$$
$$(1.26)$$

$$F = (c \, \bar{s}, \, 0^-)$$

and bubble chamber data, to be discussed later, suggest for the lowest baryon state:

$$m_{C_0^+} \simeq (2.2 - 3.5) \text{ GeV}, \quad C_0^+ = (c \, u \, d \, , \, 1/2^+).$$
$$(1.27)$$

If m_c is replaced by W_c in Eq. (1.24), one obtains a smooth extrapolation [9] to the scaling region with the condition $x' < 1$ ensuring the correct threshold cut off.

1.3 DIFFRACTIVE EFFECTS

So far we have discussed the region of validity of the parton model with respect to W, the invariant mass of the final state hadron system. Although the squared momentum transfer $q^2 = -Q^2$ is spacelike, the vector (axial) meson pole diagram of Fig. 4 is expected to be important for $Q^2 \sim m_V^2$. This contribution peaks broadly in Q^2 at $Q^2 \simeq m_V^2$ and most of its contribution lies in the region $0 \leq Q^2 \lesssim 2 \, m_V^2$. The cross section

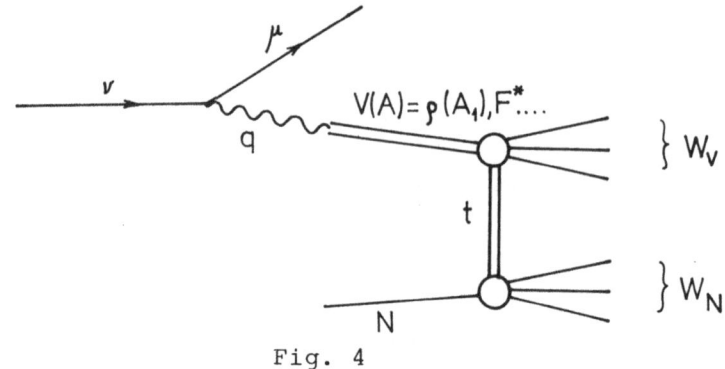

Fig. 4

has the dependence [lo]:

$$\frac{d\sigma^{diff}}{dt\ dx\ dy} \sim \frac{x(1-x)}{(Q^2 + m_V^2)^2}\ y\ \{y^2\frac{d\sigma_T}{dt}\ +$$

$$+\ (\frac{d\sigma_T}{dt} + \frac{d\sigma_L}{dt})\ (\frac{1-y-mxy/2E}{1+2mx/2Ey})\}\qquad(1.28)$$

where σ_T (σ_L) is the cross section for scattering of a transversely (longitudinally) polarized vector meson from a nucleon. This expression clearly does not have the simple x,y dependence characteristic of the parton model[+]. The point is that the parton model has a field

---

[+]One can simulate [11] the parton model distribution by a suitable superposition of vector mesons in each channel and a suitable choice of Q^2 dependence for the couplings; we are only interested in the effects of known bound states and we keep the couplings fixed at the residue of the pole, modulo certain uncertainties associated with unphysical component of polarization.

theoretical support in an asymptotically free theory
[12], where at sufficiently high momentum transfer the
strong coupling can be treated perturbatively and, in
a first approximation,neglected. This treatment clearly
must fail when quark bound state effects are important,
as is the case for q^2 sufficiently near a pole.

If the t-dependence of $\sigma_{T,L}$ is neglected, the x-
dependence is rather broad; however it is more likely
that the scattering cross section falls with decreas-
ing momentum transfer $|t|$ $(t < 0)$

$$\sigma_{T,L} \sim e^{bt} \ .$$

(1.29)

For elastic VN scattering [13] $b \approx 8$ GeV^{-2} for the ρ,
$b \sim 2$ GeV^{-2} for the ψ/J and probably has intermediate
values for intermediate masses. This has the effect
of pushing the distribution to the high y and particularly to the low x region since t is bounded by:

$$t < t_{min.}^{el} = - m^2 \ (\frac{x+m_V^2/2mEy}{1-x})^2$$

or

(1.30)

$$y > m_V^2/2E\sqrt{-t}, \quad x < \sqrt{-t}/m - m_V^2/2mEy \ .$$

For inelastic processes, the slope in t is presumably
less strong, but $|t_{min}|$ increases. If the vector meson
vertex is inelastic $m_V \to W_V$ in Eq. (1.30); if the nucleon
vertex is also significantly inelastic

$$t_{min} \simeq -w_N^2 \left(\frac{x + w_V^2/2M \; Ey}{1-x} \right) \; ,$$

so   (1.31)

$$y > \frac{w_V^2}{-t} \frac{w_N^2}{2Em}, \quad x < -\frac{t}{w_N^2} - \frac{w_V^2}{2mEy} \quad .$$

Therefore, a reasonable approximation [6] is probably to replace the total cross section by

$$\sigma_{T,L} = \int dt \; \frac{\sigma_{T,L}}{dt} \to e^{bt_{min}^{el}} \; \sigma_{T,L} \quad .$$

For diffractive production of charmed particles (via $F^* = (c\bar{s},1^-)$) this gives the distributions shown in Fig. 5. The total VN transverse cross sections for ρ, ϕ and ψ/J can be extracted from photoproduction data on $\gamma + N \to VN$ using the optical theorem:

$$\frac{d\sigma}{dt} (\gamma_N \to VN) = e^{bt} \; \frac{e^2}{g_V^2} \; \frac{1}{16\pi} \; \sigma_{tot}^2 (VN) \quad .$$ (1.32)

One finds [13]

$$\sigma_\rho \simeq 26 \; mb \; , \quad \sigma_\phi \simeq 10 \; mb, \quad \sigma_{\psi/J} \simeq 1 \; mb \; ;$$ (1.33)

to estimate the F^*N cross section the additive quark model is invoked

$$\sigma_{F^*} = \frac{1}{2} (\sigma_\phi + \sigma_{\psi/J}) \cong 5.5 \text{ mb .} \tag{1.34}$$

The couplings to the weak current are determined by CVC for the ρ, and for the F^* by the empirical rule [14] (satisfied by observed couplings to the electromagnetic current):

$$\frac{1}{g_V^2} \simeq \frac{\text{const.}}{m_V} \times \text{C.G.} \tag{1.35}$$

where C-G is the appropriate Clebsch-Gordan (or quark counting) coefficient. Then one obtains, for example, the elastic vector meson production cross sections shown in fig. 6, where they are plotted relative to the total neutrino cross section, assumed to be rising linearly with energy. (The constant in Eq. (1.35) has been taken equal for vector and axial vector couplings in accordance with pole dominated spectral function sum rules [15]. However this cannot be understood in terms of a non relativistic quark model, so the predictions for axial mesons should probably be viewed with more caution). The total diffractive contribution from each vector meson is related to the elastic contribution by

$$\frac{\sigma^{\text{el.}}}{\sigma_{\text{tot}}} = \frac{\sigma_{\text{tot}}}{16\pi b} \simeq 2.5 \frac{\sigma_{\text{tot}}(\text{mb})}{16\pi b(\text{GeV}^{-2})} \sim \begin{cases} 0.2 \\ 0.02 \text{ for} \\ 0.07 \end{cases} \begin{cases} \rho \, (A_1) \\ \psi/J \\ F^*[F_A = (c\bar{s}, 1^+)] \\ \text{if } b \simeq 4 \; . \end{cases}$$

$$\tag{1.36}$$

The qualitative features of the diffractive contributions are in many ways characteristic of the parton "sea"
contributions:

a) Neglecting V,A interference, they contribute equally
to ν and $\bar{\nu}$ scattering and therefore are relatively
enhanced by a factor of about 3 in $\bar{\nu}$ interactions.

b) They are important in the low x, high y region.
Since in practice the "sea" is observed as a flattening of the $(1-y)^2$ distribution as $y \to 1$ in the low x
region of $\bar{\nu}$-scattering:

$$d\sigma^{\bar{\nu}}(x,y) \neq 0 \quad \text{for} \quad y \to 1 \quad \text{if} \quad x \lesssim 0.1 ,$$

the effect could easily by simulated by a vector meson
contribution if Q^2 is in the appropriate range. In the
Gargamelle scaling analysis a cut $Q^2 > 1$ GeV^2 is im-
posed. This presumably eliminates much of the ρ contribution. However in that experiment $E_\nu \lesssim 10$ GeV, so for
$x \leq 0.1$, $Q^2 < 2 m E x < 2$ GeV^2.

If the A_1 contribution is to be taken seriously,
it peaks at $Q^2 = 1.2$ GeV, and a 10 % contribution to
the total cross section (c.f. fig. 6, Eq. (1.36) and a)
above) for 1 GeV$^2 < Q^2 < 2$ GeV^2 does not seem unlikely.
Similarly at high energies diffractive F^* production
might contribute about 15% - or twice that much if F_A also
contributes - to the $\bar{\nu}$ total cross section, and could
be important for Q^2 as high as 8-10 GeV^2. The difference with the "sea" contribution is that the diagrams
of Fig. 4 do not scale: as E increases Q^2 increases for
fixed x,y , and the effect will disappear unless higher
mass resonances become important.

1.4 LEPTONIC BRANCHING RATIO FOR CHARMED
PARTICLE DECAY

In order to estimate dilepton yield due to charm production by neutrinos, an estimate of the leptonic branching ratio is necessary. In terms of quarks, the dominant decay processes are expected to be:

Non-leptonic $c \rightarrow s\, u\bar{d}$ Ampl. $\sim G_F \cos^2 \theta_c$ (1.37)

Leptonic $c \rightarrow s\ell^+ \nu$ Ampl. $\sim G_F \cos \theta_c$.   (1.38)

These are a priori comparable in magnitude. However from our experience with strange particle decays we expect that the non-leptonic amplitude may be effectively enhanced by the strong interactions, i.e. by gluon exchange. This can be qualitatively understood as follows. For the lepton decay (fig. 7a), the momentum transfer carried by

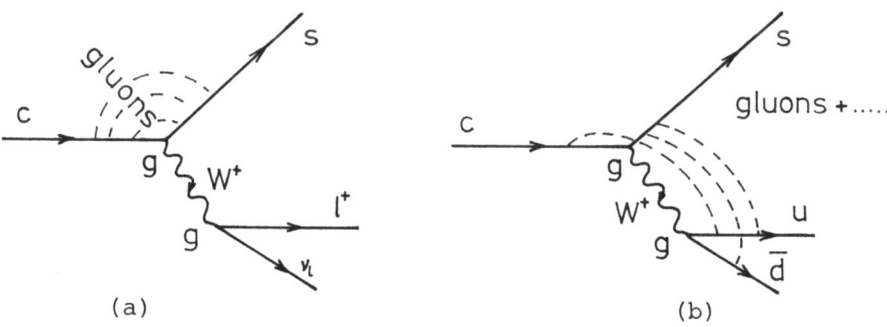

(a)   (b)

Fig. 7

the intermediate boson is always the invariant mass of the lepton pair:

$$q^2 = (p_\ell + p_\nu)^2 \ll m_W^2$$

so the effective coupling is always weak:

$$\text{Ampl.} \sim g^2/(m_W^2 - q^2) \sim G_F .$$

However, for the non-leptonic case the gluon exchange diagram of Fig. 7b allows for contributions from W's carrying high momentum, which might attenuate to some extent the mass suppression factor in the W-propagator. The effect can be calculated in an asymptotically free field theory where colored quarks are assumed to couple strongly through the exchange of an octet of colored gluons. One finds [16] that the effective fermi coupling constant is modified by a logarithmic factor:

$$G_F \rightarrow G_{eff} = G_F [1 + 6 \ln (m_W^2/\mu^2)]^y \tag{1.39}$$

where μ is a normalization mass and it turns out that

$$\left. \begin{array}{ll} y > 0, & G_{eff} > G_F \\ \\ y < 0, & G_{eff} < G_F \end{array} \right\} \quad \text{for the } \Delta I = \left\{ \begin{array}{l} 1/2 \\ \\ 3/2 \end{array} \right. \tag{1.40}$$

part of the non-leptonic current-current operator. In the GIM model the magnitude of this effect is not sufficient to account for the accuracy of the ob-served $\Delta I = 1/2$ rule. However there are various ar-guments leading one to expect a further enhancement of

the (1/2)/(3/2) amplitude ratio in the matrix elements
of the current-current operator. If there are more than
four quarks, the effect in (1.40) is somewhat increased.
The same analysis holds for the $\Delta C = 1$ current-current
operator responsible for charmed particle decay, with
$\Delta I = 1/2$ (3/2) replaced by $\Delta V = 0(1)$, V-spin being the
u $\leftrightarrow$ s symmetry group. However, the effect may be weakened in this case [17]. The only change in Eq. (1.39) in
going from strangeness to charm decay is in the choice
of the renormalization mass μ, which should be characterized by:

a) the effective masses of the external quarks in the
hadron wave function, and

b) the onset of scaling for processes involving the
appropriate quantum numbers.

Both criteria suggest $\mu^2_{\Delta C} > \mu^2_{\Delta S}$, thus decreasing the
factor in brackets in (1.39) for the charm case. For
strangeness changing decays leptonic branching ratios
are typically

$$B^S_\ell \sim 10^{-3} \quad .$$

This is due to a phase space suppression of 3-body-
final states as well as to the effective enhancement
of non-leptonic rates which is empirically of order

$$A_s \simeq \sin \theta_c^{-2} \simeq 20 \quad .$$

For the decays of massive charmed particles, 3-body
phase space is irrelevant, so we expect

$$B^C_\mu = B^C_e = (2 + A_c)^{-1} \ , \qquad\qquad (1.41)$$

and from the above discussion we expect

$$A_c < A_s \simeq 20 \ ,$$

while the minimum non-leptonic ratio expected is that
for a freely decaying quark, i.e. no enhancement. Since
the (ud) quark pair in fig. 7b can carry three colours
we obtain

$$A_c > A_{free} = 3 \ .$$

Thus we obtain

$$5\% \lesssim B_\mu = B_e \lesssim 20 \ \% \qquad\qquad (1.42)$$

as the theoretically reasonable range for the leptonic
branching ratio.

2. A SCRUTINY OF THE DATA; CHARM AND ALTERATIONS

In this lecture we shall review the neutrino data
to see to what extent various phenomena observed over
the last two years support the picture described in
lecture 1. We shall also briefly comment on proposed
modifications of the GIM model.

2.1 DIMUON EVENTS

The HPWF collaboration at Fermilab have observed neutrino induced dimuon events [18],[19] with an apparent hadronic invariant mass threshold

$$W \gtrsim 4 \text{ GeV}$$

and with relative cross sections for $E > 40$ GeV

$$\sigma_\nu(\mu^-\mu^+)/\sigma_\nu(\mu^-) \simeq 10^{-2}, \sigma_{\bar{\nu}}(\mu^+\mu^-)/\sigma_\nu(\mu^-\mu^+) \simeq 0.8 \qquad (2.1)$$

$$\sigma_\nu(\mu^-\mu^-)/\sigma_\nu(\mu^-\mu^+) \sim 0.1, \sigma_\nu(3\mu)/\sigma_\nu(2\mu) \gtrsim \frac{1}{50} \quad . \qquad (2.2)$$

As the semi-leptonic charm changing interaction respects a $\Delta C = \Delta Q$ rule, the production and semi-leptonic decay of a charmed particle will necessarily lead to a $(\mu^-\mu^+)$ state. Let us first consider the "opposite sign" dimuon events. Their spread in invariant mass rules out their interpretation in terms of a decay $X \to \mu^+ \mu^-$, and sources such as a heavy lepton or an intermediate boson appear unlikely, particularly because of the observed large energy asymmetry [20] in favor of the $\mu^-(\mu^+)$ for $\nu(\bar{\nu})$ induced events. So we shall compare the data with the naive parton model predictions for charm as dis-cussed previously.

2.2.1 Opposite sign dimuons and charm

The predictions are:

$$\sigma_\nu \, (\mu^- \mu^+) = B_\mu \, (\sigma_v + \sigma_s)$$

$$\sigma_{\bar{\nu}} \, (\mu^+ \mu^-) = B_\mu \, \sigma_s \qquad\qquad (2.3)$$

where B_μ is the muonic branching ratio and σ_v and σ_s are the valence and sea quark contributions, respectively, with

$$\sigma_s \lesssim 2\sigma_v, \ \sigma_v/\sigma_\nu^{tot} \approx \sin^2\theta_c \approx 0,05, \ B_\mu = (5\text{-}20)\% \quad .$$

Therefore we predict:

$$\sigma_{\bar{\nu}} \, (\mu^+ \mu^-)/\sigma_\nu (\mu^- \mu^+) \lesssim 2/3 \qquad\qquad (2.4)$$

and

$$B_\mu \, \sigma_v = B_\mu [\sigma_\nu - \sigma_{\bar{\nu}}] = (0,003 \text{ to } 0,01) \quad . \qquad\qquad (2.5)$$

These expectations are in agreement with the data, Eq. (2.1), which are compatible with parameters in the range (assuming always $r_s \lesssim 10 \%$):

$$(r_s, \ B_\mu) = (10\%, \ 7\%) \quad \text{to} \quad (1\%, \ 16\%); \qquad\qquad (2.6)$$

r_s is the relative strange quark content of the nucleon:

$$r_s = 2s/(u+d) = 2\bar{s}/(u+d) \quad . \qquad\qquad (2.7)$$

If this interpretation is correct, we expect the $\bar{\nu}$ distribution to be confined to small x, while the ν distribution will contain a component characteristic of the valence quark distribution ($< x > \simeq 0.2$). However if threshold effects are important at Fermilab energies, these distributions can be distorted towards the high y, low x region (see Eqs. (1.20)). In fig. 8 the data are compared with the calculations of Barger et al. [21], where the threshold effect has been included simply by a θ-function. Analyses have also been performed using smoother threshold corrections as discussed in sect. 1.2, but at the present level of the data the comparison is not significantly different, and there is no apparent discrepancy with the predictions. In particular:

a) The x-distributions are just as expected.

b) The y-distributions are compatible with predictions, but it has been pointed out [21],[8] that the fit to the ν-data is somewhat improved if there is a $(1-y)^2$ component.

The latter is not present in the conventional GIM V-A couplings. We shall comment later on possible modifications of the theory; present data is certainly not conclusive. It should also be remarked that the theoretical curves are plotted against the true x,y variables defined in Eqs. (1.4), which are not actually measured because the incident neutrino energy is unknown, and there is a presumed decay neutrino contributing to the total "hadronic" final state energy. The data are therefore plotted against quantities derived from the observed final state energy such that

$$y_{obs} < y, \qquad x_{obs} > x . \qquad\qquad (2.8)$$

The variable

$$v = xy = x_{obs} \, y_{obs} \qquad\qquad (2.9)$$

is determined by the scattered lepton energy and angle and is not subject to this difficulty. The expected and observed v distributions are shown in Fig. 9.

One may also attempt to include the decay distributions of the second muon [22], [9]. Results are again in qualitative agreement with the GIM model. In particular the E_-/E_+ asymmetry is well described [22], as opposed to the case for a heavy lepton or W hypothesis. Another interesting comparison is shown in Fig. 10, where $k_\perp$ is the component of the μ^+ momentum relative to the (v, μ^-) plane. The theoretical curves are for $V \pm A$ induced decays

$$\text{charm} \rightarrow \text{hadrons} + \mu^+ v \quad \text{(n-body)} , \qquad (2.10a)$$

and

$$P_c \rightarrow \mu^+ v \quad \text{(2-body)} , \qquad\qquad (2.10b)$$

where P_c is a charmed pseudoscalar. The distribution for the general case, Eq. (2.10a), is derived from the free quark decay distribution for

$$c \rightarrow s \, \ell \, v$$

(or equivalently the muon decay distribution). The decay (2.10a) is "allowed", while in a $V \pm A$ theory (2.10b) is forbidden by helicity conservation. The ratio of allowed to forbidden decays scales as the fourth power of the parent mass. From the measured ratio of K_{ℓ_2} to K_{ℓ_3} one expects [23] for $m_F = 2$ GeV,

$$\Gamma(F \to \ell\nu)/\Gamma \; (F \to \ell\nu + \text{all}) < 1.5 \; \% \; . \qquad (2.11)$$

The $D \to \ell\nu$ decay is further suppressed by the Cabibbo angle and should be negligible.

2.2.2 Same sign dimuons

The $\Delta C = \Delta Q$ rule led us to predict only opposite sign dimuons in charm decay. However, in spite of the analogous $\Delta S = \Delta Q$ rule for strangeness changing processes, the reaction

$$\bar{\nu} + N \to \mu^+ + K + X$$
$$\qquad\quad \hookrightarrow \mu^+ + X'$$

can occur because a decaying neutral kaon looses the memory of its initial strangeness via $K^0 \leftrightarrow \bar{K}^0$ mixing. In the GIM model, the analogous $D^0 \leftrightarrow \bar{D}^0$ mixing is expected to be negligible because [24]:

a) The $\Delta C = \Delta S$ rule for Cabibbo favored transitions implies that mixing can occur only through strangeness zero intermediate states:

$$D^0 \to (\text{hadrons})_{S=0} \to \bar{D}^0, \; \text{Ampl.} \sim \sin^2\theta_c \; .$$

b) Just as the GIM mechanism insures that the $K^0 \leftrightarrow \bar{K}^0$

transition vanishes in the limit of u,c mass degeneracy the $D^O \leftrightarrow \overline{D}^O$ transition vanishes for d,s degeneracy, which is a much better approximation.

c) because of the many channels available we expect

$$\Gamma \;\; >> \;\; \Delta\Gamma, \; \Delta m$$

for the two combinations of D^O and $\overline{D}^O$ which are decay eigenstates. Putting these factorstogether, one obtains an estimate [25] of about 10^{-4} for "wrong sign" dimuons in semi-leptonic decays.

Then the only mechanism available for the $\mu^- \bar{\mu}^-$ events in the GIM model is associated production of charm via a charm conserving reaction:

$$\nu + N \rightarrow \mu^- + C + \bar{C} + X$$

$$\mu^- + \nu + \text{hadrons} \tag{2.12}$$

$$\text{hadrons} \; .$$

Since the events occur at a level of about 10^{-3}, this requires:

$$\sigma\,(\nu N \rightarrow C\bar{C}+X)/\sigma_\nu^{tot} \sim 10^{-3}/B_\mu \simeq (0.6-1.5) \times 10^{-2} \; . \tag{2.13}$$

Is this level unreasonably high? The extrapolation of fits to ordinary hadron production leads to an esti-mate of $D\bar{D}$ production at ISR - where much more energy is available - of the order [26]

$$\sigma\,(pp \rightarrow D\bar{D} + X)/\sigma\,(pp)^{tot} \sim 5 \times 10^{-4} \; . \tag{2.14}$$

However, much of the cross section and much of the
available energy in pp collisions go into leading
particle effects which have no analogue in deep inelastic νp reactions. The large $p_\perp$ sector in pp
scattering might provide a better means of comparison
and the relative D production is expected to increase
significantly with $p_\perp$; for example [26]:

$$(D/\pi)_{tot} \approx 10^{-4}, \quad (D/\pi)_{p_\perp > 1GeV} \approx 3 \times 10^{-3} . \qquad (2.15)$$

In any case the associated production interpretation
may be tested, as it implies two unambiguous predictions:

a) Trimuon events should occur at a level

$$\sigma_\nu(3\mu) = B_\mu \sigma_\nu(\mu^-\mu^-) \approx (0.7 \text{ to } 1.6) \times 10^{-2} \sigma(\mu^-\mu^+) . \qquad (2.16)$$

The limit in (2.2) is not yet sensitive to this prediction.

b) Dimuon events in leptoproduction

$$\ell + N \to \quad \ell + C + \bar{C} + X$$
$$\ell\nu + \text{hadrons}$$
$$\text{hadrons}$$

should occur at a level of roughly

$$\sigma_\ell(\ell\ell')/\sigma_\ell(\ell) \sim 2\sigma_\nu(\mu^-\mu^-)/\sigma_\nu(\mu^-) \sim 2 \times 10^{-3} \qquad (2.17)$$

with

$$\sigma_\ell (\ell\ell^+) = \sigma_\ell (\ell\ell^-) \ .$$

2.2 ANOMALOUS x AND y DISTRIBUTIONS

If we are correct in attributing the $(\mu^-\mu^+)$ signal to the production and decay of charmed particles which have predominantly non-leptonic decays, we expect accordingly an increment in the total single muon event rate;

$$\sigma_\nu \rightarrow \sigma_\nu + \sigma_\nu^c:$$

$$\sigma_\nu^c(\mu^-)/\sigma_\nu(\mu^-) = (B_h/B_\mu)\sigma_\nu(\mu^-\mu^+)/\sigma_\nu(\mu^-) \approx (4 \text{ to } 13)\% \qquad (2.18)$$

for $B_\mu \approx (16 \text{ to } 7)\%$, where B_h is the non-leptonic branching ratio. Recalling that the uncertainty in the presumed value of B_μ is correlated with the uncertainty in

$$\sigma_{\bar\nu}(\mu^+\mu^-)/\sigma_\nu(\mu^-\mu^+) \approx (1/5 \text{ to } 2/3) \ ,$$

and that

$$\sigma_{\bar\nu}^{tot}/\sigma_\nu^{tot} = 1/3$$

we find accordingly,

$$\sigma_{\bar\nu}^c(\mu^+)/\sigma_\nu^c(\mu^+) \approx (2.5 - 25) \% \ . \qquad (2.19)$$

For ν-production, the charm contribution will not give

distributions which are strikingly different from the dominant contribution, except for a possible enhance-ment at small x if the sea contribution is important. Such an effect is apparent in the data [27] of Fig.11, although it appears somewhat excessive. The sea contri-bution to Eq. (2.18) is $\lesssim$ 8 % of σ_ν^{tot}; however the small x enhancement could be increased by a threshold dist-ortion of the valence contribution, giving still $\lesssim$ 13 %. In the y distributions no significant effect is expected, and none is observed.

Effects of charm production by anti-neutrinos can be much more dramatic, what we expect is a flattening of the y-distribution at low x, and this indeed is what is observed, as shown in Fig. 12. To make more quantitative comparisons, we assume that:

a) (20 $\pm$ 2) % of the u + d distribution lies below x = 0.1 (recall that <x> $\simeq$ 0.2 for ν-production).

b) the $\bar{u}$ + $\bar{d}$ content is $(\bar{u} + \bar{d})/(u + d)$ = (10 $\pm$ 1)%, and lies entirely below x = 0.1.

c) For charm production we take two simple hypotheses consistent with the dilepton data (see Eqs. (2.6) and (2.7)).

i) (r_s, B_h) = (0.1, 0.36) with a threshold fall-off in y such that $<y>_c$ = 0.56,

ii) (r_s, B_h) = (0.05, 0,80) with $<y>_c$ = 0.59.

The data of Fig. 12 appear consistent with naive parton ideas applied to charm production (Fig. 13). How-ever, some discrepancies appear in the comparison of different x and/or energy regions; specifically the total contribution of the events filling up the high y,

low x, region for $E_{\bar{\nu}}$ > 30 GeV seems excessively large
relative to the total cross-section. An indication of
this is in the average value [27] of y shown in Fig. 14
(where the data have not been corrected for acceptance).
For a V-A theory and no anti-partons, <y> = 0.5 for neutrinos and <y> = 0.25 for anti-neutrinos. The parameter

$$B = (q-\bar{q})/(q+\bar{q})$$

gives a measure of the anti-parton content and is related
to <y> by

$$B^{\nu} = 8 <y>^{\nu} - 3$$

$$B^{\bar{\nu}} = 3 - 8 <y>^{\bar{\nu}} \quad .$$

Since only d and s contribute to charm production by neutrinos, and only $\bar{s}$ contributes for anti-neutrinos, above
charm threshold we expect $B^{\nu} < B^{\bar{\nu}}$. Specifically, in the
model defined by a) - c) above we find

$$B^{\nu} = B^{\bar{\nu}} = 0.82 \pm 0.02 \qquad\qquad (2.20a)$$

below charm threshold, and

$$B^{\nu} = \begin{cases} 0.88 \pm 0.03 \\ 0.85 \pm 0.02 \end{cases}$$

$$\qquad\qquad (2.20b)$$

$$B^{\bar{\nu}} = \begin{cases} 0.67 \pm 0.03 \\ 0.73 \pm 0.02 \end{cases}$$

above charm threshold. These values may be compared with the values shown in Fig. 14 and the corresponding expected (corrected for acceptance) values of <y>. Note that the threshold effect, which enhances the high y contribution, has the effect of raising B^ν and lowering B^ν. For very high energies when this effect has become unimportant one would expect B^ν and $B^{\bar\nu}$ to approach each other slightly (in the absence of any other scaling violation; see below).

Our analysis depends of course on the assumption of a 10% $\bar u + \bar d$ sea, which is probably compatible with, but not required, by the data of Ref.[27] below charm threshold. In fact, the data below 30 GeV suggest a smaller anti-parton content.

If there is no $\bar u + \bar d$ sea, one hardly expects a strange quark sea. As discussed in the proceeding lecture, the sea observed in Gargamelle is conceivably a Q^2-dependent diffractive effect. If that is the case, one might attribute $\bar\nu$-induced charm production to a similar diffractive effect. From Fig. 6 and Eq. (1.36) we expect charm production via the F^* at a level of about 20% of the total $\bar\nu$ cross section (perhaps twice that if F_A also contributes - however a charm production rate higher than 25% would imply a lower leptonic branching ratio, rendering more acute the same sign dilepton problem). A careful study of diffractive production distributions, taking into account both the experimental energy spectrum and the t_{min} effects would be useful.

Aside from the fact that the data should probably not be taken too literally at this exploratory stage, we have also ignored any scaling violation which might

appear at high energy. A violation is in fact expected
in asymptotically free field theories [28], which has
the effect of depleting the high x and filling the low
x regions. Such an effect has been observed in [29] μ-p
scattering at Fermilab at a level of about a 20 % deviation (which is more than predicted; the predicted
violation is expected to be unobservable at present
energies). If the high x μp effect is ignored the rise
at low x can probably be accounted for by charm production: via a $(c\bar{c})$ sea [30] and/or diffractive production
of charm via the ψ/J. However, if there is a "true" (in
other words not understood!) scaling violation in μ-
production, this must also be taken into account in
attempting to draw conclusions from neutrino data.

2.3 EXCLUSIVE CHANNELS

$\Delta S = - \Delta Q$ and μe V^0 events. Independently of any
hypothesis concerning scaling or the parton model, the
GIM hypothesis makes a clear prediction concerning the
nature of final states arising from charm production and
decay, namely a predominance of strange particles from
the Cabibbo favored $\Delta C = \Delta S$ decays. For non-leptonic
decays the final state would be apparent associated production if the ν is scattered by an s quark, ($\Delta S = \Delta C \pm 1$
at both production and decay), and an apparent $\Delta S = -\Delta Q$
transition if the production occurs from a valence quark.
The latter case potentially provides a very striking sig-
nature, but in practice it is difficult to identify an
event of this type with certainty. One event, apparently
the reaction

$$\nu_\mu p \to \mu^- \Lambda^o \pi^+ \pi^+ \pi^+ \pi^- \qquad\qquad (2.21)$$

was in fact found by a Brookhaven bubble chamber group [31] about a year ago. Assuming the interpretation (2.21) is correct, there is an ambiguity in the μ^-/π^- identification. Identifying the faster (E = 9.79 GeV) negative track as a pion yields

$$W = (2426 \pm 12) \text{ MeV}, \qquad y \simeq 0,3, \qquad x \simeq 0.3. \quad (2.22)$$

If we assume that the reaction (2.21) is elastic, it must be

$$\nu_\mu + p \to \mu^- + C_1^{++} \ (2.4), \ C_1^{++} = (cuu, 1/2^+ \text{ or } 3/2^+). (2.23)$$

However, the octet and decuplet mass spectra lead us to expect that the doubly charged states should be heavier than the $C_o^+ = (cud, \ 1/2^+)$. The kinematics of the event are in fact compatible with a cascade decay:

$$C_1^{++}(2.4) \to C_o^+(2.1 \text{ or } 2.26) + \pi^+ \qquad\qquad (2.24)$$
$$\phantom{C_1^{++}(2.4) \to }\llcorner_{\to \pi^+ \pi^- \Lambda^o} \ .$$

The cross section for elastic C_1^{++} production has been estimated [32] to be about 1.2 % (2.2%) for $1/2^+$ ($3/2^+$) at $E_\nu = 10$ GeV; the observed event has a total energy of 13.5 GeV and 4 other events at an energy ≥ 13.5 GeV were observed in the same run.

The muon identification (2.22) is the favored one, since very fast pions are unlikely. However, if we make the alternate identification we obtain

$$W = (4604 \pm 54) \text{ MeV} \qquad y = 0.9, \qquad x = 0.1 . \qquad (2.25)$$

Then other interpretations are possible, for example [33]:

$$\nu + p \rightarrow F^+ + \Lambda + \pi$$
$$\qquad\qquad \downarrow$$
$$\qquad\qquad \pi^+ \pi^+ \pi^- \qquad\qquad\qquad (2.26)$$

which might occur via a D^* N scattering [$D^* = (c\bar{d}, 1^-)$]. However because of the Cabibbo suppression total charm production at higher energies via the D^* is not expected to exceed a percent.

Similar events have been the object of an unsuccessful search by the hydrogen bubble chamber group [34] at Fermi lab. On the basis of the BNL event, they expected three identifiable $\Delta S = - \Delta Q$ events and found none. Consequently they set a limit

$$\sigma(\nu_\mu + p \rightarrow C + X)/\gamma_\nu \lesssim 3.6 \% . \qquad (2.27)$$
$$\qquad\qquad\quad \downarrow$$
$$\qquad\qquad\quad \Lambda^0 + X'$$

For masses indicated by the interpretation (2.23) of the BNL event, and for energies in the range 10 GeV $\lesssim$ $E_\nu \lesssim$ 30 GeV appropriate to the Fermilab bubble chamber experiment, estimates [32] for the total charmed baryon production are at a level of about 4 %; however the baryons are not expected to give a Λ^0 in all decays. One should also bear in mind the possibility of the interpretation (2.25), which leaves open the question of the baryon masses. For semi-leptonic decays of charmed particles, their production by neutrinos should be signed by a dilepton and one or more strange particles in the final state. A number of events of this type have in fact been observed.

In a scan of 1800 neutrino interactions in Gargamelle two events apparently of the type

$$\nu_\mu + Z \to \mu^- + e^+ + Z^* + (V^o = 1^o \text{ or } K_s) +$$

$$1 \text{ charged track } (+ \text{ neutrals?}) \qquad (2.28)$$

were observed [35]. The estimated background contamination is less than 0.05 event. Further, if the μe events were background induced, they would have the same proportion of identified V^o's as the total event sample. One then expects 104 events of the type

$$\nu_\mu + Z \to \mu^- + e^+ + Z^* + X \text{ (no identified } V^o) \qquad (2.29)$$

whereas only 4 or 5 have been observed, and on the basis of two ($\mu^- e^+ V^o$) events one expects the sample (2.29) to contain 2 or 3 events with an unidentified neutral kaon or Λ^o. A third event of the type (2.28) was subsequently identified in a second scanning sample. In addition, the Wisconsin-Berkeley-Hawaii group working with the Fermilab 15 foot neon bubble chamber has observed four events [36] of the type (2.28) where the V^o is always identified as a K_s, and in the same sample no events of the type (2.29). The conclusion appears inescapable that new physics is involved and that the dilepton signal - at least at neutrino energies appropriate to bubble chamber exposures - are strongly correlated with strange particles. How do these results compare with the GIM model?

2.2.1. Low energy ($\mu e V^o$) events

In the first two events observed at Gargamelle [35] the invariant masses of the eV^o system are

$$\left.\begin{array}{c} (1.24 \pm 0.02) \\ (10.65 \pm 0.03) \end{array}\right\} \text{GeV and} \quad \left\{\begin{array}{c} 1.91 \pm 0.20 \\ 1.56 \pm 0.20 \end{array}\right. \quad \text{assuming}$$

$$V^o \equiv \left\{\begin{array}{c} \Lambda^o \\ K_s \end{array}\right. . \tag{2.30}$$

Then the interpretation

$$\nu_\mu + Z \to Z^* + C + \ldots$$
$$\qquad \qquad \quad \rightharpoondown e^+ \nu V^o + \ldots \tag{2.31}$$

is compatible with a charmed particle mass

$$m_C \gtrsim \left\{\begin{array}{c} 1.9 \\ 1.6 \end{array}\right. \text{GeV} \quad \text{for} \quad C = \left\{\begin{array}{c} \text{baryon} \\ \text{meson} \end{array}\right. . \tag{2.32}$$

Using theoretical cross section estimates [32], the conclusion is made that the observed event rate is

compatible with the production and decay of a charmed
baryon with [35]

$$m_C \simeq (2.5 \text{ to } 4) \text{ GeV}, \qquad B_e \gtrsim 10\% \quad . \qquad (2.33)$$

Provided the charmed baryons are not excessively heavy
as compared with charmed mesons, one expects that at
low energies the dominant charm changing process should
be elastic baryon production. For equal baryon masses
the production of the $I = 0$, $1/2^+$ state C_O^+ is favored;
the production ratios are roughly: [32]

$$C_O^+ : C_1^+ : C_1^{+*} : C_1^{+**} \simeq 1 : 0.2 : 0.4 : 0.35 : 0.7 ,$$
$$(2.34)$$

were $0,1$ is the total isospin and the asterisk denotes
a $3/2^+$ state. Moreover, the C_O^+ is expected to be the
lightest state, which will further enhance its relative
production rate near threshold. Semi-leptonic decays
with $\Delta S = \Delta C = \pm 1$ satisfy the isospin selection rule
$\Delta I = 0$, so the decay of the C_O^+ must be to an $I = 0$
hadronic state

$$C_O^+ \rightarrow \ell^+ \nu \quad + \quad \begin{cases} \Lambda^0 \\ Y_1^0 \rightarrow \Sigma \ \pi \\ \cdot \\ \cdot \\ \cdot \end{cases} \qquad (2.35)$$

The lowest available state is significantly favored by
phase space which has an $(m_C - m_Y)^5$ dependence; for
$m_C \simeq 2.5$ GeV we find

$$(m_C - m_\Lambda)^5 / (m_C - m_{Y_1^0})^5 \simeq 3 \quad . \qquad (2.36)$$

There is no reliable way to estimate the Y_1^0 coupling, but in the static SU(8) (= spin and SU(4)) limit, all couplings vanish for transitions to states other than the ground state $1/2^+$ and $3/2^+$ baryons. As the Λ^0 is the only I = 0 member of the 56-plet, a preponderance of Λ^0's in the final state for charm production near threshold would not be surprising. On the other hand, the $(\Delta m)^5$ phase space factor enhances the otherwise Cabibbo suppressed $\Delta S = 0$ decay; one estimates [37]:

$$\Gamma (C_0^+ \to n\ell\nu)/\Gamma (C_0^+ \to \Lambda\ell\nu) \simeq 1/4 \ . \tag{2.37}$$

Calculations of the Λe invariant mass spectrum [37] for $C_0^+ \to \Lambda e\nu$ show a peaking toward the lower end of the spectrum, which when compared with the observed values, Eq. (2.30) suggest that $m_{C_0^+} > 2.2$ GeV, in contrast with possible interpretations of the $\Delta S=-\Delta Q$ event discussed above.

2.2.2. High energy ($\mu e K_s$) events

The four published WBH events [36] show a high multiplicity except for one event in which only ($\mu^- e^+ K_s$) is identified in the final state. The visible energies in 34, 11, 26, and 28 GeV - as opposed to 3.1 and 3.7 GeV for the published Gargamelle events. The WBH events should thus be more characteristic of deep inelastic scattering, with the scattered charmed parton being kicked out of the nucleon (fig. 2). In this case one expects that the charm quantum number is more likely to be carried by a meson in the final state; then an abundance of kaons can be understood in terms of meson decay

$$\nu + N \to \mu^- + D + \dots$$

$$\bar{K} e^+ \nu + \dots \qquad . \qquad (2.38)$$

What is puzzling in these events however is the absence of $\mu^- e^+$ events without an identified K_s. If the ν is scattered from a valence quark one expects $\bar{K}^0$ and K^- to occur in roughly equal ratios. Since only half of the $\bar{K}^0$ are K_s, and only 2/3 of these have visible decays, we expect that on the average 1/6 of the events will contain an identified K_s. For charm production from a sea quark, there will be an additional kaon (K^+ or K^0) at the production, and one estimates that about a third of the events will contain a K_s. Allowing for 15% associated production of kaons in the recoiling hadron system still only gives:

$$\sigma(K_s)/\sigma(\text{charm}) \approx (27-29)\% \text{ for } r_s = 0.05 - 0.1 . \qquad (2.39)$$

Hopefully the present data is subject to a statistical or scanning fluke; otherwise they present a puzzle for any interpretation.

The observed event rate [36] corresponds to a relative cross section of about 1%. The transverse momenta of the e^+ are consistent with a three body decay of a particle with mass about 2 GeV. Assuming the K_s and e^+ both come from a decay, their mean invariant mass:

$$m(K_s e^+) = (1.3, 0.4, 1.1 \text{ and } 0.7) \text{ GeV}$$

suggests a mass of 1.7 for a three body decay and
m > 1.7 for a more complicated decay.

2.4 CONCLUSIONS

Other quarks? Other currents? We have seen that
the gross features of the neutrino data are in qualitative agreement with the GIM model. In particular,
dilepton data are compatible with

i) sea production for $\bar{\nu}$ and valence + sea production
for ν,

ii) a semi-leptonic branching ratio in the theoretically acceptable range

iii) a preponderance of associated strange particles
in the final state.

However a variety of motivations have led to proposals [38],[39] for extending the number of quarks,
as well as proposals that the additional quarks [39]
and/or the charmed quark [40] couple via their right
handed components to ordinary quarks. Some possible
right handed couplings are:

$$(\bar{c}d)_R, \ (\bar{c}s)_R, \ (\bar{t}d)_R, \ (\bar{u}b)_R \qquad (2.40)$$

where t ("top") and b ("bottom") are new quarks of
charge +2/3 and -1/3, respectively. We shall not discuss the merits of these models, but simply comment on
their possible relevance to the preceeding analyses.

a) If the $(\bar{c}d)_R$ coupling is comparable to the $(\bar{c}s)_L$

coupling, a dominant $(1-y)^2$ component would contribute
to the valence dilepton distribution, which probably
is consistent with the data discussed above. However,
semi-leptonic decays of charmed particles would occur
equally often via $|\Delta S| = 1$ and $|\Delta S| = 0$ transitions;
this possibility seems disfavored by the $\mu^- e^+$ data.
It predicts a much larger $D^0 \leftrightarrow \bar{D}^0$ mixing which could
account for the same sign dimuon events, but by the
same token, it leads to too larger a value for the K_L,
K_S mass difference.

b) The coupling $(\bar{c}s)_R$ would introduce a $(1-y)^2$ component into the sea contribution to both ν and $\bar{\nu}$ dilepton
distributions. This is certainly compatible with the
data, improving the fit to the ν dilepton y distribution, although it is in the wrong direction to help
the high $<y>$ problem in single muon $\bar{\nu}$ data.

c) The coupling $(\bar{t}d)_R$ has the same features as $(\bar{c}d)_R$
except that the $\mu^- e^+$ difficulty is avoided if the
t threshold is above 30 GeV.

d) The $(\bar{u}b)_R$ coupling introduces a valence component,
with a constant y dependence, into the $\bar{\nu}$ distribution,
and sufficiently above threshold $\sigma^{\bar{\nu}} \simeq \sigma^{\nu}$ if the coupling strength is the usual one. These effects have not
been observed, but again can be avoided if the threshold is sufficiently high, with threshold cut offs
confining b production to the high y, low x region.

Quantiative analyses [8], [21] of these models
have shown improved fits as compared with GIM (but the
data are far from conclusive). For example Barnett [8]
concludes that a model including the couplings b)-d)
fit the inclusive data well with

$$m_t \simeq (2-4) \text{ GeV}, \ m_b \simeq (4-5) \text{ GeV} \qquad (2.41)$$

and assuming $m_c \simeq 1.5$ for the threshold modified distributions (Eq. (1.24)). It is not clear that this t mass is sufficiently high to escape significant top production in the Fermilab bubble chamber (there is no Cabibbo or sea suppression factor) but perhaps this difficulty is no more acute than the absence of K_s-less $\mu^- e^+$ events already encountered in the simple GIM model. On the other hand, negative searches [41] for further narrow resonances at SLAC appear to rule out a $(\bar{t}t)$ state in the SPEAR energy range, implying:

$$m(\bar{t}t) \gtrsim 7.6 \text{ GeV, i.e. } m(t) \gtrsim 4 \text{ GeV} . \qquad (2.42)$$

Since the $b\bar{b}$ quark coupling to the electromagnetic current is reduced by a factor 4 in cross section, a $(b\bar{b})$ state with mass above about 5 GeV cannot be ruled out by the present data. There is in fact data [42] on e^+e^- pair production by **protons** which possibly indicates the presence of a narrow state at about 6 GeV. This could be a possible candidate for $b\bar{b}$ giving a mass $m_b \simeq 3$ GeV, m (bottom 0^- states) $\simeq 3-4$ GeV. One might worry that this gives a threshold too low to allow for a (non-Cabibbo suppressed) right handed coupling to the u quark (see d)).

If there are new, heavier quarks, but exclusively V-A weak couplings, then in the context of the Weinberg-Salam-GIM model they must couple only weakly (with mixing angle [25] $\lesssim \theta_c$) to ordinary quarks and present data are not sensitive to their effects.

FIGURES

1. Kinematic variables in deep inelastic scattering.

2. Parton model picture of deep inelastic scattering.

3. Expected x and y distributions for neutrino production of charmed particles [6].

4. Diffractive mechanism for neutrino scattering.

5. (x,y) and (E_ν, W) distributions for diffractive
   production of charmed particles [6].

6. Relative ν cross sections for elastic diffractive
   production of vector mesons. The contribution to
   the total diffractive cross sections for each vector
   meson is given by

   $$\sigma_{tot}^{diffr.}/\sigma_{el.}^{diffr.} \simeq 0.4\pi b \text{ mb GeV}^2/\sigma_{tot} \quad (VN)$$

   (Gaillard et al. [10]).

7. Strong interaction corrections to a) semi-leptonic
   and b) non-leptonic quark decay.

8. Comparison [21] of GIM predictions in the parton
   model (using θ-function threshold correction) for
   x and y distributions with experimental $x_{obs.}$ and
   $y_{obs.}$ distributions for dilepton events.

9. Some comparison [21] for $v^\pm$ and $p^\pm$ distributions.

10. Predicted and experimental distributions [22] of μ^+
    transverse momentum in ν dilepton events. L and R
    are for V-A and V+A coupling respectively.

11. Experimental x distributions [27] for Fermilab neu-
    trino scattering compared with expected distributions
    based on SLAC electroproduction data [3]. (Fig. taken
    from D. Cline, Proc. "La physique du neutrino à haute
    énergie" (CNRS, Paris, 1975), p. 107). The dashed
    lines reflect the uncertainty in the antiparton (low
    x) distribution.

12. Experimental [27] y distributions for different reg-
    ions in x and in the neutrino energy. (Fig. taken
    from C. Rubbia, paper presented at Inst. of Particle
    Physics Int'l Summer School (Montreal, 1975)).

13. Expected y distributions above charm threshold under
    the assumptions a)-c) for case i) (dots) and case
    ii) (squares).

14. Observed [27] values for <y> as a function of neu-
    trino energy compared with expected curves for fixed
    values of $B = (q-\bar{q})/(q+\bar{q})$.

REFERENCES

1. S.L. Glashow, J. Iliopoulos and L. Maiani, Phys. Rev.
   D2, 1285 (1970).

2. G. Miller et al., Phys. Rev. D5, 528 (1972);
   J.S. Poucher et al., Phys. Rev. Letters 32, 118 (1974).

3. Gargamelle Collaboration; see D.C. Cundy, Proc. XVII
   Int'l Conf. on High Energy Physics, p. IV-131 (London,
   1974).

4. For a review, see: C.H. Llewellyn Smith, Phys. Reports
   3C, 261 (1972).

5. A. De Rujula, H. Georgi, S.L. Glashow and H.R. Quinn,
   Revs. Modern Phys. 46, 391 (1974);
   G. Altarelli, N. Cabibbo and L. Maiani, Phys. Letters
   48B, 435 (1974);
   M.K. Gaillard, B.W. Lee and J.L. Rosner, Revs. Modern
   Phys. 47, 277 (1975);
   V. Barger, T. Weiler and R.J.N. Phillips, "Charm pro-
   duction by partons in neutrino scattering", University
   of Wisconsin-Madison report COO-441 (1975).

6. B.W. Lee, "Dimuon events", Fermilab-Conf-75/78-THY
   (1975).

7. H. Georgi and H.D. Politzer, "Freedom at moderate
   energies: masses in color dynamics", Harvard preprint,
   January 1976.

8. R.M. Barnett, "Evidence in neutrino scattering for
   right-handed currents associated with heavy quarks",
   Harvard preprint, January 1976.

9. E. Dermin, "Dimuon distributions due to deep inelastic
   neutrino production of new hadrons", Oxford preprint
   (1976).

10. C.-A. Piketty and L. Stodolsky, Nuclear Phys. B15,
   571 (1970);
   T. Inami, Phys. Letters 56B, 291 (1975);
   B.A. Arbuzov, S.S. Gerstein and V.N. Folomeshkin,
   preprint IHEP 75-11 Serpukhov (1975);
   B.A. Arbuzov, S.S. Gerstein, V.V. Lapin and V.N.
   Folomeshkin, preprint IHEP 75-25, Serpukhov (1975);
   J. Pumplin and W. Repko, Phys. Rev. D12, 1376 (1975);
   V. Barger, T. Weiler and R.J.N. Phillips, "Vector
   meson dominance calculations for the $\bar{\nu}$ anomaly and
   dimuon production", Wisconsin preprint COO-456 (1975);

and Phys. Rev. (to be published);
M.K. Gaillard, S.A. Jackson and D.V. Nanopoulos,
Nuclear Phys. B102, 326 (1976);
M.B. Einhorn and B.W. Lee, "Contributions of vector-
meson dominance to charmed meson production in in-
elastic neutrino and antineutrino interactions",
FERMILAB-PUB-75/76-THY (1975).

11. J.J. Sakurai and D. Schildknecht, Phys. Letters 40B,
121 (1972);
41B, 489 (1972);
42B, 216 (1972).

12. G.'t Hooft, unpublished;
D. Gross and F. Wilczeck, Phys. Rev. D8, 3633 (1973);
H.D. Politzer, Phys. Rev. Letters 30, 1346 (1973);
M. Gell-Mann and H.Leutwyler, Caltech preprint CALT-
68-409 (1973).

13. K. Berkelman, Proc. 1971 Int'l Symp. on electron and
photon interactions at high energies, Cornell Univ.,
Ithaca, Ed. N.B. Misty, p. 263 (1971);
B. Knapp et al., Phys. Rev. Letters 34, 1040 (1975).

14. D.R. Yennie, Phys. Rev. Letters 34, 239 (1975).

15. S. Weinberg, Phys. Rev. Letters 18, 507 (1967).

16. M.K. Gaillard and B.W. Lee, Phys. Rev. Letters 33,
108 (1974);
G. Altarelli and L. Maiani, Phys. Letters 52, 351
(1974).

17. J. Ellis, M.K. Gaillard and D.V. Nanopoulos, Nuclear
Physics B100, 313 (1975).

18. A. Benvenuti et al., Phys. Rev. Letters 34, 419 (1975);
ibid 35, 1199, 1203 and 1249 (1975).

19. Dimuon events also have been reported by the Cal-Tech-Fermilab group: B.C. Barish, Proc. la physique du neutrino à haute énergie, 18-2o March 1975, Ecole Polytechnique, Paris.

20. A. Pais and S.B. Treiman, Phys. Rev. Letters $\underline{35}$, 1206 (1975).

21. V. Barger, R.J. N. Phillips and T. Weiler, "Dimuon production by neutrinos: test of weak current models", SLAC-PUB-1688 (1976).

22. L.M. Sehgal and P.M. Zerwas, Phys. Rev. Letters $\underline{36}$, 399 (1976).

23. M.K. Gaillard, et al., Ref. 5).

24. R.L. Kingsley, S.B. Treiman, F. Wilczeck and A. Zee, Phys. Rev. $\underline{D11}$, 1919 (1975).

25. J. Ellis, M.K. Gaillard and D.V. Nanopoulos, Ref. TH.2116-CERN (1976).

26. M. Bourquin and J.-M. Gaillard, to be published, see also Ref. 30).

27. B. Aubert et al., Phys. Rev. Letters $\underline{33}$, 984 (1974); A. Benvenuti et al., ibid $\underline{34}$, 597 (1975); A. Benvenuti et al., "Further data on the high-y anomaly in inelastic antineutrino scattering", preprint HPWF-76/1.

28. For a recent review, see: C.H. Llewellyn Smith, Proc. 1975 Int'l Symposium on lepton and photon interactions at high energies, Ed. W.T. Kirk, p. 709 (SLAC, Stanford, 1976).

29. Y. Watanabe et al., Phys. Rev. Letters $\underline{35}$, 898 (1976); C. Chang et al., ibid, 901 (1976).

412

30. D. Sivers, "Estimating cross-sections for the pro-
    duction of new particles", Rutherford Lab. preprint
    RL-75-171, T. 142 (1975).

31. E.G. Cazzoli et al., Phys. Rev. Letters $\underline{34}$, 1125
    (1975).

32. R.E. Shrock and B.W. Lee, Fermilab-PUB-75/80-THY
    (1975);
    see also J. Finjord and F. Ravndal, Phys. Letters
    $\underline{56B}$, 61 (1975).

33. B.W. Lee, "A comment on the BNL event $\nu p \to$
    $\mu^- \Lambda^0 \pi^+ \pi^+ \pi^+ \pi^-$", Fermilab-75/38-THY (un-
    published).

34. J.P. Berge et al., Phys. Rev. Letters $\underline{36}$, 127
    (1976).

35. G. Blietzhau et al., Phys. Letters $\underline{60B}$, 207 (1976),
    and private communications from members of the Gar-
    gamelle collaboration.

36. J. von Krogh et al., "Observation of $\mu^- e^+ K_S^0$ events
    produced by a neutrino beam", Wisconsin-Berkeley-
    CERN-Honolulu preprint (1975).

37. A.J. Buras, CERN preprint TH.2142-CERN (1976).

38. The observation of μe events in $e^+ e^-$ annihilation at
    SLAC (M.L. Perl et al., Phys. Rev. Letters $\underline{35}$, 1489
    (1975)), and their most probable interpretation in
    terms of heavy lepton production and decay, require
    the introduction of a new quark pair if the cancellat-
    ion of triangle anomalies (see, for example, C.Bouchiat,
    J. Iliopoulos and Ph. Meyer, Phys. Letters $\underline{38B}$, 519
    (1975)) is to be maintained in the context of the
    minimal Weinberg-Salam model (S. Weinberg, Phys. Rev.

Letters 19, 1264 (1967); A. Salam, Proc. 8th Novel
Symposium, Stockholm p.367 (1968) (Almqvist and
Wiksells, Stockholm, 1968)) with only V-A couplings.
The extension of this model to more than four quarks
also allows for the introduction of CP violation into
the theory (M. Kobayashi and K. Maskawa, Progr.Theor.
Phys. 49, 652 (1973). See also S. Pakvasa and H.
Sugawara, University of Hawaii preprint UH-511-204-
75 (1975); L. Maiani, Rome preprint P. 75/10 (1975)
and Ref. 25).

39. F.A. Wilczek, A. Zee, R.L. Kingsley and S.B. Treiman,
Phys. Rev. D12, 2768 (1975);
H. Fritzsch, M. Gell-Mann and P. Minkowski, Phys.
Letters 59B, 239 (1975);
S. Pakavasa, W.A. Simmons and S.F. Tuan, Phys. Rev.
Letters 35, 702 (1975);
A. De Rujula, H. Georgi and S.L. Glashow, "Vector
model of weak interactions", Harvard preprint
(1975).
The primary motivation for introducing new quarks
with right-handed couplings is that one can construct
a theory which is parity conserving in the limit of
quark mass degeneracy. Parity violation arises spontaneously through the quark mass matrix generated by
the Higgs coupling. The essential prediction of these
theories is that the hadronic neutral current is parity
conserving.

40. A. De Rujula, H. Georgi and S.L. Glashow, Phys. Rev.
Letters 35, 69 (1975).

41. R.F. Schwitters, Proc. 1975 Int'l Symposium on lepton
and photon interactions at high energies, Ed. W.T.

Kirk (SLAC, Stanford, 1976), p. 5.

42. D.C. Hom et al., "Observation of high mass dilepton pairs in hadron collisions at 400 GeV", Columbia-Fermilab-Stony Brook preprint.

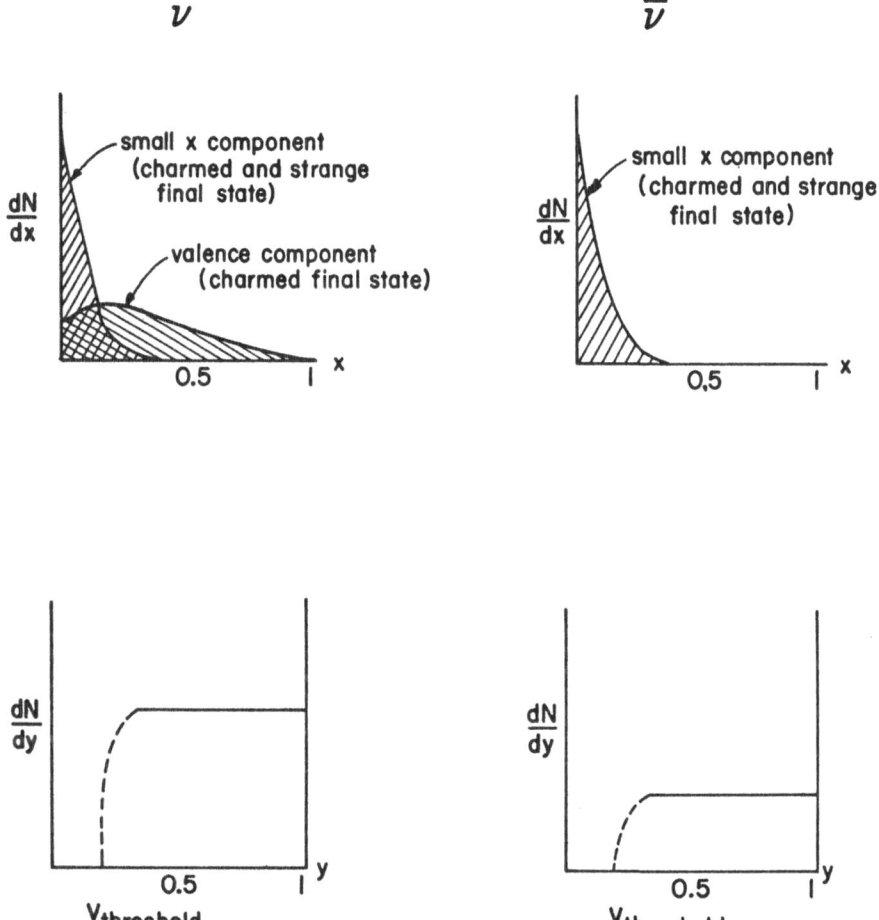

FIG. 3

416

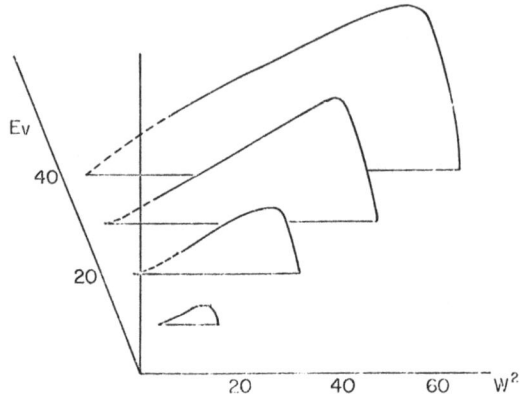

FIG. 5a

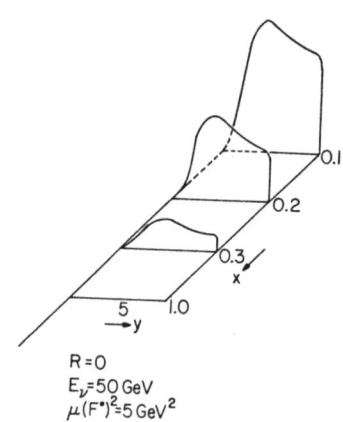

R = 0
E_ν = 50 GeV
$\mu(F^*)^2$ = 5 GeV^2

FIG. 5b

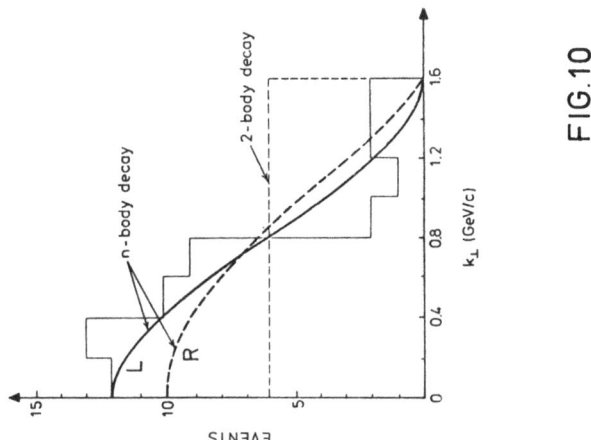

FIG.10

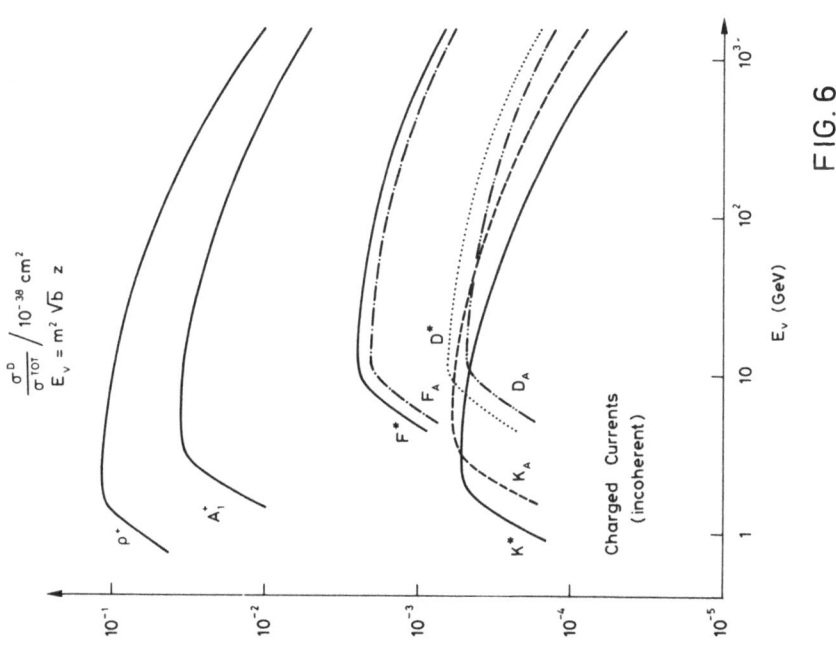

FIG. 6

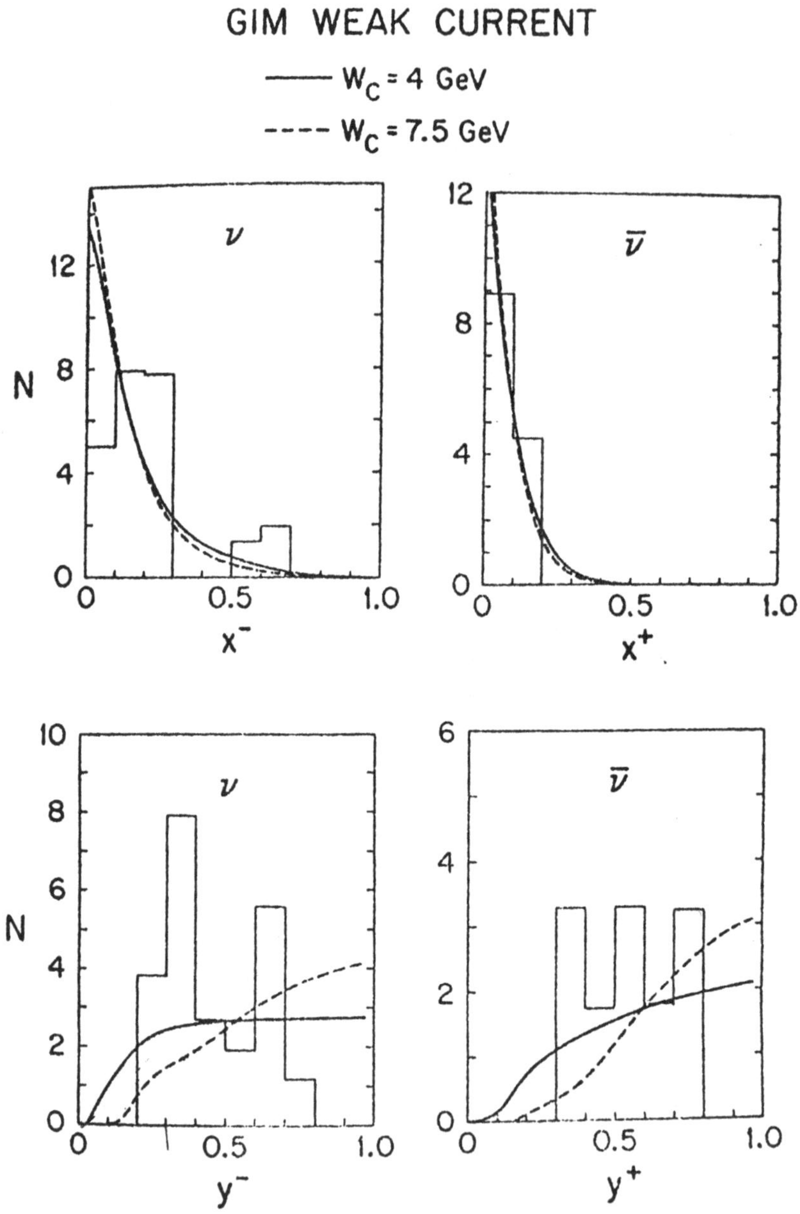

GIM WEAK CURRENT

—— $W_C = 4$ GeV

---- $W_C = 7.5$ GeV

FIG.8

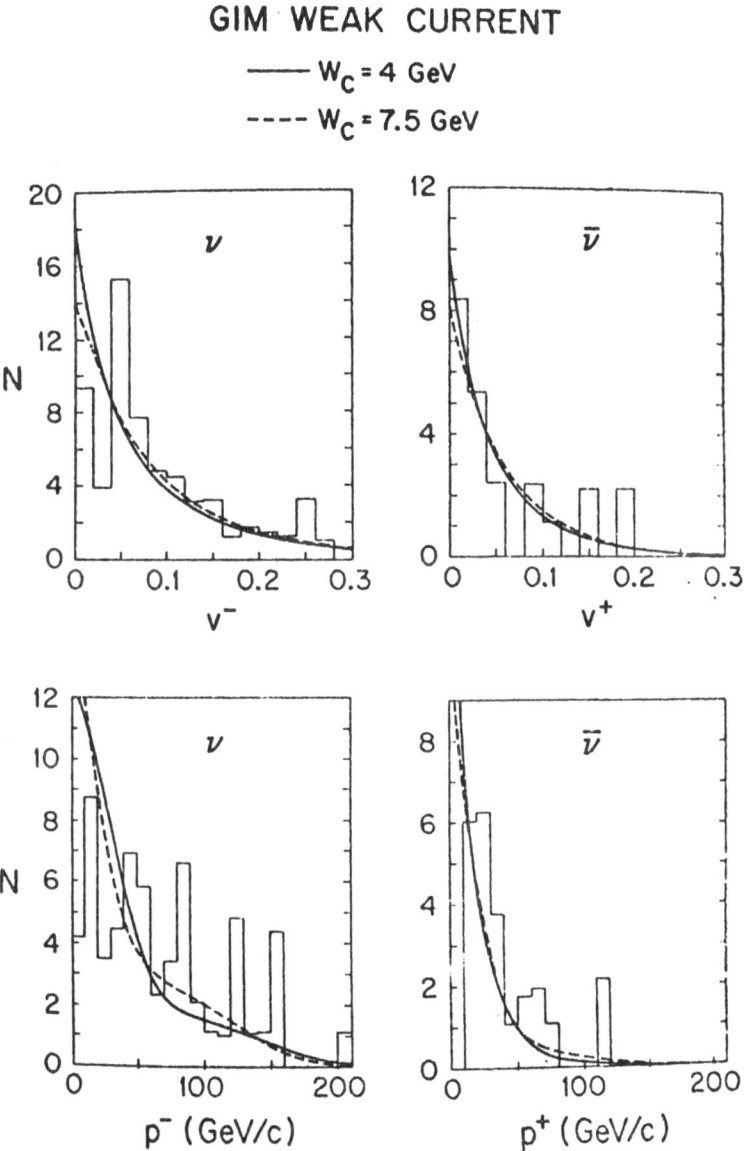

FIG. 9

420

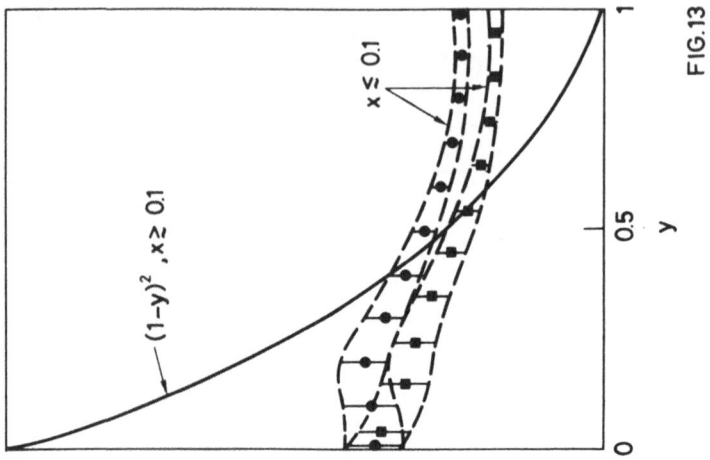

FIG.13

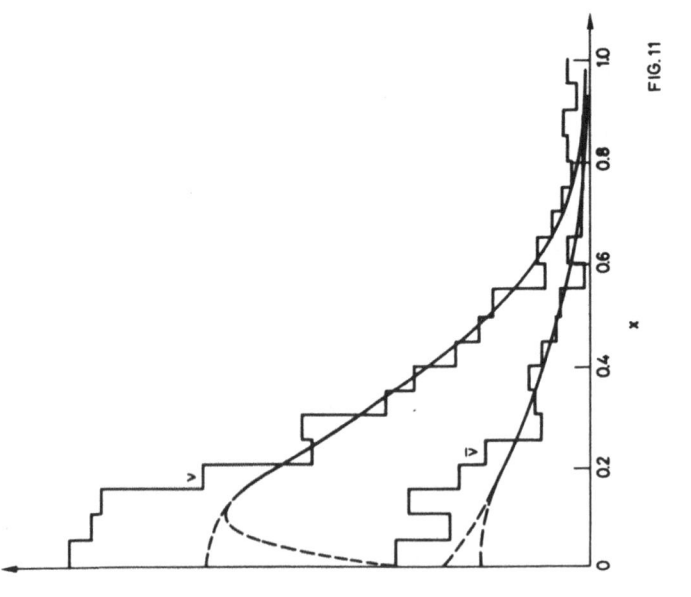

FIG.11

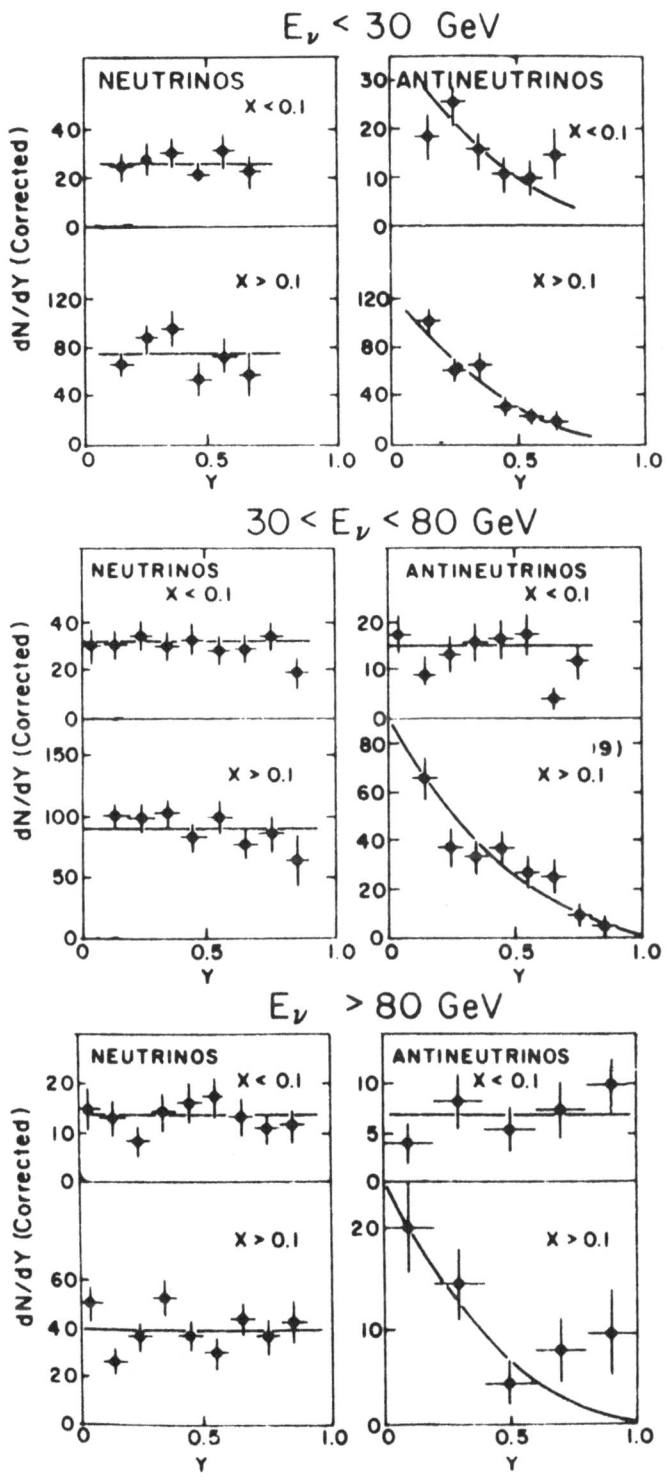

FIG.12

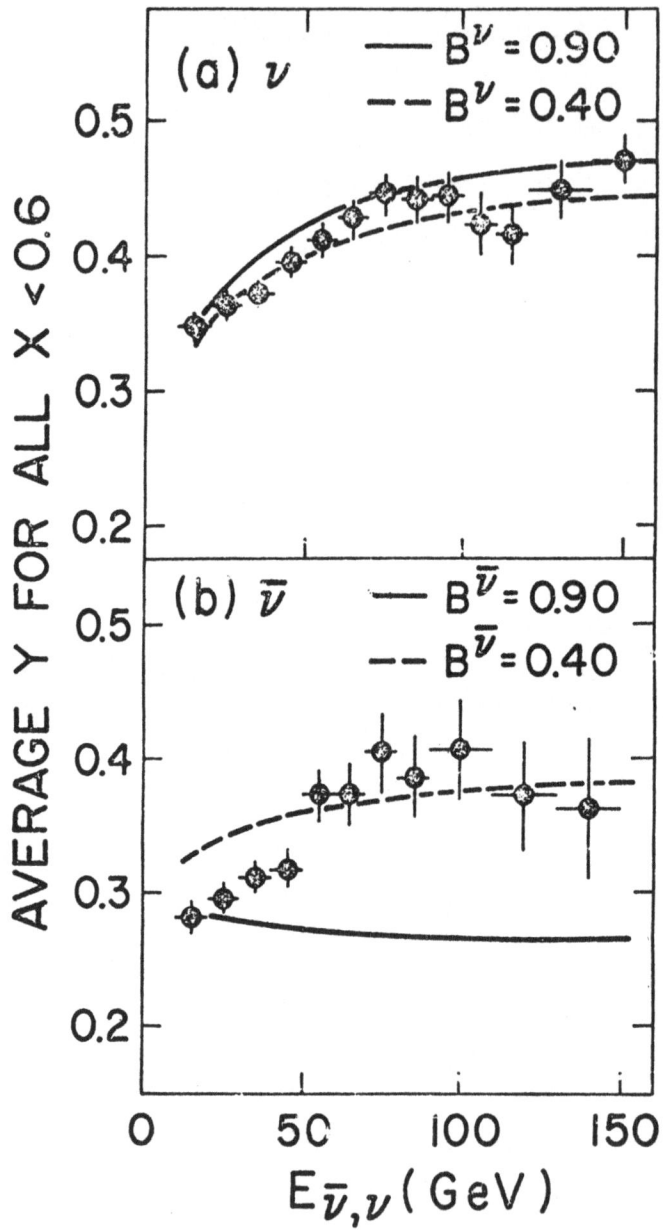

FIG. 14

Acta Physica Austriaca, Suppl. XV, 423–473 (1976)
© by Springer-Verlag 1976

RENORMALIZATION OF NONABELIAN GAUGE FIELDS[+]

by

W. KUMMER
Institut für Theoretische Physik
Technische Universität Wien

1. INTRODUCTION

Since about 1974 gauge-theories become something
like a "party-line" in theoretical elementary particle
physics. This means that a wide-spread belief links most
of our basic ideas with the concept of a local gauge-invariance of nature, although no definite proof exists
so far that gauge fields are really there - except in
quantum electrodynamics (and gravity?) Admittedly the
amount of supporting facts is quite impressive. It ranges
from a possible unification of weak and electromagnetic
interactions in such a scheme [1] to a number of qualitative, but rather convincing arguments in the field of
strong interaction physics, if the latter is based on a
field theory of gluon gauge fields. We cite "asymptotic
freedom", which could explain the almost perfect scaling

[+] Lecture given at XV. Internationale Universitätswochen
für Kernphysik,Schladming,Austria,February 16-27,1976.

of deep electroproduction [2], the enhancement of
$\Delta I = \frac{1}{2}$ - amplitudes in weak nonleptonic decays [3] and,
as an example for a more theoretical result, the persist-
ent ultraviolet divergence of radiative corrections to
all orders in the coupling of a gluon-field to matter
[4].

At the present time it has become a rather hopeless
task review within a lecture of three hours the models of
weak interactions and electromagnetism, let alone the
hadronic applications. This is typical for a situation
in which theory, unimpeded by too much precise experimental information, is stampeding into still somewhat
undetermined directions, as far as specific models of
nature are concerned.

Therefore, this report is restricted to that
aspect of the new developments which certainly will
survive: a fundamental enrichment of our knowledge, how
to handle field-theory, brought about by the complexity
of the quantization problem for nonabelian gauge-fields.
The latter is characterized by the appearance of ghosts,
unphysical "particle" states, which are an immediate consequence of the fact that the quantization must be done
in terms of a fixed gauge [5]. Most of the extensive work
of the last years refers to this general case. Already
in ref. [2], but in an even more crucial situation in
ref. [4], it was recognized that a peculiar "axial"
gauge [6] which is ghost-free [7] leads to remarkable
simplifications. Because of its "singular" nature, this
gauge has been held in low esteem for a long time. Therefore, only very recently in refs. [8], [9], [10] the
quantization and renormalization of nonabelian gauge
theories has been carried through by W. Konetschny and

the author in this gauge using modern methods. It turns out that - due to the absence of "Fadeev - Popov" (F.P.) ghosts - the steps are simplified appreciably, the logical development, however, remaining identical to the general case. This greatly facilitates the pedagogical presentation which is attempted here.

2. CLASSICAL NONABELIAN GAUGE THEORY

A local gauge transformation is defined for a vector field A^α_μ (μ refers to space and time, α is the index of the internal symmetry in the regular representation) as

$$A^\alpha_\mu \to A'^\alpha_\mu = A^\alpha_\mu + D^{\alpha\beta}_\mu \, \delta\omega^\beta(x) \qquad (2.1)$$

in terms of a gauge function $\delta\omega^\beta(x)$ and a "covariant derivative"

$$D^{\alpha\beta}_\mu = \delta_{\alpha\beta}\partial_\mu - g \, f_{\alpha\gamma\beta} \, A^\gamma_\mu \qquad (2.2)$$

with the coupling constant g and the structure constants $f_{\alpha\gamma\beta}$ of the group [11]. Matter fields (ψ = fermions, S = scalar) transform homogeneously, e.g.

$$S^\alpha \to - g \, t^\gamma_{\alpha\beta} \, \delta\omega^\gamma(x) \, S_\beta \qquad (2.3)$$

where t denotes the generator of the group in the appropriate representation of the matter field (e.g. t^γ are the Pauli-matrices for an isospinor in SU(2) etc.). Then in the covariant derivative (2.2) also $f_{\alpha\beta\gamma}$ is replaced by t. The renormalizable Lagrangian, which is invariant under (2.1) and (2.3), reads [12]

$$L_{inv} = -\frac{1}{4} F^{\alpha}_{\mu\nu} F^{\mu\nu\alpha} + \bar{\psi}(i\not{\partial} - m)\psi + (DS)^{\dagger}(DS) -$$

$$- V(S) - V_Y(S,\psi) \quad . \tag{2.4}$$

Here

$$F^{\alpha}_{\mu\nu} = \partial_{\mu} A^{\alpha}_{\nu} - \partial_{\nu} A^{\alpha}_{\mu} - g f_{\alpha\gamma\beta} A^{\beta}_{\mu} A^{\gamma}_{\nu} \tag{2.5}$$

and V and V_Y are a gauge-invariant polynomial of degree ≤ 4 in the scalar fields and the Yukawa-coupling, respectively. Formal manipulations are greatly facilitated by the compact notation [13]

$$\phi_i = \begin{cases} A^{\alpha}_{\mu}(x) \\ S^{\alpha}(x) \\ \psi^{\alpha}(x) \end{cases}$$

where i refers to the internal symmetry, the spin index and the space-time variable at the same time [14]. Thus, (2.2) and (2.3) become

$$\phi_i \rightarrow \phi'_i = \phi_i + D^{\alpha}_i(\phi) \, \delta\omega^{\alpha} \tag{2.6}$$

$$D^{\alpha}_i(\phi) = \Delta^{\alpha}_i + g \, t^{\alpha}_{ij} \, \phi_j \tag{2.7}$$

with Δ_i^α non zero for i = vector-field only, explicitely

$$\Delta_i^\alpha = \partial_\mu^{(x)} \delta(x-y)\delta_{\alpha\beta} \qquad\qquad i = (\beta,\mu,x)$$

$$\alpha = (\alpha,y) \ . \qquad\qquad (2.8)$$

In a similar way, t_{ij}^α is proportional to the matrix of
the generator in the representation i and j and to two
delta-functions, which make the space-time variables of
α, i and j coincide. A short-hand notation for the action
from (2.4) is correspondingly

$$L_{inv} = \int L_{inv} \ d^4x = \sum_{n=2}^{4} \frac{1}{n!} \ \Gamma_{i_1 \ldots i_n}^{(o)} \ \phi_{i_1} \ldots \phi_{i_n} \ . \ (2.9)$$

Each $\Gamma_{i_1 \ldots i_n}^{(o)}$ contains n-1 δ-functions so that the action
becomes the integral of a linear combination of the local
expressions in the fields (2.4).

Spontaneous symmetry breaking "distorts" the original gauge-symmetry, if V(S) has an absolute minimum
at S = $\varepsilon \neq 0$; then some or all of the vector fields
acquire a mass [15]. In order to avoid complications
from infra-red divergencies we assume that all vector-
particles acquire masses except for those related to
an abelian subgroup [16]. Quantum-mechanically this
means that the field ϕ_i has be taken in relation to
the new ground-state:

$$\phi_i' = \phi_i - \varepsilon_i \ , \qquad\qquad <0|\phi_i'|0> = 0 \ . \qquad\qquad (2.10)$$

The changes in the covariant derivatives can be simply

absorbed in a redefined

$$\Delta_i'^{\alpha} = \Delta_i^{\alpha} + \mu_i^{\alpha} \qquad (2.11)$$

with a nonzero

$$\mu_i^{\alpha} = g \; t_{ij}^{\alpha} \; \varepsilon_j \qquad (2.12)$$

for i = scalar fields only. The "direction" i in the space of scalar fields corresponds to the scalar field which will be absorbed as the longitudinal component in the vector field α [15]. We shall call it a Higgs-ghost.

3. CANONICAL QUANTIZATION, GAUGES

Whereas - in the ordinary notation - the space parts A_i^{α} of the vector field A_{μ}^{α} possess a canonically conjugate variable

$$\Pi_i^{\alpha} = \frac{\partial L}{\partial \dot{A}^{i\alpha}} = - F_{oi}^{\alpha} \; , \qquad (3.1)$$

one finds $\Pi_o^{\alpha} = 0$ and that A_o^{α} is not expressible in terms of (A_i, Π_i). Therefore a gauge-breaking term L_b must be added to L_{inv} in order to make the quantization possible. We already need it in the free part of L, therefore L_b should be at most quadratic to zero order in the coupling. The only covariant gauge with this properties, involving A_{μ}^{α} alone, is the

Fermi-gauge

$$L_b^{(1)} = - \frac{1}{2\alpha} (\partial^\mu A_\mu^\alpha)^2 \qquad\qquad (3.2)$$

with the Landau-gauge $\alpha = 0$ and the Feynman-gauge $\alpha = 1$ as special cases.

The radiation-(Coulomb-) gauge results for

$$L_b^{(2)} = - \frac{1}{2\alpha'} (\partial^i A_i^\beta)^2 \ . \qquad\qquad (3.3)$$

For a noncovariant gauge, characterized by an "axis" in R_4 [6], [7]

$$L_b^{(3a)} = - \frac{1}{2\beta^{(a)}} (n^\mu A_\mu^\alpha)^2 \qquad\qquad (3.4a)$$

or

$$L_b^{(3b)} = - \frac{1}{2\beta^{(b)}} (\partial^\nu n^\mu A_\mu^\alpha)^2 \qquad\qquad (3.4b)$$

or

$$L_b^{(3c)} = - \frac{1}{2\beta^{(c)}} [(n\partial)(n^\mu A_\mu^\alpha)]^2 \ . \qquad\qquad (3.4c)$$

(3.4b) or (3.4c) are more useful than (3.4a): Because the latter has dimension 2 in the fields, it behaves like a mass-term of a vector field and, therefore, leads to a propagator which does not converge in the ultra-violet limit.

It is clear that all these gauges are included in the compact expression

$$L_b = \int L_b \, d^4x = -\frac{1}{2} (F_i^\alpha \, \phi_i)^2 \, , \qquad (3.5)$$

where also scalar fields may appear among the ϕ_i. In the canonical procedure, e.g. in the case (3.2) at $\alpha = 1$, now

$$\Pi_o = \dot{A}_o - \partial_i A_i$$

follows and $\dot{A}_i$ and $\dot{A}_o$ can be eliminated in favour of Π_i and Π_o in the Hamiltonian density

$$H = A^\mu \Pi_\mu - L \, .$$

In the interaction representation

$$\underline{\Pi}_i = -\underline{A}_i + \partial_i \underline{A}_o$$

one may go back to $\dot{\underline{A}}_i$, $\dot{\underline{A}}_o$. Even in a covariant gauge the resulting interaction Hamiltonian is not Lorentz-covariant and in general even nonpolynomial! The free propagator of a vector-field also contains noncovariant terms. As we know from examples of field-theories, where such calculations actually can be performed, a complicated cancellation leads to an effective

$$- \underline{H}^{eff}_{int} = \underline{L}_{int} + \underline{L}'_{int} \qquad .$$

and covariant ("naive") vector propagators. In this way L'_{-int} is determined.

4. PATH INTEGRAL, FADEEV-POPOV-GHOSTS

Feynman's path integral [19] in its field-theoretic version [20] consists in replacing the operator T exp $\{-i\int H_{-int} \, d^4x\}$ by a path-integral with respect to the fields ϕ_i involving the action L,

$$T \exp [-i\int H_{-int} \, d^4x] \rightarrow \int (d\phi) \, e^{iL(\phi)} \, , \qquad (4.1)$$

where, crudely speaking, $(d\phi)$ is the limit of $\prod_r d\phi(x_r)$ for a discrete lattice in space and time. The fields are c-numbers and to be integrated upon for each box separately.

The generating functional

$$W = e^{iZ} = <0|T \, (e^{-i[\int H_{-int} \, d^4x \, - \, L_s]}) \, |0> \qquad (4.2)$$

with the source-part of the action

$$L_s = \phi_i \, j_i = \int (A^\mu j_\mu + S\rho + \bar{\psi}\eta + \bar{\eta}\psi) \, d^4x \qquad (4.3)$$

for the Green's functions

$$\frac{(-i)^n \, \delta^n W}{W \, \delta j_{i_1} \cdots \delta j_{i_n}} \, \Bigg|_{j=0} = G_{i_1 \cdots i_n} \qquad (4.4)$$

432

becomes formally

$$W' = \int (d\phi) \; e^{i(L_{inv} + L_s)}$$
(4.5)

for a gauge theory. L_s breaks the gauge-invariance, but the derivatives (4.4) contain

$$\int (d\phi) \phi_{i_1} \cdots \phi_{i_n} \; e^{i \, L_{inv}}$$

which diverges on an "orbit", where L_{inv} does not vary. In this way the gauge-difficulty is equally present in the path-integral formalism. We fix the gauge most simply at $\delta \omega^\alpha = 0$ [5], relating the fields and the gauge-transformation by a (linear) gauge-fixing condition

$$f^\alpha (\phi (\delta \omega)) = F_i^\alpha \; \phi_i (\delta \omega) = c^\alpha =$$

$$= F_i^\alpha \; \phi_i + K^{\alpha \beta} \; \delta \omega^\beta$$
(4.6)

$$K^{\alpha \beta} = F_i^\alpha \; D_i^\beta (\phi) \; ,$$
(4.7)

i.e.

$$W_c = \int (d\phi) \; \prod_\alpha \delta (\delta \omega^\alpha) \; e^{i(L + L_s)} \; .$$
(4.8)

From (4.6)

$$\prod_\alpha \delta (\delta \omega^\alpha) = \prod_\alpha \delta (F_i^\alpha \; \phi_i - c^\alpha) \; (\det K) \; .$$
(4.9)

If the observable quantities are independent of the

gauge, it must be permissible to use

$$W = \int (dc)\, e^{iL_b(c)} \qquad W_c = \int (d\phi)\, \det K\, e^{i[L_b(F_i^\alpha \phi_i) + L_{inv} + L_s]}$$

$$(4.10)$$

instead of (4.8). det K can be included in the exponential via [21]

$$\det K \propto \int e^{i\bar{u}_\alpha K_{\alpha\beta} u_\beta} (d\bar{u})\,(du) \qquad (4.11)$$

if u and $\bar{u}$ are anticommuting c-number-functions ("fields") [20]. Hence the auxiliary fields u and $\bar{u}$, the Fadeev-Popov (F.P.) ghosts, behave like scalars, <u>except</u> for a factor (-1) for each closed loop. In physical processes they only appear inside a Feynman graph.

With (4.11), (4.10) becomes (up to an irrelevant constant factor)

$$W = \int (d\phi)\,(du)\,(d\bar{u})\, e^{iL}$$

$$L = L_{inv} + L_b(F_i^\alpha \phi_i) + L_{F.P.} + L_s \,. \qquad (4.12)$$

From (3.2) with (2.8) in the Fermi-gauge $F_i^\alpha = \Delta_i^\alpha$

$$K_{\alpha\beta} = \Delta_i^\alpha D_i^\beta(\phi)$$

and thus

$$L_{F.P.} = \bar{u}_\alpha \partial^\mu (\delta_{\alpha\beta} \partial_\mu - g\, f_{\alpha\gamma\beta} A_\mu^\gamma)\, u_\beta \,, \qquad (4.13)$$

whereas in the axial gauge

$$F_i^\alpha = N_i^\alpha \tag{4.14}$$

(nonzero for (i,β) = vector particle, so that

$$N_i^\alpha \rightarrow n_\mu \, \delta_{\alpha\beta} \, \delta(x_{(i)} - x_{(\alpha)}))$$

$$L_{F.P.}^{(ax)} = \bar{u}_\alpha \, n^\mu \, (\partial_\mu - g \, f_{\alpha\gamma\beta} \, A_\mu^\gamma) \, u_\beta \quad . \tag{4.15}$$

If we choose

$$L_b(c) = - \frac{c^\beta c^\beta}{2\alpha} \tag{4.16}$$

e.g. in the Fermi-gauge (3.2) is obtained.

For "homogeneous" gauges like $\alpha \rightarrow 0$ in (4.16) or $c = 0$ in (4.8),

$$W = \int \delta(c) \, (dc) \, W_c \quad , \tag{4.17}$$

it is convenient to introduce an auxiliary field C^α via [21]

$$\delta(c^\alpha) \propto \int e^{iC^\beta c^\beta} (dC^\alpha) \quad . \tag{4.18}$$

In this case the homogeneous axial gauge has the generating functional

$$W = \int (dC) \, (d\phi) \, e^{iL}$$

$$L = L_{inv} + L_b + L_{F.P.}^{(ax)} + L_s \tag{4.19}$$

$$L_b = C^\alpha N_i^\alpha \phi_i \quad .$$

We note that the interaction term of $L_{F.P.}^{(ax)}$ can be absorbed by a "translation" in the C^α-integration:

$$C^{,\beta} = C^\beta - g \ f_{\alpha\beta\gamma} \ \bar{u}_\alpha \ u_\gamma \quad . \qquad (4.20)$$

Omitting the prime and the <u>field-independent</u> factor from the F.P. ghosts and with sources K^α for C^α we finally arrive at the generating functional for the <u>ghost-free</u> axial gauge [8]

$$W = \int (dC) \ (d\phi) \ e^{iL} \qquad (4.21)$$

$$L = L_{inv} + C^\alpha \ N_i^\alpha \ \phi_i + L_s \qquad (4.22)$$

$$L_s = j_i \ \phi_i + C^\alpha K^\alpha \quad . \qquad (4.23)$$

From the free part of L in (4.22) the propagator of the vector field is easily derived [22],

$$k^{-2} \ [g_{\mu\nu} + \frac{k_\mu k_\nu}{(kn)^2} - \frac{n_\mu k_\nu + n_\nu k_\mu}{(kn)}] \ , \qquad (4.24)$$

there is also a mixed C-A_μ-propagator,

$$k_\mu/(kn) \ , \qquad (4.25)$$

but no C-C-propagation and no C-vertex to the fields ϕ_i.

Whereas the absence of ghosts in the homogeneous **axial gauge has been known for some time** [7] only very

436

recently [47] the observation was made that even in the
inhomogeneous gauge (4.6) with (4.14) the Fadeev-Popov
part (4.15) can be eliminated as well. Using instead of
(4.10) the equally acceptable

$$W = \int (dc) e^{iL_b(c)} \det^{-1} (N_i^\alpha \Delta_i^\beta - g\, f_{\alpha\beta\gamma} c^\gamma) W_c \qquad (4.10')$$

eliminates the det K. A choice like (3.4b) or (3.4c) for
L_b leads to a propagator of type (4.24) with an additional
term

$$k_\mu k_\nu \beta^{(b)}/(kn)^2 \qquad \text{or} \qquad \beta^{(b)} k^2 k_\mu k_\nu/(kn)^4$$

in the square bracket. In this way besides n_μ another
gauge parameter appears. Here we shall follow refs.
[8],[9] and [10] and consider (4.21), (4.22) and (4.23),
because the complications from the additional auxiliary
field are essentially trivial, whereas this gauge has
still a simpler propagator for the vector particle (cf.
chapter 10.), than the one for the general axial gauge.

From (4.24) and (4.25) we notice that a ghost-pole
is still present at (kn) = 0, although the explicit
ghost-fields have been eliminated. It even remains,
if all integrals are done in the Euclidean region
("singular" gauge), but its position depends on n_μ.
Because we are going to show below that all observable
S-matrix elements are independent of n_μ, we anticipate
that the prescription, how to surround this pole in a
complex integration, is quite arbitrary. The most convenient one is a principal value prescription [8]

$$\frac{1}{(kn)} \rightarrow \frac{1}{2} \left(\frac{1}{(kn)+i\varepsilon'} + \frac{1}{(kn)-i\varepsilon'} \right) \ , \ \varepsilon' > 0 \ . \qquad (4.26)$$

This singularity at $(kn) = 0$ must be dealt with at different stages of our approach. First we turn to power-counting: (4.24) does not decrease at $k^2 \rightarrow \infty$, $(kn) = 0$.

However, from the form

$$k^{-2} (g_{\mu\nu} - N_{\mu\nu} \log (\overline{kn}))$$

for the vector-propagator, with

$$N_{\mu\nu} = n^2 \frac{\partial^2}{\partial n_\mu \partial n_\nu} + n_\mu \frac{\partial}{\partial n_\nu} + n_\nu \frac{\partial}{\partial n_\mu}$$

$$(\overline{kn})^2 = (kn)^2 + \varepsilon'^2$$

in any internal line of a Feynman graph - the n_μ can be taken to be different for each vector line, for the moment - in Euclidean space

$$k \rightarrow (i \ K_o, \ \vec{K})$$

$$n \rightarrow (i \ N_o, \ \vec{N})$$

and from

$$\log (KN)^2 \leq \log K^2 + \log N^2 \qquad (4.27)$$

it is clear that Feynman-integrals will contain at worst propagators of the form $\log K^2/K^2$, a situation which is covered by Weinberg's general proof of power

counting [23]. Thus an essential prerequisite for renormalization is available [8].

In order to give a well-defined meaning to all
integrals without spoiling the symmetry of the theory
we shall assume that all quantities are calculated at
a complex dimension d [24]

$$\int d^4k \rightarrow \int d^dk \quad .$$

In the limit d → 4 a superficially divergent integral
with r internal integrations develops a pole of degree
r (at most) at d - 4 = 0.

Then the BPH theorem [25] tells us that power-
counting, together with the fact that all terms in L
are of dimension ≤ 4, guarantee that the addition of
a finite number of counter-terms to L

$$L \rightarrow L' = L^{ren} + L^{counter}$$

is sufficient to make all Green's functions of the theory
finite. However, in the case of a gauge theory, where the
gauge was broken right at the start, it is not obvious
that the renormalized S-matrix is gauge-invariant and
unitary.

5. RELATIONS BETWEEN GREEN'S FUNCTIONS FROM
GAUGE INVARIANCE

5a) Ghostfree case

Unrenormalized, but dimensionally regularized
Green's functions are treated first in the case of the

ghost-free axial gauge. (4.21) is unchanged under any
transformation of the field-integrations. This is
especially true for a gauge-transformation (2.6) which
affects L_b and L_s in (4.22) alone:

$$\frac{\delta L}{\delta \omega^\alpha} = (j_i + N_i^\beta\, c^\beta)\, D_i^\alpha (\phi) \; . \tag{5.1}$$

The integration measure is invariant under such a transformation. Replacing

$$\phi_i \to \frac{1}{i}\frac{\delta}{\delta j_i} \; , \qquad\qquad c^\alpha \to \frac{1}{i}\frac{\delta}{\delta K^\alpha}$$

$\delta W = 0$ becomes

$$N_i^\alpha\, D_i^\beta\, \left(\frac{1\delta}{i\delta_j}\right)\, \frac{\delta W}{i\delta K^\alpha} + j_i\, D_i^\beta\, \left(\frac{1\delta}{i\delta}_j\right)\, W = 0 \; , \tag{5.2}$$

which is an identity of the Slavnow-Taylor-type [26],
but much simpler in this F.P.-ghost-free case [8].
It contains all the information on the guage-invariance
of the original theory; the "equation of motion" of the
auxiliary field c^α is ($c^\alpha \to c^\alpha + \delta c^\alpha$ in W):

$$\int (K^\alpha + N_i^\alpha \phi_i)\, (d\phi)\, (dC)\, e^{iL} =$$

$$= (K^\alpha + N_i^\alpha\, \frac{1}{i}\frac{\delta}{\delta j_i})\, W = 0 \; . \tag{5.3}$$

Differentiating (5.3) at vanishing sources j_i according
to (4.4) we find the expected result

$$N_i^{\alpha} \, G_{i,i_1 \ldots i_n} = 0 \quad . \tag{5.4}$$

This "transversality" with respect to n_{μ} ceases to hold for external C-legs. This is summarized in the derivative of (5.3) with respect to K^{β}

$$N_i^{\alpha} \, \frac{1}{i} \, \frac{\delta^2 W}{\delta j_i \delta K^{\beta}} = - \, \delta_{\alpha\beta} \, W - K^{\alpha} \, \frac{\delta W}{\delta K^{\beta}} \tag{5.5}$$

(5.5), (2.11), (4.14) and the antisymmetry of the structure constants may be used to simplify the first term in (5.2),

$$N_i^{\alpha} \, (\Delta_i^{!\beta} + g \, t_{ij}^{\beta} \, \frac{1}{i} \, \frac{\delta}{\delta j_j}) \, \frac{\delta W}{i\delta K^{\alpha}} \Big|_{K=0} =$$

$$= (N_i^{\alpha} \, \Delta_i^{\beta} + g \, f_{\alpha\gamma\beta} \, N_j^{\gamma} \, \frac{1}{i} \, \frac{\delta}{\delta j_i}) \, \frac{\delta W}{i\delta K^{\alpha}} \Big|_{K=0} =$$

$$= (N\Delta) \, \frac{\delta W}{i\delta K^{\beta}} \Big|_{K=0} \quad .$$

If we divide by $W_0(N\Delta) = W_0(n\partial)$ and differentiate as in (4.4), we arrive at the set of generalized Ward-identities (cf. fig. 1, greek indices refer to external C-legs)

$$G_{\alpha,i_1 \ldots i_n} = \sum_{s=1}^{n} (H_{\alpha,i_s} \, G_{i_1 \ldots i_{s-1} \, i_{s+1} \ldots i_n} + \tag{5.6}$$

$$+ g \, t_{ij}^{\beta} \, h_{\alpha\beta} \, G_{i_1 \ldots i_{s-1}j \, i_{s+1} \ldots i_n})$$

with the "propagators"

$$H_{\alpha,i} = i \ (N\Delta)^{-1}_{\alpha\beta} \ \Delta^{,\beta}_i \qquad (5.7)$$

and

$$h_{\alpha\beta} = i \ (N\Delta)^{-1}_{\alpha\beta} \qquad . \qquad (5.8)$$

For $n = 1$ the result

$$G_{\alpha,i} = H_{\alpha,i} \qquad (5.9)$$

simply means that the "mixed" C-A_μ-propagator - for spontaneous symmetry-breaking the propagator from C to either a vector-field or to a Higgs-ghost (cf. (5.7) and (2.11) - is the same to all orders in perturbation theory.

Renormalization is more readily discussed in terms of one-particle-irreducible (1-p-i) vertex-functions. Their generating functional is a Legendre-transform of $Z = -i\log W$ [27].

Expressing j_i and K^α in terms of new field-variables

$$a_i = \frac{\delta Z}{\delta j_i} \ , \quad z^\alpha = \frac{\delta Z}{\delta K^\alpha} \qquad (5.10)$$

the quantity

$$\Gamma(a,z) = Z(j(a,z), K(a,z)) - K^\alpha(a,z)z^\alpha - j_i(a,z)a_i$$
$$(5.11)$$

obeys

$$\frac{\delta \Gamma}{\delta a_i} = - j_i \ , \qquad \frac{\delta \Gamma}{\delta z^\alpha} = - K^\alpha \ .$$
(5.12)

If we call

$$\phi_A = (\phi_i, \ c^\alpha)$$

$$a_A = (a_i, \ z^\alpha) \ ,$$
(5.13)

(5.10) and (5.12) become

$$\frac{\delta z}{\delta j_A} = a_A \ , \qquad \frac{\delta \Gamma}{\delta a_A} = - j_A \ .$$
(5.14)

Differentiating the eqs. (5.14) with respect to a_B one sees that

$$\Gamma_{AB} = \frac{\delta^2 \Gamma}{\delta a_A \delta a_B} \bigg|_{a = 0}$$

is the inverse of the propagator $\delta^2 z / \delta j_A \delta j_B \big|_{j=0}$.

Thus Γ_{AB} is the (1-p-i) self-energy. In a similar way for the general 1-p-i vertex

$$\Gamma_{A_1 \ldots A_n} = \frac{\delta^n \Gamma}{\delta a_{A_1} \ldots \delta a_{A_n}} \bigg|_{a = 0}$$
(5.15)

can be shown.

Using (5.12) and

$$\frac{1}{i} \frac{\delta W}{\delta j_i} = a_i \; W, \qquad \frac{1}{i} \frac{\delta W}{\delta K^{\alpha}} = z^{\alpha} \; W \tag{5.16}$$

(5.2) may be rewritten immediately in terms of Γ [8]

$$D_i^{\beta}(a) \; [\frac{\delta \Gamma}{\delta a_i} - z_{\alpha} \; N_i^{\alpha}] = D_i^{\beta}(a) \; \frac{\delta \hat{\Gamma}}{\delta a_i} = 0 \tag{5.17}$$

$$\hat{\Gamma} = \Gamma - z_{\alpha} \; N_i^{\alpha} \; a_i = \Gamma - L_b \; \Big|_{K=Z, \; \phi=a} \qquad . \tag{5.18}$$

(5.17) is an identity of the "Lee-type" [28]. In our simple case it just states the gauge-invariance of $\hat{\Gamma}$ or, equivalently, the gauge-invariance of all 1-p-i vertices, except for $\Gamma_{\alpha, i}$, the vertex with one C-leg and one field-leg, which can be the vector field only (cf. (4.14)).

Also the "equation of motion", (5.3), can be translated into

$$\frac{\delta \hat{\Gamma}}{\delta z^{\alpha}} = 0 \qquad . \tag{5.19}$$

This means that all 1-p-i vertices involving C-legs vanish, except for

$$\Gamma_{\alpha, i} = N_i^{\alpha} \tag{5.20}$$

which is unchanged to all orders.

It will be fundamental for the discussion of re-normalization that $\hat{\Gamma}$ and $L_{inv}(a)$ fulfill the same re-lations (5.17) and (5.19), i.e. that both are gauge-invariant and z-independent.

5b) Identities with Fadeev-Popov-Ghosts

Transforming the integration-variable ϕ_i in (4.12) as (2.6) leads to a Slavnov-type identity which is very difficult to handle. The trick of Becchi, Stora and Rouet [29] generalizes the gauge-transformation in a non-linear way, so as to include the F.-P.-ghosts,

$$\delta\omega_\alpha = u_\alpha \, \delta\lambda$$

$$\delta\bar{u}_\alpha = -F_i^\alpha \, \phi_i \, \delta\lambda$$

$$\delta u_\alpha = -\frac{g}{2} \, f_{\alpha\beta\gamma} \, u_\beta \, u_\gamma \, \delta\lambda \qquad\qquad (5.21)$$

where $\delta\lambda$ is anticommuting with u and $\bar{u}$. (2.6) and (5.21) leave $L_{inv} + L_b + L_{F.P.}$ invariant, as can be seen from (4.12), (4.16); $1/\alpha$ is absorbed in the definition of F_i now:

$$L_b = -\frac{1}{2}(F_i^\alpha \phi_i)^2, \quad L_{F.P.} = \bar{u}_\alpha F_i^\alpha D_i^\beta(\phi) u_\beta \ . \qquad (5.22)$$

Also the Jacobi identity and the relation

$$t_{ij}^\alpha \, D_j^\beta - t_{ij}^\beta \, D_j^\alpha = f_{\alpha\beta\gamma} \, D_i^\gamma \qquad\qquad (5.23)$$

is used, which in turn, follows from the group property, i.e. the fact that

$$\left(\frac{\delta^2}{\delta\omega_\alpha^1 \delta\omega_\beta^2} - \frac{\delta^2}{\delta\omega_\beta^2 \delta\omega_\alpha^1}\right) \phi_i$$

is again a group transformation on ϕ_i.

The Jacobian is 1 to $O(\delta\lambda)^2$ so that from (4.12)

$$\delta\lambda \int (d\phi)(d\bar{u})(du) \{D_i^\alpha u_\alpha j_i + \bar{\xi}_\alpha \frac{g}{2} f_{\alpha\beta\gamma} u_\beta u_\gamma +$$

$$+ F_i^\alpha \phi_i \xi_\alpha\} e^{iL} = 0 \qquad (5.24)$$

is obtained, if L_S is extended to contain sources for the F.P.-ghosts as well:

$$L_S = j_i \phi_i + \bar{\xi}_\alpha u_\alpha + \bar{u}_\alpha \xi_\alpha . \qquad (5.25)$$

Replacing (differentiation of anticommuting variables is always defined from the left)

$$u_\alpha \to \frac{1}{i} \frac{\delta}{\delta\bar{\xi}^\alpha}$$

$$\phi_i \to \frac{1}{i} \frac{\delta}{\delta j_i}$$

and differentiating once with respect to ξ_β at $\xi = \bar{\xi} = 0$ leads to the original identity by Taylor and Slavnov [26]

$$j_i D_i^\alpha (\frac{1}{i} \frac{\delta}{\delta j}) \frac{1}{i} \frac{\delta W}{\delta\xi_\beta \delta\bar{\xi}_\alpha} \bigg|_{\xi=0} + F_i^\beta \frac{1}{i} \frac{\delta W}{\delta j_i} = 0 . \qquad (5.26)$$

The first term contains the Green's functions with two F.-P.-legs and an arbitrary number of others. Taking n derivatives with respect to j_i procedures fig. 2. As compared to the ghost-free case (fig. 1) the appearance of an additional closed loop (involving the F.-P.-ghost) is emphasized. It leads to a divergence, which is not compensated by counter-terms in G and which must be taken care of by a renormalization of the gauge-transformation.

(5.26) can be written in terms of a generating functional for 1-p-i-vertices, but this is quite complicated [28]. It is much easier to follow J. Zinn-Justin [30] and to introduce separate sources for all new composite operators in (2.6) with (5.21), i.e.

$$L_s = j_i \phi_i + \bar{\xi}_\alpha u_\alpha + \bar{u}_\alpha \xi_\alpha + k_i D_i^\alpha u_\alpha - \frac{g}{2} f_{\alpha\beta\gamma} u_\beta u_\gamma \ell_\alpha . \qquad (5.27)$$

The composite operators $D_i^\alpha u_\alpha$ and $f_{\alpha\beta\gamma} u_\beta u_\gamma$ are invariant under (5.21) and (2.6). Thus the only change in L again comes from the first three terms in (5.27)

$$\int (d\phi)(du)(d\bar{u}) \{ j_i D_i^\alpha u_\alpha - \frac{g}{2} \bar{\xi}_\alpha f_{\alpha\beta\gamma} u_\beta u_\gamma + F_i^\alpha \phi_i \xi_\alpha \} e^{iL} = 0$$

or using derivatives with respect to the new sources k_i and ℓ_α

$$j_i \frac{\delta W}{\delta k_i} + \bar{\xi}_\alpha \frac{\delta W}{\delta \ell_\alpha} + F_i^\alpha \frac{\delta W}{\delta j_i} \xi^\alpha = 0 . \qquad (5.28)$$

Also the "equation of motion", obtained by a change $\bar{u} \to \bar{u} + \delta\bar{u}$ in W

$$\int (d\phi)(d\bar{u})(du) \ [F_i^\alpha D_i^\beta u_\beta + \xi_\alpha] \ e^{iL} = 0$$

can be simplified in a similar way

$$(F_i^\alpha \frac{1}{i} \frac{\delta}{\delta k_i} + \xi_\alpha) \ W = 0 . \qquad (5.29)$$

In the form (5.28) and (5.29) the transcription into

identities for l-p-i vertices is straightforward. In analogy to (5.11) and (5.12) one defines

$$\Gamma = Z - j_i a_i - \bar{\xi}_\alpha \Omega_\alpha - \bar{\Omega}_\alpha \xi_\alpha = \Gamma(a,\Omega,\bar{\Omega},k,\ell) , \qquad (5.30)$$

keeping k and ℓ as parameters, with

$$a_i = \frac{\delta Z}{\delta j_i} , \quad \Omega_\alpha = \frac{\delta Z}{\delta \bar{\xi}_\alpha} , \quad \bar{\Omega}_\alpha = - \frac{\delta Z}{\delta \xi_\alpha}$$

$$\frac{\delta \Gamma}{\delta a_i} = - j_i , \quad \frac{\delta \Gamma}{\delta \bar{\Omega}_\alpha} = - \xi_\alpha , \quad \frac{\delta \Gamma}{\delta \Omega_\alpha} = \bar{\xi}_\alpha$$

$$\frac{\delta \Gamma}{\delta k_i} = \frac{\delta Z}{\delta k_i} , \quad \frac{\delta \Gamma}{\delta \ell_\alpha} = \frac{\delta Z}{\delta \ell_\alpha} . \qquad (5.31)$$

Then (5.28) and (5.29) become

$$F_i^\alpha a_i \frac{\delta \Gamma}{\delta \bar{\Omega}_\alpha} - \frac{\delta \Gamma}{\delta \Omega_\alpha} \frac{\delta \Gamma}{\delta \ell_\alpha} + \frac{\delta \Gamma}{\delta a_i} \frac{\delta \Gamma}{\delta k_i} = 0 \qquad (5.32)$$

$$F_i^\alpha \frac{\delta \Gamma}{\delta k_i} - \frac{\delta \Gamma}{\delta \bar{\Omega}_\alpha} = 0 . \qquad (5.33)$$

As in (5.18) the gauge-breaking term $L_b(a)$ can be subtracted out

$$\hat{\Gamma} = \Gamma + \frac{1}{2} (F_i^\alpha a_i)^2 \qquad (5.34)$$

so that instead of (5.32), (5.33)

$$\frac{\delta \hat{\Gamma}}{\delta \Omega_\alpha} \frac{\delta \hat{\Gamma}}{\delta \ell_\alpha} - \frac{\delta \hat{\Gamma}}{\delta a_i} \frac{\delta \hat{\Gamma}}{\delta k_i} = 0 \qquad (5.35)$$

$$F_i^\alpha \frac{\delta \hat{\Gamma}}{\delta k_i} - \frac{\delta \hat{\Gamma}}{\delta \bar{\Omega}_\alpha} = 0 \ . \tag{5.36}$$

One can check by explicit computation that

$$\hat{L} = L - j_i \phi_i - \bar{\xi}_\alpha u_\alpha - \bar{u}_\alpha \xi_\alpha + \frac{1}{2}(F_i^\alpha \phi_i)^2 \tag{5.37}$$

i.e. the action except for the ordinary source-terms and L_b fulfils the <u>same relations (5.35) and (5.36)</u> if $u_\alpha \to \Omega_\alpha$, $\bar{u}_\alpha \to \bar{\Omega}_\alpha$, $\phi_i \to a_i$ just as in the ghost-free gauge (5.17) and (5.18).

We may note that for the linear **gauges** (4.6) considered here, it would have been sufficient to consider an "amputated" Slavnov-transform

$$\delta \bar{u}_\alpha = 0$$

$$\delta u_\alpha = \frac{g}{2} f_{\alpha\beta\gamma} u_\beta u_\gamma \, \delta\lambda \tag{5.38}$$

instead of (5.21). It leaves $L_{inv} + L_{F.P.}$ invariant and in this sense it is more similar to the simple gauge-transformation in the ghost-free case. Still the composite operators in (5.27) are invariant under (5.38). Then the last term in (5.28) is changed into the contribution of L_b,

$$- F_j^\alpha \frac{\delta}{i\delta j_i} F_i^\alpha \frac{\delta}{\delta k_i}$$

but the Lee-identities (5.35) and (5.36) turn out to be unchanged.

6. RENORMALIZED ACTION

The number of loop-integrations is exhibited clearly by Nambu's method [31], where one replaces

$$L \to \frac{1}{\eta} L \quad .$$

This introduces a factor η for each propagator and η^{-1} for each vertex. The power of η in a 1-p-i Feynman-graph counts the number of loop-integrations. Then in the dimensionally regularized scheme we are able to isolate the divergent parts as $\tau = d-4 \to 0$ in the 1-p-i vertices

$$\Gamma'_{i_1 \ldots i_n} = \Gamma^{(0)}_{i_1 \ldots} + \frac{\eta}{\tau} \Gamma^{(1)}_{i_1} + \ldots \quad . \tag{6.1}$$

Let us take the terms up to $O(\eta)$ first. The symmetry breaking parameter ε_i is calculable in order η^0, i.e. in the tree-approximation ($\varepsilon_i = \varepsilon_i^{(0)}$). Obviously in terms of $\varepsilon_i^{(0)}$ the one point-functions to $O(\eta)$

$$\Gamma'_i = O(\eta) \neq 0$$

$$W'_i = O(\eta) \neq 0 \tag{6.2}$$

which must be taken care of in writing down explicit Ward-identities from the general Lee-identity. From power-counting $\Gamma^{(1)}_{i_1 \ldots i_n} \neq 0$ for $n \leq 4$ only.

The discussion of renormalization is again very simple in the ghost-free case. From (5.20) the only vertex involving an external C-leg is $\Gamma_{i,\alpha}$ which is not changed to all orders and thus certainly yields no contribution to $\Gamma^{(1)}$. The 1-p-i vertices with the other fields are gauge-invariant from (5.17) and so are the $\Gamma^{(1)}$. Thus the new action

$$L_{inv} \rightarrow L_{inv}^{ren} + L^{counter}$$

$$L^{counter} = -\frac{\eta}{\tau} \sum_{n=2}^{4} \frac{1}{n!} \Gamma_{i_1 \ldots i_n} \phi_{i_1} \ldots \phi_{i_n} \qquad (6.3)$$

will remain gauge-invariant to $O(\eta)$. Using (6.3) in all possible ways inside a vertex with two loop-integrations isolates $\Gamma^{(2)}$, which then is subtracted in L too. It also allows the computation of ε_i to $O(\eta)$ – and so on. In this way L_{inv} together with all counter-terms retains its gauge-invariance to all orders. Note that L_b and L_s are not affected. Although the theory is made finite in this way, clearly a final finite normalization (e.g. to make the residue of the propagators at the pole equal to one etc.) is still left open.

With F.-P.-ghosts the steps are essentially the same. Instead of (5.17) and (5.18) for $\hat{\Gamma}$ and L_{inv}, now (5.35) and (5.36) are used for $\hat{\Gamma}$ and $\hat{L}$. By complete induction in the loop expansion parameter $\hat{L}^{ren} + L^{counter}$ is the most general local polynomial of degree 4 which obeys (5.35) and (5.36). This can only have the form of (5.37) again with

$$f_{\alpha\beta\gamma} \rightarrow f_{\alpha\beta\gamma}^{ren}$$

$$D_i^{\alpha} \rightarrow D_i^{\alpha\ ren} \quad .$$

The latter quantities obey the same commutator relations as f and D. By a continuity argument the identification of the "renormalized" group transformations with the original one can be made [30].

7. PHYSICAL SOURCES

The sources j_i, K^α etc., used so far, must be completely arbitrary in order to allow taking the functional derivatives from the generating functional. The sources of <u>physical</u> particles are a subset of all j_i, where the j_i are submitted to further restrictions. In the literature on gauges with F.-P.-ghosts the approach is to <u>choose</u> a certain condition (involving the F.-P.-ghosts also) and to show from the gauge-invariance and unitarity of the S-matrix as obtained by projecting the $G_{i_1.....i_n}$ onto this subspace of "physical" sources that such a condition is adequate and consistent [32].

However, physical sources are also the eigenvectors of the self-energy-matrix and it is from that property that the condition on physical sources must be really <u>derived</u>. Of course, it can be expected that the result of such a derivation will coincide with the assumption mentioned above. At present, an analysis of this problem seems to exist for the ghost-free axial gauge alone [9]. In the notation (5.13) a physical source is among the eigenvectors [34]

$$e_A = (e_i, e_\alpha) \tag{7.1}$$

of the 1-p-i two-point vertex $\tilde{\Gamma}_{AB}$, which denotes the Fourier-transform of Γ_{AB}:

$$\overset{\sim}{\Gamma}_{AB} \; e_B \;\; = \lambda \; e_A \quad . \tag{7.2}$$

For a "physical" source near the mass shell $m^2_{(r)}$

$$\lambda_{(r)} \sim - (p^2 + m^2_{(r)}) \; z^2_{(r)} \tag{7.3}$$

must hold, where $z^2_{(r)}$ is a certain proportionality constant.

From the last section we conclude that also the renormalized Γ-s fulfil a relation like (5.17). Differentiating once with respect to a_j at $a = 0$ ($\Gamma_i = 0$) yields the Ward-identity

$$\Delta_i^{!\alpha} \; \Gamma_{ij} \;\; = \; 0 \quad . \tag{7.4}$$

Using (7.4) in (7.2) for $A = i$ and (5.20) gives

$$\Delta_i^{!\alpha} \; N_i^\beta e_\beta \; = \; (N\Delta) \; e_\alpha \; = \; \lambda \Delta_i^{!\alpha} e_i \tag{7.5}$$

for any eigenvector, whereas in the special case $A = \alpha$ (C-leg) (7.2) becomes (cf. (5.20))

$$N_i^\alpha \; e_i \; = \; \lambda e_\alpha \quad . \tag{7.6}$$

It is clear that for an eigenstate e_A with vanishing C-field-component e_α

$$\Delta_i^{!\alpha} \; e_i \; = \; N_i^\alpha \; e_i \; = \; 0 \quad . \tag{7.7}$$

We now show that eigenvectors with the property (7.3) for their eigenvalues obey in fact (7.7).

(7.7) is trivially fulfilled for i = fermion or scalar, other than a Higgs-ghost (cf. (2.11), (4.14)). For the vector fields and the Higgs-ghost Γ_{ij} is restricted by (7.4). Near the mass-shell (7.3) it is sufficient to consider the Γ_{ij} which is at most quadratic in p_μ and μ_i^α, the latter appearing in the combination $\tilde{\Delta}_i^{,\alpha}$, (the momentum space form of $\Delta_i^{,\alpha}$).

We simplify by choosing a linear combination of Higgs-ghosts and of the would-be massive Yang-Mills-fields such that in terms of the new fields the quadratic matrix μ_i^α becomes diagonal. Let us denote by μ the parameter of the only Higgs-ghost χ which in this basis belongs to a given (would-be) massive vector field. Then we may suppress the (diagonal) internal symmetry index and use instead of (7.4) and (4.14)

$$\tilde{\Delta}_i^{,*} \, \tilde{\Gamma}_{ij} = 0 \tag{7.8}$$

$$\tilde{\Delta}_i^{,} = (p_\mu, \, -i_\mu) \tag{7.9}$$

$\tilde{\Gamma}_{ij}$ depends on $\tilde{\Delta}_i^{,}$ and $\tilde{N}_i = (n_\mu, \, 0)$.

In terms of the orthogonal "unit-vectors"

$$f_i^{(1)} = \tilde{\Delta}_i^{,} / \sqrt{|\tilde{\Delta}_i^{,}|^2}$$

$$f_i^{(2)} = (\tilde{\Delta}_i^{,} - \tilde{N}_i |\Delta^{,}|^2 /_{(N\Delta)}) \, [\, |\tilde{\Delta}^{,}|^2 (\tilde{\gamma}-1) \,]^{-1/2}$$

$$\tilde{\gamma} = \frac{N^2 |\Delta'|^2}{(N\Delta)^2}$$

(7.10)

the projection operators

$$T_{ij} = f_i^{(2)} f_j^{(2)*}$$

(7.11)

and

$$L_{ij} = \delta_{ij} - f_i^{(1)} f_j^{(1)*}$$

(7.12)

span the space orthogonal to $\tilde{\Delta}_i'$. Hence [8]

$$\tilde{\Gamma}_{ij} = a L_{ij} |\tilde{\Delta}'|^2 + b |\tilde{\Delta}'|^2 (\tilde{\gamma}-1) T_{ij} \quad .$$

(7.13)

For the calculation of the eigenvalues (7.13) near the mass-shell $|\tilde{\Delta}'|^2 \sim 0$ terms at most quadratic in $\tilde{\Delta}_i$ (but with arbitrary powers of $(\widetilde{N\Delta})$) suffice in (7.13). Then a and b are constants. From the renormalization procedure the expression coincides with the renormalized free Lagrangian plus the $L^{counter}$ (cf. (6.3)). In momentum space the latter must be a polynomial in $\tilde{\Delta}_i'$ of second degree, because of power-counting. We observe from the explicit form of $f_i^{(2)}$ that this quantity cannot appear in the counter-terms.

Therefore

$a = z^2$

$b = 0$

at $|\tilde{\Delta}'|^2 \sim 0$ and with (5.20) the 6 x 6 self-energy matrix

$$\tilde{r}_{AB} \approx z^2 \left(\begin{matrix} \tilde{\lambda}_i' \, \tilde{\lambda}_j'^* - \delta_{ij} |\lambda'|^2 & \tilde{N}_i' \\ \tilde{N}_i' & 0 \end{matrix} \right)$$

$$\tilde{N}_i' = \tilde{N}_i / z^2 \qquad\qquad (7.14)$$

depends on n_μ in the $C-\phi_i$-parts only. Expressing $\tilde{N}_i'$ in terms of $f^{(1)}$ and $f^{(2)}$ (cf. (7.10) and writing (7.14) in an orthonormal basis $f_i^{(k)}$, (k) = 1,2,...5, $f_i^{(6)} = 0$ $f_{C-field} = 1$ one easily finds the three-fold eigenvalue $z^2 |\tilde{\lambda}'|^2$ with eigenvectors in the directions (3), (4) and (5) respectively. They correspond to the polarizations of a massive vector-particle with mass μ and have no C-components, i.e. (7.7) holds. The remaining three eigenvectors correspond to unphysical states. Their eigenvalues are n-dependent.

We know that the polarization of a massive vector particle must obey

$$p^\mu e_\mu^{(v)} = 0 \qquad\qquad (7.15)$$

which disagrees with (7.7),

$$p^\mu e_\mu = i\mu \, e^{(x)}$$

$$n^\mu e_\mu = 0 \quad . \qquad\qquad (7.16)$$

Thus a redefinition

$$e_\mu^{(v)} = e_\mu + \frac{i}{\mu} p_\mu \, e^{(x)} \qquad\qquad (7.17)$$

is necessary or

$$A_\mu \rightarrow A_\mu - \frac{1}{\mu} \partial_\mu \chi \qquad\qquad (7.18)$$

in terms of fields. (7.18) is nothing else but the transformation into the "unitary" gauge [35] in which A_μ and
χ decouple in the free Lagrangian. With (7.17) eq. (7.16)
does no longer represent any restriction on $e_\mu^{(v)}$, except
at (pn) = 0.

In order to exclude the latter case, it is sufficient to take n time-like here.

As mentioned above, in the presence of F.-P.-
ghosts one usually assumes a relation like the first
one in (7.7),

$$(\Delta_i^{\prime \alpha} + \eta \gamma_i^\alpha + \ldots) \, e_i = 0 \qquad\qquad (7.19)$$

where, as indicated in (7.19) order by order in the loop
expansion rather complicated vertices involving the
F.-P.-ghosts are added [33].

8. GAUGE DEPENDENCE OF $G_{i_1 \ldots i_n}$

Obviously Green's functions depend on the gauge-
breaking term in the Lagrangian and are gauge-dependent
[36]. For the axial gauge this means that the $G_{i_1 \ldots i_n}$
are, in general, even not Lorentz-invariant. Let us
change n_μ by δn_μ in L_b only, remembering that in the
renormalized equ. (4.22) $L_{inv} + L_{counter}$ is still gauge-
invariant. For the renormalized generating functional
(4.21)

$$\delta W = - i \quad \delta N_i^\alpha \quad \frac{\delta^2 W}{\delta k^\alpha \delta j_i} \tag{8.1}$$

is obtained. In this special change of δn_μ clearly a compensation of the infinities in the $(G_{i_1 \ldots i_n})_{n + \delta n}$ via the counter-terms can no longer be expected. Nevertheless, it will turn out to be sufficient for the proof of the gauge-invariance of the S-matrix in the next chapter to consider this special change of n_μ. (5.6) holds for the renormalized $G_{i_1 \ldots i_n}$ too. For (8.1) we need the l.h.s. of fig. 1 "bent together" via δN_i^α to one of the ϕ_i-legs. By differentiation of (8.1) with

$$\delta G_{i_1 \ldots i_n} = \frac{\delta W_{i_1 \ldots i_n}}{W_0} - G_{i_1 \ldots i_n} \frac{\delta W_0}{W_0} \tag{8.2}$$

and

$$\frac{\delta W_0}{W_0} = i \, \delta N_i^\alpha \, G_{\alpha, i} \tag{8.3}$$

in the summation of the first term on the r.h.s. of (5.6) the disconnected term from closing i_s with α drops out. The result for $\delta G_{i_1 \ldots i_n}$ is shown in fig. 3.

A special case is the change in the propagator, fig. 4. It yields important information on the propagator near the mass-shell of a physical particle state e_i

$$\tilde{G}_{ij} \approx \frac{e_i e_j}{\lambda}$$

$$\lambda = z^2 \sigma$$

$$\sigma = - (p^2 + \mu^2) \quad . \tag{8.4}$$

Projecting

$$\delta \hat{G}_{ij} \simeq \frac{1}{\lambda} [e_i \delta e_j + (\delta e_i) e_j - 2 \frac{\delta Z}{Z} e_i e_j + \frac{\delta \mu^2}{\sigma} e_i e_j] \tag{8.5}$$

onto e_i, e_j with

$$e_i e_i = 1, \qquad e_i \delta e_i = 0 \tag{8.6}$$

and comparing this with a similar projection of fig. 4 we find with (7.7)

$$\delta \mu^2 = 0 \tag{8.7}$$

$$\delta Z/Z = - e_k w_{kj} e_j \quad . \tag{8.8}$$

where w is defined in fig. 4 in terms of the (hatched) 1-p-i vertex.

Also in linear gauges with F.-P.-ghosts, a change of F_i^α in L_b

$$\delta L_b = - (F_i^\alpha \phi_i) (\delta F_j^\alpha \phi_j) \tag{8.9}$$

makes the use of the identities of fig. 2 possible with the line containing F_i^α joined via δF_j^α to another leg ϕ_j [33].

Relations like (8.7) and (8.8) are produced here as well. We shall keep both relations in the form (8.7) and (8.8) in the ghost-free case, because of its similarity to the formulae in other gauges, although a further simplification is possible: Z must be homogeneous in [39]. This is possible for $\delta Z/\delta n_\mu = 0$ only.

9. GAUGE-INDEPENDENCE AND UNITARITY OF THE S MATRIX

Under the special gauge-transformation δn_μ of the last chapter the S-matrix

$$S \equiv (\prod_{s=1}^{n} \sigma_{(s)} Z_{(s)} e_{i_s}^{(s)}) \tilde{G}_{i_1 \ldots i_n} \Big|_{\sigma_{(s)} \to 0} \tag{9.1}$$

changes by the amount ($\delta\sigma = 0$, cf. (8.7))

$$\delta S = (\prod_{s=1}^{n} \sigma_{(s)} Z_{(s)} e_{i_s}^{(s)}) \delta\tilde{G}_{i_1 \ldots i_n} \Big|_{\sigma = 0} +$$

$$+ \prod_s \sigma_{(s)} Z_{(s)} [\sum_{k=1}^{n} (e_{i_1}^{(1)} \ldots \delta e_{i_k}^{(k)} \ldots e_{i_n}^{(n)} +$$

$$+ e_{i_1}^{(1)} \ldots e_{i_n}^{(n)} \frac{\delta Z_{(k)}}{Z_{(k)}})] \tilde{G}_{i_1 \ldots i_n} \Big|_{\sigma = 0} . \tag{9.2}$$

Near the mass-shell $\sigma_{(k)} \to 0$ we may amputate the propagator from the leg i_k in G:

$$\tilde{G}_{i_1 \ldots i_n} \simeq \frac{e_{i_k}^{(k)} e_j^{(k)}}{\lambda_{(k)}} \overset{\approx}{G}_{i_1 \ldots i_n}^{(k)} . \tag{9.3}$$

Thus the second term in (9.2) vanishes with (8.6), whereas the third one is reproduced with the replacement

$$\tilde{G} \to \lambda^{-1}_{(k)} \overset{\approx}{G}(k) \; .$$

In the first term of (9.2) using fig. 3 we immediately observe that the first sum on the r.h.s. of fig. 3 vanishes again because of $H_i e_i \propto \Delta'_i e_i = 0$. Near the mass-shell i_k the second term of fig. 3 can be written as indicated in fig. 5. With (8.8) its cancellation together with the last expression in (9.2) is obvious. The fact that $\delta S/\delta n_\mu = 0$ for the special change of n_μ (in L_b only) is sufficient to conclude the general gauge dependence on-shell. In reality there can be no changes at all in the counter-terms which are relevant on the mass-shell, because $S_{n+\delta n}$ is just as finite as - and even identical to - S_n.

For the proof of unitarity we believe that the only legitimate approach is the one of an "S-matrix theory of fields"; i.e. only S-matrix elements are observable, the other parts of Green's functions are only relevant to the extent that they contribute - say as part of another Green's-function - to the S-matrix. Then we cannot expect more but

$$S^\dagger S = 1$$

or with $S = 1 + iT$

$$- i \, (T_{\textcircled{a}\,\textcircled{\beta}} - T^\dagger_{\textcircled{a}\,\textcircled{\beta}}) = \sum_{\textcircled{\gamma}} T^\dagger_{\textcircled{a}\,\textcircled{\gamma}} \, T_{\textcircled{\gamma}\,\textcircled{\beta}} \tag{9.4}$$

where $\textcircled{a}$, $\textcircled{\beta}$, $\textcircled{\gamma}$ refer to states of physical particles (cf.fig. 5). For each intermediate state $\textcircled{\gamma}$ also the polarizations are summed over.

The foregoing discussion teaches us that the l.h.s. of (9.4) is independent of n_μ. It is the absorptive part at a discontinuity. Quite generally the latter is obtained, if all the Feynman-graphs in $T_{\textcircled{a}\textcircled{b}}$ are "cut" by replacing the propagators along the line cutting $T_{\textcircled{a}\textcircled{b}}$ either by their mass-shell values or those lines are contracted to a point. This is the content of the Landau-rules, which according to the general analysis [40] may be readily extended to include the "abnormal" terms $(pn)^{-1}$, $(pn)^{-2}$ in the propagators of the ghost-free axial gauge (cf. (4.24) and below).

For the study of the propagator it is sufficient to consider the free (renormalized) Lagrangian. For the latter a diagonalization of μ_i^α as in chapter 7 leads to an expression which exactly coincides with the free Lagrangian of the Abelian Higgs-model [15] for each massive vector meson A_μ and its Higgs-ghost χ. The propagators simply follow by the replacements $p_\mu \to \tilde{\Delta}_i'$, $n_\mu \to \tilde{N}_j$ in (4.24), i.e.

$$\tilde{G}_{ij} = -\frac{1}{|\Delta'|^2}[\delta_{ij} - \frac{\tilde{\Delta}_i'\tilde{N}_j + \tilde{N}_i\tilde{\Delta}_j'^*}{(\tilde{N}\tilde{\Delta})} + \frac{\tilde{\Delta}_i'\tilde{\Delta}_j'^*\tilde{N}^2}{(\tilde{N}\tilde{\Delta})^2}] , \qquad (9.5)$$

or explicitely

$$\Delta_{\mu\nu}^{(A-A)} = -(p^2+\mu^2)^{-1}[g_{\mu\nu} - \frac{p_\mu n_\nu + p_\nu n_\mu}{(pn)} + \frac{p_\mu p_\nu n^2}{(pn)^2}] \qquad (9.6)$$

$$\Delta_\mu^{(A-\chi)} = -i\mu(p^2+\mu^2)^{-1}[n^2 p_\mu - (np)n_\mu](np)^{-2} \qquad (9.7)$$

$$\Delta^{(\chi-\chi)} = -(p^2+\mu^2)^{-1}[1 + \mu^2 n^2 (pn)^{-2}] \qquad (9.8)$$

According to [40] the Landau-rules generalize to

$$\alpha_i \ (p_i^2 + \mu^2) = 0 \tag{9.9}$$

$$\beta_i \ (p_i n) = 0 \tag{9.10}$$

for each vector - or Higgs-ghost line with momentum p_i
and "line-parameters" α_i, β_i. From the n_μ-independence
of the S-matrix, the l.h.s. of (9.4) is n-independent.
As far as the "abnormal" pole-terms (9.10) is concerned
this can be achieved, if all the graphs, where the solution $(pn) = 0$ of (9.10) is taken, cancel somehow.
Also the positions of the cuts belonging to the absorptive parts must be n-independent. Then in (9.10)
only $\beta_i = 0$ must be considered [41]. In this manner
the "normal" poles of (9.5) at $p^2 + \mu^2 = |\overset{\sim}{\Delta}{}'|^2 = 0$
contribute alone. We conclude that the l.h.s. is of
the form

$$\sum_{j'k'} \ T^+_{\textcircled{α} \ j'} \ T_{j'k'} \ T_{k' \ \textcircled{β}} \tag{9.11}$$

where in the sum over all intermediate particles, $T_{j'k'}$
for each massive vector particle is a contribution like
(9.5) with the replacement $|\overset{\sim}{\Delta}{}'|^{-2} \to i\pi\delta(|\Delta'|^2)$. In order
to show the equality of (9.11) with the r.h.s. of (9.4)
we use the Ward-identities (5.6) (cf.fig.1) with all
external lines (including the C-leg) at their respective mass-shells. By this we mean that all lines except
the C-leg are multiplied by $(p^2_{(i)} + \mu^2_{(i)})$ in the limit
$(p^2_{(i)} + \mu^2_{(i)}) \to 0$.

There is no pole-term in the line joining α and i_s in the first sum and only at most one in the second one (by amputation of the j-leg the propagator has a pole at $p^2_{(j)} = - \mu^2_{(j)}$), which would require the existence of an unstable state by

$$p^2_{(\alpha)} + \mu^2_{(\alpha)} = 0$$

$$p^2_{(i_s)} + \mu^2_{(s)} = 0$$

$$p^2_{(j)} + \mu^2_{(j)} = 0$$

$$p_{(\alpha)} + p_{(i_s)} + p_{(j)} = 0$$

We exclude the latter by assumption and thus amputating the C-A-propagator in the C-leg we arrive at (cf. (5.7) in momentum space)

$$\tilde{\Delta}_{i'} \ T_{i'} \ \textcircled{δ} \ = \ 0 \ . \tag{9.12}$$

The general state $\textcircled{$\delta$}$ contains the initial state $\textcircled{$\beta$}$ of $T_{j'} \ \textcircled{$\beta$}$ in (9.11), as well as intermediate lines (scalars, fermions, abelian gauge-fields) on the mass-shell. According to (9.12) all terms in T_{ij} (cf. (9.5)) may be dropped except δ_{ij}, i.e. for each vector-particle + Higgs-ghost (i = 0,1,2,3 for the vector A_μ, i = 5 for the Higgs-particle χ) we have

$$T^*_{\textcircled{ρ} \ i} \ \delta_{ij} \ T_{j \ \textcircled{σ}} \ = \ T^{*(A)}_{\textcircled{ρ} \ \mu} \ g_{\mu\nu} \ T^A_{\nu \ \textcircled{σ}} \ + \ T^{*(\chi)}_{\textcircled{ρ}} \ T^{(\chi)}_{\textcircled{σ}} \ =$$

$$= \ T^{(A)*}_{\textcircled{ρ} \ \mu} \ (g_{\mu\nu} + \frac{p_\mu p_\nu}{\mu^2}) \ T^{(A)}_{\nu \ \textcircled{σ}} \ , \tag{9.13}$$

if (9.12) in its explicit form

$$p_\mu \; T_\mu^{(A)} {}_{\textcircled{\scriptsize σ}} \; + \; i\mu \; T^{(\chi)} {}_{\textcircled{\scriptsize σ}} \; = \; 0$$

is used to eliminate the amplitudes with the χ-line.
Thus the internal line contains the sum over the polarizations of a physical, massive vector particle
$e_\mu^{(v)}$ (cf.(7.15)) and the Higgs-ghost disappears in the
unitarity sum.

For intermediate fermions the numerator of the
propagator appearing in T_{ij} is $\not{p} + m$, i.e. the sum of
the spinor-polarizations too. In this way the r.h.s.
of (9.4) is reproduced and unitarity holds.

Again the basic ingredients in the proof are
the same in the presence of F.-P.-ghosts. This proof,
which is more involved, may be found in ref. [43].

10. CONCLUSIONS

It is clear from the development of Yang-Mills
theories in the ghost-free case that the "naive" expectations to obtain a complete liberation from the
F.-P.-ghost-problem is almost true. The nontrivial
problem is always the "abnormal" singularity at (pn)=0,
but it can be dealt with in the power- counting argument
as well as in the other steps of the argument leading
to a unitary S-matrix. Thus it can be expected that
the ghost-free axial gauge works equally well in other
more general considerations involving nonabelian Yang-
Mills fields.

On the other hand, for explicit low order calculations in perturbation theory, e.g. in some unified
theory of weak- and electromagnetic interactions, usually a gauge like (3.2) with F.-P.-loops may be more
advantageous. The first reason is the large number of
terms in each propagator, although the square bracket
in (4.24) may be simplified appreciably in the special
case $n_0 = 1$, $\vec{n} = 0$ to contain a contribution from the
space-part ($\underline{i} = 1,2,3$) only:

$$-(\delta_{\underline{ij}} - k_{\underline{i}} k_{\underline{j}}/k_o^2)k^{-2} \quad . \tag{10.1}$$

Still, the additional pole-term at $k_o = (kn) = 0$ is
difficult to handle. In the usual Feynman-trick with
parameter-integrals it may be treated just as another
propagator besides $1/k^2$; this increases very much the
number of such integrals.

It must be stressed also that for a light-like
axial gauge, $n^2 = 0$, which has been advocated for its
simplicity in certain applications [44], the whole
proof of renormalization, unitarity etc. breaks down
right from the start together with the disappearance
of power-counting [45]. Possibly the difficulties are
even of a basic nature [46].

FIGURE CAPTIONS

Fig. 1. Ward-Identity in the ghost-free axial gauge;
        an open circle denotes a Green's function G.

Fig. 2. Ward-identity with two F.-P.-ghosts (dotted
        line).

Fig. 3. Gauge-dependence of $G_{i_1 \ldots i_n}$ in the ghost-free gauge.

Fig. 4. Change in the propagator.

Fig. 5. The last term of fig. 3 near the mass shell (k).

Fig. 6. The unitarity relation.

REFERENCES

1. S. Weinberg, Phys. Rev. Lett. 19, 1264 (1967);
   A. Salam, Elementary Particle Physics, N.Svartholm
   ed. (Stockholm 1968), p.367.

2. D.J. Gross and F. Wilczek, Phys. Rev. Letters 30,
   1343 (1973), Phys. Rev. D8, 3633 (1973), D9, 980
   (1974);
   N. D. Politzer, Phys. Rev. Lett. 30, 1346 (1973).

3. M.K. Gaillard and B.W. Lee, Phys. Rev. Lett. 33,
   108 (1974);
   G. Altarelli and L. Maiani, Phys. Lett. 52, 351
   (1974).

4. W. Kainz, W. Kummer and M. Schweda, Nucl. Phys.
   B79, 484 (1974).

5. B.S. de Witt, Phys. Rev. 162, 1195 (1967);
   L.D. Fadeev and V.N. Popov, Phys. Lett. 25B,
   29 (1967).

6. W. Kummer, Acta Phys. Austr. 14, 149 (1961).

7. R.L. Arnowitt and S.I. Fickler, Phys. Rev. 127,
   1821 (1962);
   J. Schwinger, Phys. Rev. 130, 402 (1963);
   Y.P. Yao, Journ. of Math. Phys. 5, 1319 (1964);
   E.S. Fradkin and I.V. Tyutin, Phys. Rev. D2, 2841 (1970).

8. W. Kummer, Acta Phys. Austr. $\underline{41}$, 315 (1975).

9. W. Konetschny and W. Kummer, Nucl. Phys. $\underline{B100}$, 106 (1975).

10. W. Konetschny and W. Kummer, "Unitarity in the Ghost-free axial gauge", prep. TU Vienna, Jan.1976.

11. The extension to more gauge-fields with different couplings is straightforward.

12. C.N. Yang and R.L. Mills, Phys. Rev. $\underline{96}$, 191 (1954).

13. B.S. de Witt, Phys. Rev. $\underline{162}$, 1195 (1967).

14. Our metric is $g_{00} = -g_{11} = -g_{22} = -g_{33} = -1$.

15. P. Higgs, Physics $\underline{12}$, 132 (1966); T.W.B. Kibble, Phys. Rev. $\underline{155}$, 1554 (1967).

16. In the latter case in each order of perturbation theory the soft-photon technique can be used along the lines of ref.[17]. Now it seems that this can be extended to nonabelian gauge-theories as well [18] by simply averaging over the internal symmetry of the nonabelian vector fields.

17. F.Bloch and A. Nordsieck, Phys. Rev. $\underline{52}$, 54 (1937).

18. Y.P. Yao "On the infrared problem in nonabelian gauge theories", Michigan prep. UMHE 75-38; Th. Appelquist, J. Carazzone, H. Kluberg - Stern and M. Roth "Infrared finiteness in Yang-Mills theories" FNAL Pub 76/16-THY.

19. R.P. Feynman, Rev. Mod. Phys. $\underline{20}$, 367 (1948).

20. Cf.e.g. F.A. Berezin, The method of second quantization, Academic Press, New York, London 1966.

21. A field-independent factor is irrelevant. It cancels in (4.4).

22. For simplicity we write it without symmetry breaking, because this is irrelevant for the following
argument. The propagators for the spontaneously
broken case are very similar, cf. (9.6) below.

23. S. Weinberg, Phys. Rev. $\underline{118}$, 838 (1960).

24. G. t'Hooft and M. Veltman, Nucl. Phys. $\underline{B44}$, 189
(1972).

25. N.N. Bogoliubov and O.S. Parasiuk, Acta Math. $\underline{97}$,
227 (1957);
K. Hepp, Commun. Math. Phys. $\underline{1}$, 95 (1965).

26. A. Slavnov, Theor. and Math. Phys. $\underline{10}$, 99 (1972);
J.C. Taylor, Nucl. Phys. $\underline{B33}$, 436 (1971).

27. G. Jona - Lasinio, Nuovo Cim. $\underline{34}$, 1790 (1964).

28. B.W. Lee, Phys. Lett. $\underline{46B}$, 214 (1973).

29. C. Becchi, A. Rouet and R. Stora, Phys. Lett. $\underline{52B}$,
344 (1974).

30. J. Zinn-Justin, Lectures at the International Summer
Institute for Theoretical Physics, Bonn 1974.

31. Y. Nambu, Phys. Lett. $\underline{26B}$, 626 (1966).

32. For an excellent review of the whole subject of
the renormalization of gauge - fields cf. ref. [33].

33. G. Costa and M. Tonin, Rivista del Nuovo Cim. $\underline{5}$,
29 (1975).

34. e_i refers to the ordinary fields, e_α to the C-field
in the ghost-free axial gauge.

35. S. Weinberg, Phys. Rev. Lett. $\underline{27}$, 1688 (1970).

36. In the so called "back-ground-field" method [37] one
does not introduce external sources as in (4.2)

and (4.3). Instead, the field is replaced by a quantum field $\phi_i^{(Q)}$ plus a classical field $\phi_i^{(c)}$. Despite a gauge-breaking term for $\phi_i^{(Q)}$ the Green's function (in terms of external $\phi^{(c)}$-legs) may retain a gauge-invariance. A short and clear introduction into the complicated literature can be found in [38].

37. B.S. DeWitt, Phys. Rev. 160, 1113 (1967), 162, 1195 (1967), 162, 1239 (1967);
    J. Honerkamp, Nucl. Phys. B48, 269 (1972);
    R. Kallosh, Nucl. Phys. B78, 293 (1974).

38. M.T. Grisaru, P. van Nieuwenhuizen and C.C. Wu, "Background field method vs. normal field theory in examples...", Brandeis prep. 1975.

39. The propagator (4.24) in Feynman-graphs with no external C-legs leads to expressions homogeneous in n_μ!

40. L.D. Landau, Nucl. Phys. 13, 181 (1959), R.J. Eden, P.V. Landshoff, D.I. Olive and J.C. Polkinghorne, The Analytic S-matrix, Cambridge Univ. press 1966.

41. A more elaborate argument can be found in ref. [10]. It is based on Veltman's version of the cutting rule [42].

42. R.E. Cutkosky, Rev. Mod. Phys. 33, 448 (1961); M. Veltman, Physica 29, 186 (1963).

43. G. t'Hooft, Nucl. Phys. B38, 173 (1971) and B35, 161 (1971);
    B.W. Lee and J. Zinn - Justin, Phys. Rev. D5, 3137 (1972).

44. Cf. the second and third ref. [2] and J.M. Cornwall, Phys. Rev. D10, 500 (1974).

45. Cf. eq. (4.27).

46. Cf. Cornwall, ref. (44), end of the "Appendix",
    and ref. (4).

47. J. Frenkel, A class of ghost-free nonabelian gauge
    theories, prep. Sao Paulo Univ., IFUSP/P-71,
    Dec. 1975.

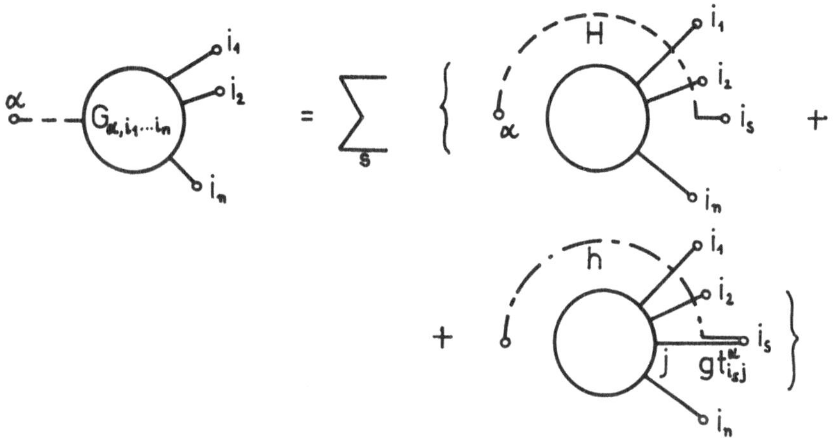

Fig. 1

Fig. 2

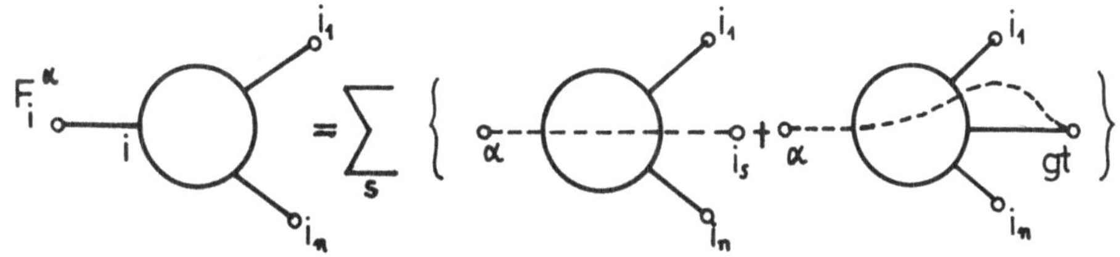

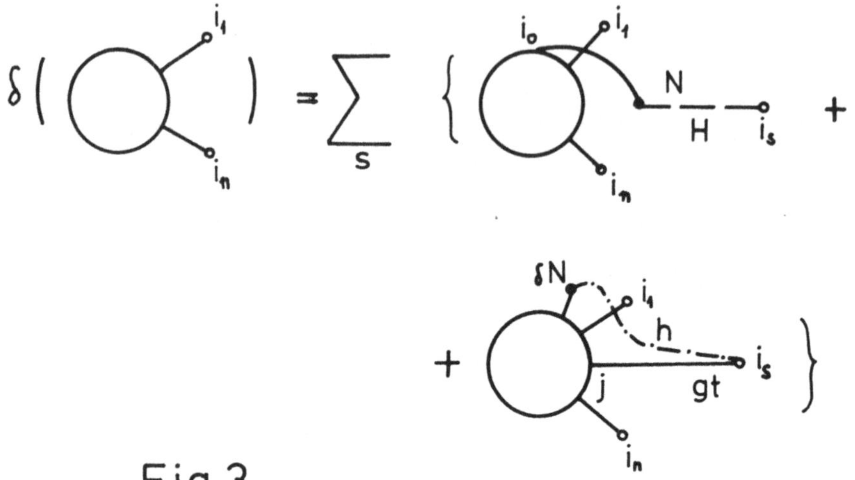

472

Fig. 3

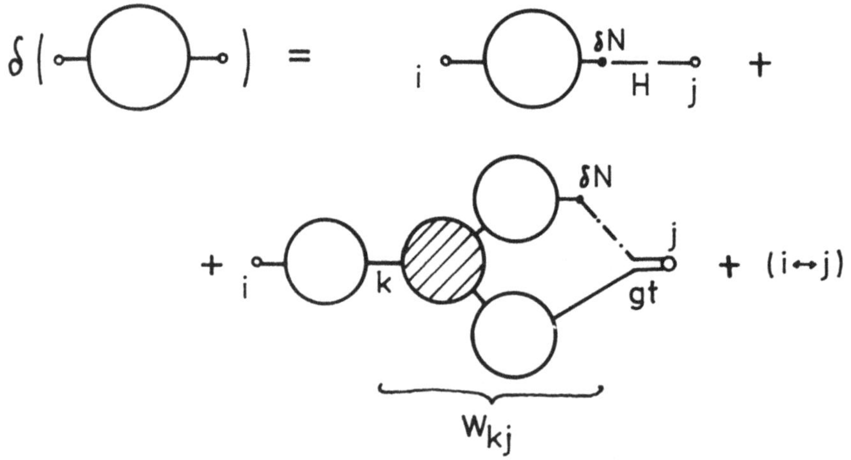

Fig. 4

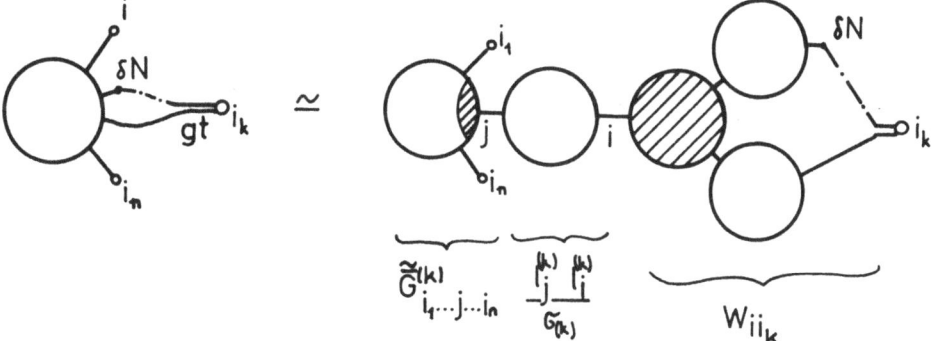

Fig. 5

Fig. 6

Acta Physica Austriaca, Suppl. XV, 475–498 (1976)
© by Springer-Verlag 1976

SUPERSYMMETRY[+]

by

J. WESS
Institut für Theoretische Physik
Universität Karlsruhe

INTRODUCTION

Supersymmetry is an attempt to enlarge the concept
of Lie-Algebras. This attempt has been successful [1]:
conserved currents have been constructed in the framework
of renormalizable field theories such that the conserved
charges form a closed algebra only if commutators or anticommutators are used respectively. Theories which possess
such a symmetry have interesting properties, there is a
remarkable cancellation of divergencies [2]. Fields with
different spin (and different statistic) are in the same
multiplet, supersymmetries contain the Poincaré group in
a non-trivial way.

[+] Lecture given at XV. Internationale Universitätswochen
für Kernphysik,Schladming,Austria, February 16-27,1976.

Supersymmetry restricts the possible field theoretical models very much - therefore it has not been
possible to find a model in reasonable agreement with
present day particle phenomenology. It has been possible,
however, to combine supersymmetry with all the current
concepts used in renormalizable Lagrangian field theories such as internal symmetries, gauging of internal
symmetries, spontaneous breaking of supersymmetry as
well as of an internal symmetry, Hipp mechanism. To
demonstrate this is the purpose of these lectures. We
start with an introduction to the supersymmetry formalism.

I.SUPERSYMMETRY

Supersymmetry is based on the following algebra [3]:

$$[P_\mu, Q_\alpha^i]_- = [P_\mu, \bar{Q}_{\dot\alpha i}]_- = [P_\mu, P_\nu]_- = 0$$

$$\{Q_\alpha^i, Q_\beta^j\}_+ = \{\bar{Q}_{\dot\alpha i}, \bar{Q}_{\dot\beta j}\}_+ = 0$$

$$\{Q_\alpha^i, \bar{Q}_{\dot\beta j}\}_+ = 2\sigma_{\mu\alpha\beta} P^\mu \delta_j^i \qquad . \qquad (1)$$

P^μ is the energy momentum operator, which generates four-
dimensional translations. Q_α^i are constant Weyl spinors-
α takes the values 1, 2 and i the values 1.. N. $\bar{Q}_{\dot\alpha i}$ is the
complex conjugate of Q_α^i.

If we introduce parameters which are elements of a
Grassmann algebra, i.e.:

$$\{\theta_i^\alpha, \theta_j^\beta\} = \{\bar\theta^{\dot\alpha i}, \bar\theta^{\dot\beta j}\} = \{\theta_i^\alpha, \bar\theta^{\dot\beta j}\} = \{\theta, Q\} = 0$$

the anticommutators of (1) can be written as commutator relations:

$$[\theta Q, \theta Q]_- = [\bar\theta\bar Q, \bar\theta\bar Q]_- = 0$$

$$[\theta Q, \bar\theta\bar Q]_- = 2\theta\sigma_\mu \bar\theta \ P^\mu$$

$$\theta Q = \theta^\alpha_i Q^i_\alpha, \quad \bar\theta\bar Q = \bar\theta^{\dot\alpha i} \bar Q_{\dot\alpha i} \qquad . \tag{2}$$

Formally, we can deal with the algebra (1) as with a Lie algebra. We can define a "Group element" of the corresponding "Lie group":

$$G(\theta,\bar\theta,x) = \exp i\{\theta Q + \bar\theta\bar Q - x_\mu P^\mu\} \quad . \tag{3}$$

Using Hausdorff's formula it is easy to multiply two group-elements:

$$G(\zeta,\bar\zeta,y)G(\theta,\bar\theta,x)=G(\theta+\zeta,\bar\theta+\bar\zeta,x+y - i\zeta\sigma\bar\theta + i\theta\sigma\bar\zeta) . \tag{4}$$

Instead of (3) we could have chosen the following para-metrisation:

$$G_1(\theta,\bar\theta,x) = e^{i\{\theta Q-xP\}} \ e^{i\bar\theta\bar Q}$$

$$G_2(\theta,\bar\theta,x) = e^{i\{\bar\theta\bar Q-xP\}} \ e^{i\theta Q}$$

$$G(\theta,\bar\theta,x) = G_1(\theta,\bar\theta, \ x + i\theta\sigma\bar\theta)$$

$$= G_2(\theta,\bar\theta, \ x - i\theta\sigma\bar\theta) . \tag{5}$$

This yields the following multiplication law:

$$G(\zeta,\bar{\zeta},y) \cdot G_1(\theta,\bar{\theta},x) = G_1(\theta+\zeta,\bar{\theta}+\bar{\zeta},x+y+2i\theta\sigma\bar{\zeta}+i\zeta\sigma\bar{\zeta})$$

$$G(\zeta,\bar{\zeta},y) \cdot G_2(\theta,\bar{\theta},x) = G_2(\theta+\zeta,\bar{\theta}+\bar{\zeta},x+y-2i\zeta\sigma\bar{\theta}-i\zeta\sigma\bar{\zeta}). \qquad (6)$$

The "Lie group" can be formally looked upon as a "Lie transformation group" in the space spanned by $\{x^\mu, \theta_i^\alpha:, \bar{\theta}^{\dot\alpha i}\}$

$$G(\zeta,\bar{\zeta},y) : (x,\theta,\bar{\theta}) \rightarrow (x+y-i\zeta\sigma\bar{\theta}+i\theta\sigma\bar{\zeta},\theta+\zeta, \bar{\theta}+\bar{\zeta})$$

$$\text{or} \quad : (x,\theta,\bar{\theta}) \rightarrow (x+y+2i\theta\sigma\bar{\zeta}+i\zeta\sigma\bar{\zeta}, \theta+\zeta, \bar{\theta}+\bar{\zeta})$$

$$\text{or} \quad : (x,\theta,\bar{\theta}) \rightarrow (x+y-2i\zeta\sigma\bar{\theta}-i\zeta\sigma\bar{\zeta},\theta+\zeta, \bar{\theta}+\bar{\zeta}) \qquad . \qquad (7)$$

The infinitesimal motion is generated by the following operators:

$$Q = \frac{\partial}{\partial\theta} - i\sigma^\mu\bar{\theta}\frac{\partial}{\partial x^\mu}, \quad \bar{Q} = -\frac{\partial}{\partial\bar{\theta}} + i\theta\sigma^\mu\frac{\partial}{\partial x^\mu}, \quad P_\mu = i\frac{\partial}{\partial x^\mu}$$

$$\text{or } Q = \frac{\partial}{\partial\theta} \qquad , \quad \bar{Q} = -\frac{\partial}{\partial\bar{\theta}} + 2i\theta\sigma^\mu\frac{\partial}{\partial x^\mu}, \quad P_\mu = i\frac{\partial}{\partial x^\mu}$$

$$\text{or } Q = \frac{\partial}{\partial\theta} - 2i\sigma_\mu\bar{\theta}\frac{\partial}{\partial x^\mu}, \quad \bar{Q} = -\frac{\partial}{\partial\bar{\theta}} \qquad , \quad P_\mu = i\frac{\partial}{\partial x^\mu}. \quad (8)$$

The differential operators, thus obtained, satisfy the commutation relations (1). They form a representation of the algebra. To obtain other representations we consider functions of these variables.

These functions were introduced by Salam and Strathdee [4] and called "superfields"

$$G(\zeta,\bar{\zeta},y) : \Phi(\theta,\bar{\theta},x) \rightarrow \Phi(\theta+\zeta,\bar{\theta}+\bar{\zeta}, x+y-i\zeta\sigma\bar{\theta}+i\theta\sigma\bar{\zeta})$$

$$\Phi_1(\theta,\bar{\theta},x) \rightarrow \Phi_1(\theta+\zeta,\bar{\theta}+\bar{\zeta},x+y+2i\theta\sigma\bar{\zeta}+i\zeta\sigma\bar{\zeta})$$

$$\Phi_2(\theta,\bar{\theta},x) \rightarrow \Phi_2(\theta+\zeta,\bar{\theta}+\bar{\zeta},x+y-2i\zeta\sigma\bar{\theta}-i\zeta\sigma\bar{\zeta}). \quad (9)$$

The product and sum of two superfields of the same type transforms again as a superfield. If we multiply or sum two superfields of a different type we first have to make a shift in the variables, according to (5) in order to obtain a superfield again; for example

$$\Phi_1(\theta,\bar{\theta},x)\ \Phi_2(\theta,\bar{\theta},x - 2i\theta\sigma\bar{\theta})$$

is again a superfield of type one.

From (9) we see that a superfield of type one, (two) which does not depend on $\bar{\theta}$ (θ) will not depend on $\bar{\theta}$ (θ) after the transformation either. The simplest superfield will therefore be of the form $\Phi_1(\theta,x)$. We shall call it a scalar superfield. If we expand $\Phi_1(\theta,x)$ in powers of θ we obtain, for the case N = 1:

$$\Phi_1(\theta,x) = A(x) + \theta^\alpha\psi_\alpha(x) + \theta^\alpha\theta^\beta\varepsilon_{\alpha\beta}F(x) . \quad (10)$$

The infinitesimal transformation law for the field components becomes according to (9):

$$\delta A = \zeta\psi$$

$$\delta\psi \;=\; 2F\zeta \;+\; 2i\sigma_\mu\,\bar\zeta\,\frac{\partial}{\partial x^\mu}\,A$$

$$\delta F \;=\; -\,i\,\frac{\partial}{\partial x^\mu}\,\psi\sigma^\mu\,\bar\zeta\;\;. \tag{11}$$

Notice that $\Phi_1(x,\theta) = a = \text{const}$ is a scalar superfield of type one and so is $\Phi_1(\theta,x) + a$.

$$\Phi_1^\dagger \;=\; A^* \;+\; \bar\psi\bar\theta \;+\; \bar\theta\bar\theta F$$

is a scalar superfield of type two.

In order to construct actions which are invariant under supersymmetry transformations we observe that the F component of Φ_1 transforms with a gradient. If the Lagrangian transforms like the F component of a multiplet, the action will be invariant.

As possible candidates for a Lagrangian we take:

$$\Phi_1\,\Phi_1^\dagger \;=\; \Phi_1(\theta,x)\,\Phi_1^\dagger\,(\bar\theta,\; x-2i\theta\sigma\bar\theta)$$

$$\sim A\,\Box\,A^* \;+\; FF^* \;-\; \frac{i}{2}\psi\sigma^\mu\partial_\mu\,\bar\psi$$

$$\Phi_1^2 \;\sim\; 2AF \;-\; \frac{1}{2}\,\psi\psi$$

$$\Phi_1^{\dagger 2} \;\sim\; 2A^* F^* \;-\; \frac{1}{2}\,\bar\psi\bar\psi$$

$$\frac{1}{3}\,\Phi^3 \;\sim\; A^2 F \;-\; \frac{1}{2}\,A\,\psi\psi$$

$$\frac{1}{3}\Phi_1^{\dagger 3} \;\sim\; A^{*2}F^* \;-\; \frac{1}{2}\,A^*\,\bar\psi\bar\psi \tag{12}$$

The Lagrangian:

$$L \sim \Phi\Phi^{\dagger} + \frac{1}{2}m(\Phi^2+\Phi^{\dagger 2}) + \frac{1}{3}g(\Phi^3+\Phi^{\dagger 3}) + \lambda\Phi + \lambda^*\Phi^{\dagger}$$

$$\sim -\frac{i}{2}\psi\sigma^{\mu}\partial_{\mu}\bar{\psi} - \partial_{\mu}A\partial^{\mu}A^* + FF^* + \lambda F + \lambda F^*$$

$$+\frac{1}{2} m (2AF + 2A^*F^* - \frac{1}{2}\psi\psi - \frac{1}{2}\bar{\psi}\bar{\psi})$$

$$+g (A^2F + A^{*2}F^* - \frac{1}{2}A\psi\psi - \frac{1}{2}A^*\bar{\psi}\bar{\psi}) \tag{13}$$

has been studied in great detail by B. Zumino and the author [5]. It is themost general renormalizable La-grangian which is renormalizable and which can be constructed in terms of one scalar superfield only.
If we introduce a set of superfields Φ_a (a = 1.. M) the Lagrangian above can be easily generalized to:

$$L \sim \Phi_a\Phi_a^{\dagger} + \{\frac{1}{2}m_{ab}\Phi_a\Phi_b - \frac{1}{3}g_{abc}\Phi_a\Phi_b\Phi_c - \lambda_a\Phi_a + h.c.\} \tag{14}$$

m_{ab} and g_{abc} are totally symmetric.

Componentwise:

$$L = - \frac{i}{2}\psi_a\sigma^{\mu}\partial_{\mu}\bar{\psi}_a - \partial_{\mu}A_a\partial^{\mu}A_a^* + F_aF_a^*$$

$$+ \{+ \frac{1}{2} m_{ab} (F_aA_b + F_bA_a - \frac{1}{2}\psi_a\psi_b) - \lambda_a F_a$$

$$- g_{abc} (A_aA_bF_c - \frac{1}{2} A_a \psi_b \psi_c) + h.c.\} \tag{15}$$

482

The subsidiary fields F_a can be eliminated, and, with the following equation of motion:

$$F^*_a = \lambda_a + m_{ab}A_b + g_{abc}A_bA_c \tag{16}$$

we obtain

$$L = -\frac{i}{2}\psi_a\sigma^\mu\partial_\mu\bar\psi_a - \partial_\mu A_a\partial^\mu A^*_a - F_aF^*_a$$

$$+ \{\frac{1}{4}m_{ab}(\psi_a\psi_b + \bar\psi_a\bar\psi_b) + \frac{1}{2}g_{abc}A_a\psi_b\psi_c + h.c.\} . \tag{17}$$

In order to restrict the class of renormalizable Lagrangians one can use R-invariance defined as follows:

$$R : \Phi(\theta,x) \to e^{2in\alpha}\Phi(e^{-i\alpha}\theta,x) \tag{18}$$

for the components:

$$R : A \to e^{2in\alpha}A$$

$$\psi \to e^{2i(n-\frac{1}{2})\alpha}\psi$$

$$F \to e^{2i(n-1)\alpha}F , \tag{19}$$

n is called the R-character of the superfield. Let us take three superfields Φ_o, Φ_2 and Φ_1 with R character 1, 1, 0 respectively. The most general, renormalizable R-invariant supersymmetric Lagrangian is [7]:

$$L \sim \Phi_0 \Phi_0^\dagger + \Phi_1 \Phi_1^\dagger + \Phi_2 \Phi_2^\dagger +$$

$$+ \{\lambda_0 \Phi_0 + \lambda_2 \Phi_2 + (m_0 \Phi_0 + m_2 \Phi_2) \Phi_1$$

$$+ (g_0 \Phi_0 + g_2 \Phi_2) \Phi_1 \Phi_1 + h.c.\} \qquad . \qquad (20)$$

II. SPONTANEOUS SYMMETRY BREAKING

This problem has been carefully investigated by L. O'Raifeartaigh [7], P. Fayet [8] and H. Nicolai [9]. In this lecture I would like to follow the work of O'Raifeartaigh. We start with the potential of equation [17].

$$V(A) = \sum_j F_j^* (A) \, F_j \, (A) \geq 0$$

$$F_j^* = \lambda_j + m_{jk} A_k + g_{jk\ell} A_k A_\ell \qquad (21)$$

We notice that the potential is always larger or equal zero. Next we notice that the invariant Lagrangian could have been constructed with the scalar superfields $\Phi_j + a_j$ which would have led to the potential

$$V(A_j + a_j) = V'(A_j) = F_j^{*'} F_j' \qquad (22)$$

where

$$F'_j = \lambda'_j + m'_{jk}A_k + g'_{jk\ell}A_kA_\ell$$

$$\lambda'_j = \lambda_j + m_{jk}a_k + g_{jk\ell}a_ka_\ell$$

$$m'_{jk} = m_{jk} + 2g_{jk\ell}a_\ell$$

$$g'_{jk\ell} = g_{jk\ell} \quad . \tag{23}$$

This tells us that if there is a supersymmetric theory which has a minimum of the potential at $A_j = -a_j$, there is also a supersymmetric theory where this point is shifted to the origin. It therefore suffices to look for a supersymmetric theory with a potential minimum at the origin which breaks the symmetry. The important terms of the potential are:

$$V(A) = \lambda_a\lambda_a^* + \lambda_a^* m_{ab} A_b + \lambda_a m_{ab}^* A_b^*$$

$$+ \lambda_a^* g_{abc} A_b A_c + \lambda_a g_{abc}^* A_b^* A_c^*$$

$$+ m_{ab} m_{ad}^* A_b A_d^* + \dots \quad . \tag{24}$$

The condition for an extremum is

$$m_{ab} \lambda_b^* = 0 \quad . \tag{25}$$

The condition for a minimum:

$$M^2 = \begin{bmatrix} \frac{1}{2} m_{ab}^* m_{bc} & g_{acb} \lambda_b^* \\ g_{acb}^* \lambda_b & \frac{1}{2} m_{ab} m_{bc}^* \end{bmatrix} \geq 0 \tag{26}$$

M^2 is the mass matrix of the boson fields ,

$$L_{Bm} \sim (A^*, A) \, M^2 \, \binom{A}{A^*} , \qquad (27)$$

condition (26) guarantees that there are not tachyons. m_{ab} is the mass matrix of the Fermions, (25) shows that there is at least one massless Fermion. The term $g_{acb} \cdot \lambda^*_b$ does not enter in the mass matrix of the Fermi fields, it therefore leads to spontaneous symmetry breaking. Spontaneous symmetry breaking will occur if and only if:

$$g_{acb} \, \lambda^*_b \neq 0 . \qquad (28)$$

To find an absolute minimum is, as usually, very diffi-cult. In our case $V \geq 0$, $V = 0$ would therefore be an absolute minimum. Unfortunately, $V = 0$ cannot be a symmetry breaking minimum because from $V = 0$ follows $F_a = 0$, but $F_a(0) = \lambda_a$. Therefore, $\lambda_a = 0$ which vio-lates (28), the condition for symmetry breaking. There-fore, we have to find a potential which has at $A = 0$ an absolute minimum such that $V(0) > 0$. First, we derive from (26) a necessary condition for $V(0)$ to be a local minimum:

Lemma 1 of O'Raifeartaigh: If $V(0)$ is a local minimum then

$$m_{bc} \, \psi^*_c = 0 \qquad \text{has as consequence} \qquad g_{acb} \lambda^*_b \psi^*_c = 0 .$$

Proof: Equ. (26) means that

$$(\psi^*, \psi) \, M^2 \, \binom{\psi}{\psi^*} \geq 0$$

for any ψ. For a ψ which satisfies $m_{ab} \psi_b^* = 0$ this condition becomes

$$\text{Re } \psi_a \, g_{abc}^* \, \lambda_b \, \psi_c \geq 0 \; .$$

With ψ, $e^{i\alpha}\psi$ for any α satisfies $m_{ab} e^{-i\alpha} \psi_b^* = 0$ as well. Therefore $\text{Re } e^{2i\alpha} \psi_a \, g_{abc}^* \, \lambda_b \, \psi_c \geq 0$ for any α. This can only be true if $\psi_a \, g_{acb}^* \, \psi_c \, \lambda_b = 0$. It follows that $(\psi^*, \psi) M^2 \binom{\psi}{\psi^*} = 0$. From (26) follows

$$M^2 \binom{\psi}{\psi^*} = 0 \qquad\qquad \text{or} \qquad\qquad g_{acb} \, \lambda_b^* \, \psi_c = 0 \; .$$

This proves the lemma.

Now we make the following observations: If $V(A)$ has a local minimum at $A = 0$, then $V(c\lambda_a^*) = V(0)$ for all c. This follows from $F_j(c\lambda_a^*) = F_j(0)$ due to lemma 1 and equations (23) and (25).

If $V(0)$ is an absolute minimum, the same is true for $V(c\lambda^*)$ because $V(c\lambda^*) = V(0)$ and $V \geq V(0)$ everywhere. If spontaneous symmetry breaking occurs at $V(0)$ it also occurs at $V(c\lambda^*)$. To show this let us shift the point $c\lambda^*$ to zero. From (26) we learn:

$$V(A_j + c\lambda_j^*) = V'(A_j)$$

where

$$\lambda_j' = \lambda_j, \; m_{jk}' = m_{jk} + 2c \, g_{jk\ell} \lambda_\ell^*, \; g_{k\ell j}' = g_{k\ell j} \; .$$

It follows that $g'_{acb} \lambda'_b = g_{acb} \lambda_b \neq 0$, the condition for spontaneous symmetry breaking. From $V(c\lambda^*) = V(0)$ and $\lambda^* \neq 0$ follows that there are at least two massless bosons in the theory. We can now formulate lemma 2. Let Ω be the subspace where the matrix m_{ab} is different from zero. Let $\tilde{m}$ and $\tilde{g}$ be the matrices m_{ab} and $g_{abc}\lambda^*_c$ restricted to this subspace.

Lemma 2: If $V(0)$ is an absolute minimum where spontaneous symmetry breaking occurs, $\tilde{m}^{-1} \tilde{g}$ is nilpotent. Proof: With λ_j, m_{jk} and $g_{jk\ell}$ also $\lambda'_j = \lambda_j$, $m'_{jk} = m_{jk} + 2cg_{jk\ell}\lambda^*_\ell$,

$$g'_{jk\ell} = g_{jk\ell}$$

are parameters for a potential with an absolute minimum with symmetry breaking at the origin. Lemma 1 tells us that $m\psi = 0$ implies $m'\psi = 0$. The reciprocity of the two sets of parameters tells us that $m'\psi = 0$ implies $m\psi = 0$.

The subspace where m is zero is identical with the subspace where m' is zero. On the subspace Ω, where m is non-zero also $m' = m + 2c\, g$ is non-zero. This means that the eigenvalue equation

$$(\tilde{m}^{-1} \tilde{g} + \frac{1}{2c})\psi = \frac{1}{2c} \tilde{m}^{-1} \tilde{m}' \psi$$

has no solution for finite c. Thus $\tilde{m}^{-1} \tilde{g}$ has only eigenvalues zero. This proves the lemma.

Using this very restrictive conditions on the parameters λ_j, m_{jk} and $g_{jk\ell}$, O'Raifeartaigh has constructed a model with spontaneous symmetry breaking. The simplest model which, according to the conditions, has to have at least three fields is defined through:

$$\lambda_a = (0, 0, \lambda) \ , \quad \lambda = \lambda^*$$

$$m_{ab} = \begin{bmatrix} 0 & m & 0 \\ m & 0 & 0 \\ 0 & 0 & 0 \end{bmatrix} \qquad g_{ab} = \begin{bmatrix} g & 0 & 0 \\ 0 & 0 & 0 \\ 0 & 0 & 0 \end{bmatrix} \qquad (29)$$

$$g_{abc} = \frac{1}{\lambda^2} (g_{ab} \lambda_c + g_{bc} \lambda_a + g_{ca} \lambda_b) .$$

This Ansatz satisfies all the conditions: (28)

$$g_{acb} \lambda_b^* = g_{ac} \neq 0,$$

the condition for spontaneous symmetry breaking. Lemma 1 is satisfied because $(0, 0, 1)$, the only eigenvector of m_{ab} with eigenvalue zero, is also eigenvector of g_{ab} with eigenvalue zero. Lemma 2 is satisfied because

$$\underset{\sim}{m}^{-1} \underset{\sim}{g} = \begin{bmatrix} 0 & & 0 \\ \dfrac{g}{m} & & 0 \end{bmatrix}$$

is nilpotent.

We are left to show that V has an absolute minimum indeed. To this end we construct the potential explicitly:

$$F_o = \lambda + g A_1^2$$

$$F_1 = mA_2 + 2g A_o A_1$$

$$F_2 = m A_1 .$$ (30)

The potential becomes:

$$V = |F_1|^2 + |F_2|^2 + |F_3|^2 =$$

$$\lambda^2 + (m^2 + 2\lambda g) a_1^2 + (m^2 - 2\lambda g) b_1^2$$

$$+ g^2 (a_1^2 + b_1^2)^2 + F_1 F_1^*$$ (31)

where $A_j = a_j + ib_j$. If $m^2 > |2\lambda g|$, the potential is al-
ways $V \geq \lambda^2$. At $A = 0$ it takes its absolute minimum
$V = \lambda^2$. The masses show spontaneous symmetry breaking.

In terms of superfields:

$$L \sim \Phi_a \Phi_a^\dagger - \lambda (\Phi + \Phi^\dagger) + m (\Phi_1 \Phi_2 + \Phi_2^\dagger \Phi_1^\dagger)$$

$$- g (\Phi_1^2 \Phi_o + \Phi_o^\dagger \Phi_1^{\dagger 2}) .$$ (32)

If we assign the R-character of these fields as in (20),
i.e. $(\Phi_o, \Phi_2, \Phi_1) \sim (1, 1, 0)$, L is a special case of
(20) and therefore R invariant. If, moreover, we impose
the following symmetry:

$$\Phi_1 \to -\Phi_1 , \qquad \Phi_2 \to -\Phi_2 , \qquad \Phi_o \to \Phi_o$$ (33)

on the Lagrangian (20), the Lagrangian (32) would be the
most general, renormalizable, R-invariant, supersymmetric
Lagrangian invariant under (33). Therefore, also after renormalization, the Lagrangian (32) should be of the same
structure and one can hope that spontaneous symmetry breaking persists to all orders.

III. GAUGE THEORIES AND HIGGS MECHANISM

It is easy to combine supersymmetry with an internal symmetry. Let S_i be a set of scalar superfields, they are assumed to belong to a representation of an internal symmetry group:

$$S_i \rightarrow (e^{-i\lambda^\ell a_\ell})_{ij} \, S_j \tag{34}$$

$S \rightarrow e^{-i\Lambda}S$ in matrix notation. λ^ℓ are the generators of the symmetry group in the representation under consideration. The individual peaces of the Lagrangian (14) (now also in matrix notation) will be invariant if the representation is orthogonal and if g_{abc} is a symmetric form-invariant tensor ($g \, d_{abc}$ in the case of SU_3, as an example).

If the representation is unitary, $S^\dagger S$ will still be invariant. To obtain an invariant mass term, and a coupling term as well, we introduce a second set of scalar superfields, subjected to the following transformation law:

$$T \rightarrow e^{-i\Lambda^\dagger} T \; . \tag{35}$$

The Lagrangian:

$$L \sim \frac{1}{2} \, (S^\dagger S + T^\dagger T) + \frac{1}{2} \, m \, (T^\dagger S + S^\dagger T) \tag{36}$$

is invariant.

Gauging this symmetry means to make the parameters

a_ℓ x-dependent. Constant parameters a_ℓ can be considered as scalar superfields, going over to x-dependent transformations we will have to replace the a_ℓ's by a set of scalar superfields ϕ_ℓ

$$\Lambda = \lambda^\ell \phi_\ell (x,\theta) \ .$$

The mass term will still be invariant under this transformation, but not the term $S^\dagger S + T^\dagger T$ because now $\Lambda^\dagger \neq \Lambda$. This had to be expected, gauging is impossible without vector fields. We have to introduce a superfield which contains, as component, at least one vector field.

A superfield that transforms like G in equ. (4) can be taken hermitean: $V^\dagger = V$. The decomposition of such a field, which we shall call a vector superfield, is as follows:

$$V(x,\theta,\bar\theta) = (1 + \frac{1}{4} \theta\theta\bar\theta\bar\theta \,\Box\,)C$$

$$+ (i\theta + \frac{1}{2} \theta\theta\sigma^\mu\bar\theta \frac{\partial}{\partial x^\mu})\chi + \frac{i}{2} \theta\theta (M + iN)$$

$$+ (-i\bar\theta + \frac{1}{2} \bar\theta\bar\theta\sigma^\mu \frac{\partial}{\partial x^\mu})\bar\chi - \frac{i}{2} \bar\theta\bar\theta (M - iN)$$

$$- \theta\sigma_\mu \bar\theta v^\mu + i\theta\theta \bar\theta\bar\lambda - i \bar\theta\bar\theta \theta\lambda$$

$$+ \frac{1}{2} \theta\theta\bar\theta\bar\theta \,D \ . \tag{37}$$

The fields C, M, N and D are real scalar fields, χ, λ are Weyl fields, v_μ is the desired vector field.

In the following it will be convenient to work in the representation of G always, for this purpose, we have to carry out the shift of the superfields of type one and two:

$$\Phi_1 + \Phi_1^\dagger = (1 + \tfrac{1}{4}\,\theta\theta\bar\theta\bar\theta)\,(A + A^*)$$

$$-i(i\theta + \tfrac{1}{2}\,\theta\theta\sigma^\mu\bar\theta\,\frac{\partial}{\partial x^\mu})\psi \;-\; 2i\,\theta\theta\,F$$

$$+\, i(-i\theta + \tfrac{1}{2}\,\bar\theta\bar\theta\sigma^\mu\frac{\partial}{\partial x^\mu})\psi \;+\; 2i\,\bar\theta\bar\theta\,F^*$$

$$+\, i\theta\sigma_\mu\bar\theta\,\frac{\partial}{\partial x^\mu}\,(A - A^*) \; . \tag{38}$$

This shows that the following functions of the components of a scalar superfield transform like the components of a vector superfield:

$$C = A + A^*$$

$$\chi = -i\psi,\; \bar\chi = i\bar\psi$$

$$v^\mu = -i\partial^\mu\,(A - A^*)$$

$$N = -\,(F + F^*),\, M = -i\,(F - F^*)$$

$$\lambda = D = 0 \; . \tag{39}$$

It is important to be able to characterise the type of a superfield in this representation. For a vector field we have $v^\dagger = V$. For a scalar superfield we have to shift the condition: does not depend on $\bar\theta$, or $\frac{\partial}{\partial\bar\theta}\,\Phi = 0$, we obtain

$$\bar{D}_{\dot{\alpha}} \Phi_1 = 0, \quad \bar{D}_{\dot{\alpha}} = -\left(\frac{\partial}{\partial \theta^{\dot{\alpha}}} + i\theta^{\alpha}\sigma^{\mu}{}_{\alpha\dot{\alpha}} \frac{\partial}{\partial x^{\mu}}\right) \tag{40}$$

and, analogously:

$$D_{\alpha} \Phi_2 = 0, \quad D = \left(\frac{\partial}{\partial \theta} + i\sigma_{\mu}\bar{\theta} \frac{\partial}{\partial x^{\mu}}\right) . \tag{41}$$

We compute:

$$\{D, \bar{D}\} = -2i\sigma_{\mu} \frac{\partial}{\partial x^{\mu}} , \quad \{D,D\} = \{\bar{D},\bar{D}\} = 0 . \tag{42}$$

Now we are ready to gauge the internal symmetry group. We follow the work of S. Ferrara and B. Zumino [10].

The kinetic term of equ. (36) transforms like

$$S^{\dagger}S \rightarrow S^{\dagger} e^{i\Lambda^{\dagger}} e^{-i\Lambda} S , \tag{43}$$

to make it invariant we have to introduce the Yang Mills potential V, it has to transform as follows:

$$e^{V} \rightarrow e^{-i\Lambda^{\dagger}} e^{V} e^{i\Lambda} \tag{44}$$

expanded:

$$V \rightarrow V + i(\Lambda - \Lambda^{\dagger}) + .. \tag{45}$$

The Lagrangian

$$\frac{1}{2}(S^{\dagger}e^{V} S + T^{\dagger} e^{-V} T) + \frac{1}{2}m(S^{\dagger}T + T^{\dagger}S) \tag{46}$$

is gauge invariant. When we expand the exponential we obtain, as first term, the Lagrangian (36).

The kinetic term for the vector field has to be constructed. First we have to find the analogon to the Yang Mills fields: We recall:

$$V = V^\dagger , \qquad \bar{D}\Lambda = D\bar{\Lambda} = 0 \quad . \tag{47}$$

From (44) follows:

$$e^{-V} D_\alpha e^V \to e^{-i\Lambda} (e^{-V} D_\alpha e^V) e^{i\Lambda} + e^{-i\Lambda} D_\alpha e^{i\Lambda} \quad .$$

Applying $\bar{D}_{\dot\beta}$ and making use of (42) yields:

$$\bar{D}_{\dot\beta} (e^{-V} D_\alpha e^V) \to e^{-i\Lambda} [\bar{D}_{\dot\beta} (e^{-V} D_\alpha e^V)] e^{i\Lambda}$$

$$- 2i \, \sigma^\mu_{\alpha\dot\beta} \, e^{-i\Lambda} \partial_\mu \, e^{i\Lambda} \quad .$$

The Yang-Mills-Superfield:

$$W_\alpha = \bar{D}_{\dot\beta} \bar{D}^{\dot\beta} (e^{-V} D_\alpha e^V) \to e^{-i\Lambda} \, W_\alpha \, e^{i\Lambda} \quad . \tag{48}$$

Because the product of three $\bar{D}$ operators is zero we obtain $\bar{D}W = 0$. W_α is a scalar superfield with a vector index. We know how to construct Lagrangians for scalar superfields. A suitable Lagrangian is:

$$L \sim \text{Tr} \, (W_\alpha W^\alpha + W^\dagger_\alpha W^{\dagger\alpha})$$

$$\sim \text{Tr} \left(-\frac{1}{2} v_{\mu\nu}^2 - \frac{i}{2} \bar{\lambda} \gamma^\mu D_\mu \lambda + \frac{1}{2} D^2 \right) \quad . \tag{49}$$

Here:

$$v_{\mu\nu} = \partial_\mu v_\nu - \partial_\nu v_\mu + i [v_\mu , v_\nu] ,$$

$$D_\mu \lambda = \partial_\mu \lambda + i [v_\mu , \lambda] ;$$

the trace is to be taken over the matrix indices. The Lagrangians (49) and (46) together solve the problem of gauging an internal symmetry consistent with super-symmetry. Due to the exponential in (46), it looks like an unrenormalizable theory. However, equs. (45) and (39) tell us that the C-component of the vector superfield is a gauge-field and, in a particular gauge, can be put equal to zero. In the decomposition (37) there is no θ independent term left, any function of the superfield is a polynomial in the component-fields. The case of SO(2) (supersymmetric quantum electrodynamics) has been thoroughly discussed in ref. [11].

Through the Higgs mechanism, a massless vector field aquires a finite mass via a nonvanishing vacuum expectation value of a scalar field. The missing degree of freedom (the longitudinal component of the vector field) is furnished by the scalar field. How does it work for super fields? Let us first study the massive vector field. A suitable Lagrangian is:

$$L \sim W_\mu W^\mu + W_\mu^\dagger W^{\mu\dagger} + \frac{1}{2} m^2 v^2$$

$$\sim \frac{1}{2} D^2 - i\lambda\sigma_\mu \partial^\mu \bar{\lambda} - \frac{1}{4} v_{\mu\nu} v^{\mu\nu}$$

$$+ \tfrac{1}{2} m^2 \ (2CD - \partial_\mu C \partial^\mu \ C - v_\mu^2 + M^2 + N^2$$

$$- 2 \ \chi\lambda - 2 \ \bar{\chi}\bar{\lambda} - 2i\chi\sigma^\mu\partial_\mu\bar{\chi}) \ . \tag{50}$$

We learn that a massive vector superfield contains as additional degrees of freedom: the longitudinal component of a vector field v_μ, a scalar field C and a Weyl spinor χ. These are exactly the degrees of freedom of a scalar superfield. Based on this observation P. Fayet [12] has constructed several models with the Higgs mechanism, we discuss the simplest one. Take the Lagrangian:

$$L \sim \tfrac{1}{2} \ S^\dagger e^V \ S + \zeta V + W_\mu W^\mu + W_\mu^\dagger \ W^{\mu\dagger} \ . \tag{51}$$

It contains one scalar superfield and one vector superfield. The Lagrangian is gauge invariant:

$$S \to e^{-i\Lambda} S$$

$$V \to V + i \ (\Lambda - \Lambda^\dagger) \ . \tag{52}$$

From (39) follows that the D component of the super vector field V does not change under gauge transformations. To demonstrate the Higgs mechanism we follow P. Fayet and choose the special gauge

$$S = i \ v \tag{53}$$

v is a constant. This is a gauge which is supersymmetric - a constant is a scalar superfield. The connection of this gauge with the previous one was thoroughly discussed by P. Fayet [13].

In this special gauge we expand the exponential in (51)

$$L \sim v^2 (1 + V + V^2 + \ldots) + \zeta V$$

$$+ W_\mu W^\mu + W_\mu^\dagger W^{\mu\dagger} \quad . \tag{54}$$

We can choose $\zeta = -v^2$ and obtain

$$L \sim W_\mu W^\mu + W_\mu^\dagger W^{\mu\dagger} + v^2 V^2 + \ldots \tag{55}$$

the Lagrangian for a vector field with mass according to (50).

The author wishes to acknowledge helpful discussions with H. Nicolai on the subject of spontaneous symmetry breaking.

REFERENCES

1. J. Wess and B. Zumino, Nucl. Phys. B70, 39 (1974); Phys. Lett. 49B, 52 (1974).

2. J. Wess and B. Zumino, Phys. Lett. 49B, 52 (1974); Nucl. Phys. B78, 1 (1974); J. Iliopoulos and B. Zumino, Nucl. Phys. B76, 310 (1974);

S. Ferrara, J. Iliopoulos and B. Zumino, Nucl. Phys. B77, 413 (1974);

W. Lang and J. Wess, Nucl. Phys. B81, 249 (1974).

3. S. Ferrara, J. Wess and B. Zumino, Phys. Lett. 51B, 239 (1974);

    J. Wess, Bonn Lecture Notes (1974);

    A. Salam and B. Strathdee, Nucl. Phys. B76, 477 (1974)

4. A. Salam and J. Strathdee, Phys. Rev. D11, 1521 (1975)

5. J. Iliopoulos and B. Zumino, Nucl. Phys. B76, 310 (197

    S. Ferrara, J. Iliopoulos and B. Zumino, Nucl. Phys. B77, 413 (1974).

6. P. Fayet, Phys. Lett. 58B, 67 (1975).

7. L. O'Raifeartaigh, Nucl. Phys. B96, 331 (1975).

8. P. Fayet, Nucl. Phys. B90, 104 (1975).

9. H. Nicolai, Diplomarbeit Karlsruhe.

10. S. Ferrara and B. Zumino, Nucl. Phys. B79, 413 (1974).

11. J. Wess and B. Zumino, Nucl. Phys. B78, 1 (1974).

12. P. Fayet, Higgs Model and Supersymmetry, Preprint Ecole Normale Supérieure Paris.

Acta Physica Austriaca, Suppl. XV, 499–519 (1976)
© by Springer-Verlag 1976

ON DYNAMICAL SYMMETRY BREAKING FOR LEPTONS[+]

by

P. BUDINI

International Centre for Theoretical Physics

Trieste, Italy

and

Istituto di Fisica Teorica

Università di Trieste, Italy

ABSTRACT

    Starting from massless Majorana-Weyl multiplets con-
sidered as limit of short distance interacting Dirac spinors, models of dynamical symmetry breaking are proposed.
In the first model (for leptons), based on SU(3) x U(1)
symmetry, it is shown how this symmetry could be reduced
to U(2) x U(1) for weak interactions by virtue of the
chiral projectors. The lepton self-masses can be made

[+] Lecture given at XV. Internationale Universitätswochen
    für Kernphysik,Schladming,Austria,February 16-27, 1976.

infra-red driven and/or small by the internal algebra.
In the second model, based on an extension of the Konopinski-Mahmoud multiplet to the SU(4) x U(1) algebra,
it is shown how, by performing the dynamical symmetry
breaking starting from the original Lagrangian, all
spinors get large self-masses of ultraviolet origin;
if the QED generating transformation of the diagonal
Lagrangian is first performed, the uncharged spinors
(neutrinos) remain massless, while the charged ones get
a self-mass, and this can be of infra-red origin (the
integral equations are of the Fredholm type) and might
be made small. The possible interplay of infra-red and
ultraviolet driven masses is discussed.

I. INTRODUCTION

Starting from massless spinors (Weyl or Majorana)
interacting either directly [1] or via vector and axial-
vector fields [2], one can generate masses. In so doing
one has the necessity of breaking the original scale invariance of the massless Lagrangian by introducing a
fundamental mass or length.

One can conceive this length as the range of the
region "inside" which the fields and particles manifest
their fundamental nature of massless Weyl or Majorana
fields and, as such, do not conserve parity, but conserve spinor number, when interacting; the original interactions will generate masses mainly from the inside
region (ultraviolet). "Outside" it the fields appear
massive (possibly Majorana) spinors and combine in
Dirac bi-spinors and in massless and massive bosons;

in the "outside" interactions they conserve parity.

In the physical world the "outside" interactions are represented by the familiar strong and electromagnetic ones among massive spinors and bosons and represent parity-conserving interactions of massive relatively loose compound systems built up from the fundamental ones. These will also give rise to self-mass effects but from the "outside" region (infra-red); the symmetry will be broken but representations still valid to classify fields.

Is it conceivable to observe the "inside" interactions without using energies (c.m.) corresponding to the fundamental mass? The answer would be yes if we supposed the existence of an agent active only in the inside region. This agent could be a boson interacting with the fundamental spinors and having a mass of the order of the fundamental one (or a force having the range of the fundamental length). Its Green function would then be representative of the fundamental length, and in all interactions mediated by it the spinor fields would appear as massless Weyl fields and parity would not be conserved. Besides, such interactions could be manifest not only in high-energy phenomena (that is for wave lengths of the order of the fundamental length) but also at low energy provided the fermion-boson coupling constant be high enough and that the spinor masses are dynamically originated, possibly in the outside region that is infra-red driven, as compared with the short distance interaction. However, these interactions will in general be very weak at low energy as compared with those strong ones among the large composite systems, because of the small volume involved by the static limit of the intermediate boson

Green function. We do have in Nature a possible good example of these interactions: the weak ones. These in fact suggest to us that Fermi coupling constant is the low-energy value of a propagator and that itself represents the fundamental length of the theory.

We will now try to present two simple models along these lines of thought: the one framed on our knowledge of leptons; the other will only draft a possible attempt to include both leptons and hadrons.

II. MODEL FOR LEPTONS

Let us suppose that the fundamental (inside) field is represented by two triplets of Weyl fields: the one having the quantum numbers of the electron and let us insert them, formally, in one multiplet:

$$\Psi_e = \begin{bmatrix} e_L \\ e_R \\ \nu_L^{(e)} \end{bmatrix} , \tag{1}$$

where $e_{L,R} = \frac{1}{2}(1 \pm \gamma_5)e(x)$ and $\nu_L^{(e)}$ refers to the electron neutrino, and the other with those of the muon

$$\Psi_\mu = \begin{bmatrix} \mu_L \\ \mu_R \\ \nu_L^{(\mu)} \end{bmatrix} . \tag{1'}$$

Let us suppose that the Lagrangian density is the formally SU(3) x U(1) invariant

$$L = L_o + L_I \, , \tag{2}$$

where L_o is the free Lagrangian for the massless spinors, while

$$L_I = g \sum_{e,\mu} \sum_{i=0}^{8} \bar{\Psi}(\ell) \gamma_\rho \lambda_i \Psi(\ell) A_i^\rho \, , \tag{3}$$

where the U(3) generators λ_i (including $\lambda_o = a \cdot 1$) act in the triplet spaces of each of the two leptons[+].

The fields $A_i^\rho(x)$ are massless vector-axial-vector fields, but they are not necessarily conceived as Yang-Mills gauge fields. In fact (3) need only be invariant for phase and chiral transformations of the spinors with constant phases such that the currents in (3) are conserved.

Let us take the Gell-Mann [3] representation of the λ_i then it is easily seen that, because of the chiral projectors, $\lambda_1 \lambda_2 \lambda_6 \lambda_7$ give zero contribution so that the interaction Lagrangian reduces to:

$$L_I = g \sum_{e,\mu} (\bar{\Psi}_\ell \lambda_3 \gamma_\rho \Psi_\ell A_3^\rho + \bar{\Psi}_\ell \lambda_{\pm} \gamma_\rho \Psi_\ell A_{\mp}^\rho +$$

$$+ \bar{\Psi}_\ell \lambda_8 \gamma_\rho \Psi_\ell A_8^\rho + a \bar{\Psi}_\ell \gamma_\rho \Psi_\ell A_o^\rho) \tag{4}$$

[+] In a more realistic theory one would start with multiplets of Dirac spinors and then show how, for very short distance interactions, the Dirac spinors may reduce to Weyl ones if the masses are of dynamical origin and infra-red driven (see Sec. IV). This will be discussed in a subsequent work.

where we have put $\lambda_\pm = \frac{1}{2}(\lambda_4 \pm i\lambda_5)$ and similarly for $A_\mp^\rho$.

If we now write L_I in terms of the leptons and neutrinos, we obtain:

$$L_I = g \sum_{e,\mu} \{ \bar{\ell} \gamma_\rho \gamma_5 \ell A_3^\rho + \bar{\ell}_L \gamma_\rho \nu_L^{(\ell)} A_-^\rho + \text{c.c.} +$$

$$+ \frac{1}{\sqrt{3}}(\bar{\ell} \gamma_\rho \ell - 2\nu_L^{(\ell)} \gamma_\rho \nu_L^{(\ell)}) A_8^\rho +$$

$$+ a(\bar{\ell} \gamma_\rho \ell + \nu_L^{(\ell)} \gamma_\rho \nu_L^{(\ell)}) A_0^\rho \} . \tag{5}$$

It is seen that in such a way the original symmetry is reduced to $U(2) \times U(1)$ and the unwanted fields A_1 A_2 A_6 A_7 are eliminated[+]. After symmetry breaking, the non-diagonal terms will give rise to μ-$e\bar{\nu}\nu$ decay, while the diagonal ones to electromagnetic and neutral currents.

Let us now allow for the generation of spinor masses by iterating L_I. As discussed in a previous paper [4], iteration of an interaction Lagrangian where the left-handed and right-handed currents appear separated, can only give two kinds of effective Lagrangians: those where products of current of the same chirality appear, that is

$$L_{\text{eff}}^I(x) = g^2 \int [j_\rho^L(x) \, j_\rho^L(x+\eta) + j_\rho^R(x) \, j_\rho^R(x+\eta)] \, \Delta(\eta) d^4\eta, \tag{6}$$

where

[+] But this happens only for the massless (inside) fields. For the dressed (outside) fields the original symmetry will be restored.

$$j_\rho^{L,R} = \frac{1}{2} \bar{\Psi} \gamma_\rho (1 \pm \gamma_5) \Psi,$$

and those where currents of opposite chirality appear,

$$L_{eff}^{II}(x) = g^2 \int [j_\rho^L(x) j_\rho^R(x+\eta) + j_\rho^R(x) j_\rho^L(x+\eta)] \Delta(\eta) d^4\eta. (7)$$

The first ones, in the case bound states are formed, will give effective vertices where a composite vector-axial-vector boson is coupled to a left-handed or right-handed current (and, under certain conditions for $\Delta(\eta)$, they will gauge [4]). The second ones can give rise to mass terms when they refer to the same fields.

In our model it is easily seen that we can only have mass terms from the diagonal terms:

$$L_I^d = g \sum_{e,\mu} \bar{\Psi} \gamma_\rho (\lambda_3 A_3^\rho + \lambda_8 A_8^\rho + a A_0^\rho) \Psi =$$

$$= g \sum_{e,\mu} [(j_\rho^L - j_\rho^R) A_3^\rho + \frac{1}{\sqrt{3}} (j_\rho^L + j_\rho^R - 2 j_\rho^L(\nu)) A_8^\rho +$$

$$+ a (j_\rho^L + j_\rho^R + j_\rho^L(\nu)) A_0^\rho] \tag{8}$$

(where now $j_\rho^{L,R}$ refers only to e and μ). Furthermore, only the leptons may acquire mass since the neutrino fields appear only with their left-handed components, and we have that (7) in our case becomes

$$L_{eff}^{(II)}(x) = \frac{ig^2}{2} \sum_{e,\mu} \int (j_\rho^L(x) j_\rho^R(x+\eta) + j_\rho^R(x) j_\rho^L(x+\eta)) \cdot$$

$$\cdot (-\Delta_3(\eta) + \tfrac{1}{3}\Delta_8(\eta) + a^2\Delta_o(\eta)) \, d^4\eta =$$

$$= -2ig^2 \int_{e,\mu} [\, \bar{\ell}(x+\eta)\, \ell(x)\, \bar{\ell}(x)\, \ell(x+\eta) \; -$$

$$-\; \ell(x+\eta)\, \gamma_5 \ell(x)\, \bar{\ell}(x)\, \gamma_5 \ell(x+\eta) \,][\, -\Delta_3(\eta) + \tfrac{1}{3}\Delta_8(\eta) + a^2\Delta_o(\eta)\,]\, d^4\eta,$$

$$(9)$$

which, apart from the propagators $\Delta(\eta)$, is of the form of the chiral invariant Lagrangian postulated by Nambu-Jona-Lasinio [1]. This will give rise to the self-energy for the leptons:

$$\Sigma^{(\ell)} = 2g^2 \int [\, T_2\, S_F(\eta) \;-\; \gamma_5\, T_2\, S_F(\eta)\, \gamma_5\,] \quad \cdot$$

$$\cdot\; [\, a^2\Delta_o(\eta) + \tfrac{1}{3}\Delta_8(\eta) - \Delta_3(\eta)\,]\, d^4\eta, \qquad (10)$$

which is only logarithmically divergent as compared with the Nambu-Jona-Lasinio case. However, we know that the $\Sigma(p)$ appearing in $S_F(\eta)$ is asymptotically vanishing for $p^2 \to \infty$ ($\eta \to 0$) and consequently the integral in (10) converges, but still its main contribution comes from the ultraviolet region ($\eta \simeq 0$).

We will have finally:

$$\Sigma^{(e,\mu)}(p^2) \neq 0 \qquad \text{while} \qquad \Sigma^{(\nu)}(p^2) \equiv 0. \qquad (11)$$

Before discussing the possible values of lepton masses, let us go back to (5) and point out that since chiral and gauge invariance are broken, the first two lepton vertices will generate a Nambu-Goldstone pole and, by the Schwinger mechanism [5], A_3, A_+ and A_- will acquire a mass. Let us briefly resume this mechanism.

Let

$$D^{\mu\nu}(q^2) = -i \ (g^{\mu\nu} - \frac{q^\mu q^\nu}{q^2}) \ \frac{1}{q^2 - q^2 \Pi(q^2)} \tag{12}$$

be the complete vector meson propagator, where $\Pi(q^2)$ represents the proper part of vacuum polarization and is given by:

$$i(g^{\mu\nu}q^2 - q^\mu q^\nu)\Pi(q^2) = -g^2 \int d^4x \ e^{iqx} <0|Tj^\mu(x) j^\nu(0)|0>_{pr.p.} \tag{13}$$

If $\Pi(q^2)$ has a pole,

$$\Pi(q^2) = \frac{g^2 \lambda^2}{q^2} \tag{14}$$

then from (12) the meson gets a mass given by

$$M^2 = g^2 \lambda^2 \ . \tag{15}$$

Now it is known that a pole for the vertex Γ^μ induced by (3) is required by the Ward-Takahashi identity

$$q_\mu \Gamma^\mu(p, p+q) = \gamma^5 G^{-1}(p+q) + G^{-1}(p)\gamma^5 \ , \tag{16}$$

where

508

$$G^{-1}(p) = -i [\not{p} - \Sigma(p)],$$

since the current must be conserved, and, since $\Sigma(p) \neq 0$ for leptons, (16) must be $\neq 0$ for $q \to 0$ and $\Gamma^{\mu}(p,p+q)$ must have a pole in q. As a consequence (see Ref.[2]) the vacuum polarization $\Pi(q^2)$ gets a pole represented by (14) where

$$g^{\mu\nu} g^2 \lambda^2 = -i \, \tilde{\Pi}^{\mu\nu}_A (0) \tag{17}$$

where $\tilde{\Pi}^{\mu\nu}_A (0)$ is the vacuum polarization tensor associated with the non-singular part of the non-conserved axial-vector current. Equivalently, the residue $g^2 \lambda^2$ can be computed from the identity (16) once the self-mass $\Sigma(p)$ is fixed.

Going back to our model, let us suppose we are computing the mass of the intermediate mesons $A_{\mp}$ in (4). We would obtain integrals of the type: [2]

$$g^{\mu\nu} \lambda^2 g^2 = ig^2 \text{Tr} \int \frac{d^4 q}{(2\pi)^4} \frac{(\hat{q}+m) \gamma^{\mu} (\hat{q}+m) \gamma^{\nu} - (\hat{q}+m) \gamma^{\mu} \gamma_5 (\hat{q}+m) \gamma^{\nu} \gamma_5}{(q^2 - m^2)^2}$$

$$= g^{\mu\nu} M^2_{\pm} , \tag{18}$$

which is logarithmically divergent and needs a cut-off unless one assumes to substitute m in (18) with an asymptotically vanishing $\Sigma(p)$.

We know that in second order the terms $g\bar{\ell}\gamma_{\rho}(1 + \gamma_5)\nu A^{\rho}_{+}$ must give rise to the Fermi Lagrangian, and in this way $M_{\mp}$ is connected with the Fermi coupling constant, or better Fermi length, by the relation:

$$G_F = \frac{g^2}{M_{\mp}^2} = \lambda^{-2} \, , \qquad (19)$$

where (15) has been taken into account.

It is seen from (18) that G_F is independent explicitly of g. It only depends on the dynamically acquired masses of the fermions which in turn might depend on g.

In this way (19) acquires the meaning of the self-consistency equation which fixes the scale of the model. The masses are determined implicitly in terms of the Fermi length which establishes the measure of the fundamental length of the theory.

Usually when computations of dynamical symmetry breaking are performed starting from conventional models, e.g. Salam-Weinberg model [6], it is found that $g^2 \sim 1$ and, consequently,

$$M_{\mp}^2 \simeq G_F^{-1} \, , \qquad (20)$$

which confirms the possible role of the Fermi length.

In these models also the fermion masses are obtained of the same order of magnitude, unless special hypotheses are invoked [6].

In our model it is seen from (10) that the fermion masses will depend on the coupling constant $a^2 g^2$ of the

U(1) invariant term of the Lagrangian; in fact supposing that the intermediate propagators $\Delta_o(\eta)$ $\Delta_3(\eta)$ $\Delta_8(\eta)$ are equal (for $\eta \sim 0$, which gives the main contribution) one would get positive fermion masses only for

$$a^2 \geq \frac{2}{3} , \tag{21}$$

and zero in the case of equality. The fermion masses will then, in general, depend on a and it will be possible to choose the value of a such as to make them small compared with those of intermediate bosons.

Once the leptons have acquired a mass, they will obey a Dirac equation and it will be possible to choose a linear combination of the fields A_3^ρ A_8^ρ A_o^ρ appearing in the diagonal Lagrangian L_I^d in such a way that one of the fields A^ρ interacts with the conserved current $\bar{\ell}\gamma_\rho \ell$ and as a consequence it may remain massless. There are many possibilities for this choice; let us take the simpler one:

$$A_3 = \frac{1}{\sqrt{3}} T - \sqrt{\frac{2}{3}} Z$$

$$A_8 = \frac{1}{\sqrt{3}} A + \frac{2}{3} T + \frac{\sqrt{2}}{3} Z$$

$$A_o = \frac{\sqrt{2}}{3} A - \frac{\sqrt{2}}{3} T - \frac{1}{3} Z . \tag{22}$$

Then (8) becomes:

$$L_I^d = g \sum_{e,\mu} [\bar{\ell}\gamma_\rho \ell \ A^\rho + \frac{1}{\sqrt{3}} \bar{\ell}\gamma_\rho \gamma_5 \ell (T^\rho - \sqrt{2} \ Z^\rho) -$$

$$- \frac{1}{\sqrt{3}} \, \bar{\nu}_L \gamma_\rho \nu_L \, (2 T^\rho + \sqrt{2} \, z^\rho) \,] \qquad (23)$$

A^ρ remain massless, B^ρ and Z^ρ massive as usual.

The Lagrangian (23) has the interesting property that, if a dynamical symmetry breaking is started from this Lagrangian, then the equations for the fermion masses become convergent (of the Fredholm type) and the self-masses are obtained mainly from low momenta in the integrals or they are "infra-red driven" according to Englert et al. [6]. In fact, the self-mass would have the form:

$$\Sigma^{(\ell)} (p^2) = \frac{-3 i g^2}{(2\pi)^4} \int d^4 k \, \frac{\Sigma(k^2)}{k^2 - \Sigma(k^2)} \, [\frac{1}{(p-k)^2} - \frac{1}{(p-k)^2 - M_3^2}] \, , \quad (24)$$

where M_3 is the mass of the boson A_3^ρ.

One would then conceive that this is also a possibility of diagonalization of the fermion masses and that, while at high energy (and low distances) the diagonalization (8) prevails, at low energies and large distances (23) is dominant.

In terms of the previous algebra, the charge operator of the model is now:

$$Q = \frac{1}{\sqrt{3}} \lambda_8 + \frac{2}{3} \lambda_o = \begin{bmatrix} 1 & & \\ & 1 & \\ & & 0 \end{bmatrix} . \qquad (25)$$

If we now substitute Dirac spinors for the massive leptons,

$$\psi^{(\ell)} = \begin{bmatrix} \psi_L^{(\ell)} \\ \\ \psi_R^{(\ell)} \end{bmatrix}, \tag{26}$$

we will obtain that the fields after dynamical symmetry breaking

$$\Psi = \begin{bmatrix} \psi^{(e)} \\ \nu_L^{(e)} \\ \psi^{(\mu)} \\ \nu_L^{(\mu)} \end{bmatrix} \tag{27}$$

constitute the basis for an $U(4)$ or $SU(4)$ symmetry algebra with an integer charge operator:

$$Q = \begin{bmatrix} 1 & & & \\ & 0 & & \\ & & 1 & \\ & & & 0 \end{bmatrix} .$$

We have only considered mass terms obtained by the iteration of the Lagrangian (5). But we know, as mentioned, that, provided g is large enough, also composite states may be generated from the effective Lagrangian (6) between fermions-anti-fermions. As discussed in a previous paper [4] these composite states will be either vector-axial-vector or scalar-pseudoscalar and will, in general, span the whole $U(3)$ algebra depending on kernels and coupling constants. The size of the composite states will, in general, be comparable to the fundamental length and will only allow spin 1 states, since higher orbital excitations will be excluded by centrifugal forces [7].

It is also interesting to point out that the spin 1
composite states, depending on not restrictive conditions
on the binding kernel (1/r dependence for r → 0), are
subject to gradient gauge transformations if the constituent spinors are subject to a non-constant phase
transformation [7]. In this way, even if we started
from non-gauge vector and axial-vector fields A_i^0 (and
neutral Majorana spinors); the composite vector-axial-
vector fields resulting from the model would gauge (and
the Dirac spinors would get charge). This could be a
motivation for considering the gauge fields we know
(e.g. the e.m. one) as composite fields.

III. THE SYMMETRIC MODEL

In the previous model there was a dominance of
left-handed components among spinors. This asymmetry
guaranteed us the neutrino masslessness.

It is difficult to accept that a basic Lagrangian
may have this asymmetry.

We know that a left-right symmetry is offered by
the Konopinski-Mahmoud [8] Lagrangian based on the
multiplet

$$
\Psi = \begin{bmatrix} e_L \\ \mu_{eR} \\ \nu_L^{(e)} \\ \nu_R^{(\mu)} \end{bmatrix} \qquad . \tag{28}
$$

33

Then we could try tu build up a model starting from the
Weyl multiplet:

$$\Psi \;=\; \begin{bmatrix} \psi_L \\ \psi_R \\ \psi_L^{(\nu)} \\ \psi_R^{(\nu)} \end{bmatrix} \tag{29}$$

where the last two components have the quantum numbers
of the neutrinos, and consider it as a basis for a U(4)
algebra. This model automatically becomes exceedingly
ambitious because one cannot insert from the beginning
the muon and electron quantum numbers as in the multiplet (28); in fact from this symmetric multiplet one
could not easily obtain mass terms of the type (9)
since the electron appears only with the left-handed
component and the muon with the right-handed one[+].

Let us nevertheless start from the SU(4) x U(1)
symmetric Lagrangian:

$$L = L_O + L_I \; ,$$

where

$$L_I = g \sum_{i=o}^{15} \overline{\Psi} \, \lambda_i \, \gamma_\rho \, \Psi \, A_i^\rho \tag{30}$$

[+] One could write coupled mass equations where the left-
handed electron field combines with the right-handed
muon and vice versa to get two mass eigenvalues; but
we will not examine this possibility here.

and

$$\lambda_o = a \cdot I \, .$$

As before, due to chiral projectors, the couplings with unwanted boson fields will be eliminated.

Let us now examine the diagonal components giving rise to possible mass terms. We obtain that the Lagrangian corresponding to (8) is

$$L_I^d = g \; [\, (j_\rho^L - j_\rho^R) A_3^\rho + \frac{1}{\sqrt{3}} \; (j_\rho^L + j_\rho^R - 2j_\rho^L(\nu)) A_8^\rho \; +$$

$$+ \; \frac{1}{\sqrt{6}} \; (j_\rho^L + j_\rho^R + j_\rho^L(\nu) \; - \; 3j_\rho^R(\nu)) A_{15}^\rho \; +$$

$$+ \; a \; (j_\rho^L + j_\rho^R + j_\rho^L(\nu) \; + \; j_\rho^R(\nu)) A_o^\rho \,] \quad . \tag{31}$$

We will suppose equality of the boson propagators and obtain ,for the Lagrangian corresponding to (9),

$$L_{eff.}^{(II)} = \frac{ig^2}{2} \int \{ \, ((j_\rho^L j_\rho^R + j_\rho^R j_\rho^L) \, (-1 + \tfrac{1}{3} + \tfrac{1}{6} + a^2) \; +$$

$$+ \; [j_\rho^L(\nu) j_\rho^R(\nu) + j_\rho^R(\nu) j_\rho^L(\nu)] (-\tfrac{3}{6} + a^2) \; +$$

$$+ \; [j_\rho^L(\nu) j_\rho^R + j_\rho^R j_\rho^L(\nu)] (\tfrac{1}{6} - \tfrac{2}{3} + a^2) \; +$$

$$+ \; [j_\rho^R(\nu) j_\rho^L + j_\rho^L j_\rho^R(\nu)] (-\tfrac{3}{6} + a^2) \} \Delta(\eta) d^4\eta \; . \tag{32}$$

It is seen that for $a^2 = \frac{1}{2}$ all terms in the integral
go simultaneously to zero, or they are all different
from zero for $a^2 \neq \frac{1}{2}$. There seems to be no possibility
in this model to leave some components of the multiplet
massless (neutrinos) while others get massive at least
for the ultraviolet driven masses.

However, let us now suppose $a^2 = \frac{1}{2}$ and let us
choose another basis for the boson fields

$$A = \frac{A_8}{\sqrt{3}} + \frac{A_{15}}{\sqrt{6}} + \frac{1}{\sqrt{2}} A_o$$

$$\frac{1}{\sqrt{2}} B = - \frac{\sqrt{2}}{\sqrt{3}} A_8 + \frac{A_{15}}{3\sqrt{2}} - \frac{1}{2} A_o$$

$$\frac{1}{\sqrt{2}} C = - \frac{\sqrt{3}}{2} A_{15} + \frac{1}{2} A_o \quad . \tag{33}$$

Then the Lagrangian (31) becomes

$$L_I^d = g \ [\bar{\Psi} \ \gamma_\rho \ \gamma_5 \ \Psi \ A_3^\rho + \bar{\Psi} \ \gamma_\rho \ \Psi \ A^\rho +$$

$$+ \frac{1}{\sqrt{2}} \ \bar{\nu} \ \gamma_\rho (1 + \gamma_5) \nu \ B^\rho \ +$$

$$+ \frac{1}{\sqrt{2}} \ \bar{\nu} \ \gamma_\rho (1 - \gamma_5) \nu \ C^\rho] \quad . \tag{34}$$

It is seen that in this basis we will obtain an infra-
red driven mass for the (charged) field Ψ, while the
neutrino field will remain massless since there is no
possible communication between its left-handed and
right-handed components and there is no ultraviolet

driven mass since $a^2 = \frac{1}{2}$ and all terms in (32) are zero (in the above hypothesis).

IV. SOME SPECULATIVE OUTLOOKS

It is amusing to think of Eq. (32) for $a^2 \neq \frac{1}{2}$ as that responsible for the ultraviolet driven masses of quarks, while for $a^2 = \frac{1}{2}$, (34) originates the infra-red driven lepton masses and leaves neutrinos massless. In this way leptons and neutrinos would represent the "out-side" dynamical solutions (infra-red driven masses: $a^2 = \frac{1}{2}$) while quarks "inside" solutions (ultraviolet driven masses: $a^2 \neq \frac{1}{2}$).

Still, among others, the problem remains how to explain the multiplicity of basic fermion fields necess-ary to give account of the complexity of the so-called elementary systems.

One possibility was suggested by Nambu and Jona-Lasinio [1] and is offered by the fact that equations like (10) admit more solutions for self-masses. For zero value of the bare mass, (10) has the solution

$$\Sigma(p) = 0 , \tag{35}$$

called "trivial" by Nambu-Jona-Lasinio, and a non-trivial one:

$$\Sigma(p^2) \neq 0 . \tag{36}$$

If we admit a non-zero mass (possibly breaking the symmetry of the original Lagrangian [1]) then the possibility arises that the self-energy equation (10) gets a solution not far from the bare one, corresponding to the trivial one of the massless case, and one or two (corresponding to $\Sigma(p^2) = \pm m$) corresponding to the non-trivial one; the last ones seem to be large - of the order of the intermediate boson mass - and could correspond to quark fields [6]. In the case examined by Nambu-Jona-Lasinio the difficulty arose that some of these masses had a negative sign. In our case, due to the negative contribution to self-mass equations, originating from the internal algebra (see (10)), perhaps the problem of negative self-masses [1] could be overcome.

These eigensolutions could correspond to different possibilities of existence for the fermions and, the transition from one eigensolution to the other being forbidden (in the massless case they correspond to orthogonal worlds [1]), they could be the origin of the conserved quantum number for the elementary fermions.

ACKNOWLEDGEMENTS

The author acknowledges with thanks useful discussions with F. Englert, and also P. Furlan who pointed out to him that the choices 2/3 and 1/2 for a^2 giving infra-red driven masses simply corresponds to the $U(3)$ and $U(4)$ symmetry for the models.

REFERENCES

1. Y. Nambu and G.Jona-Lasinio, Phys. Rev. 122, 345
   (1960) and Phys. Rev. 124, 246 (1961).

2. F. Englert and R. Brout, Phys. Rev. Letters 13,
   321 (1964);
   R. Jackiw and K. Johnson, Phys. Rev. D 8, 2386 (1973);
   J.M. Cornwall and R.E. Norton, Phys. Rev. D 8, 3338
   (1973).

3. M. Gell-Mann and Y. Ne'eman: The Eight-fold Way,
   p. 49 (Benjamin, 1964).

4. P. Budini and P. Furlan, Nuovo Cimento 30A, 63 (1975).

5. J. Schwinger, Phys. Rev. 125, 397 (1962).

6. F. Englert and R. Brout, Phys. Letters 49B, 77 (1974);
   F. Englert, J.-M.Frere, P.Nicoletopulos, Univ. of
   Bruxelles preprint (1975).

7. P. Budini and P. Furlan, ICTP, Trieste, preprint
   IC/74/27.

8. E. Konopinski and H. Mahmoud, Phys. Rev. D 92, 1045
   (1953).

Acta Physica Austriaca, Suppl. XV, 521–568 (1976)
© by Springer-Verlag 1976

SOLITON MODELS OF HADRONS[+]

by

P. VINCIARELLI

CERN, Geneve

TABLE OF CONTENTS

[+]Lecture given at XV.Internationale Universitätswochen für
Kernphysik,Schladming,Austria, February 16-27, 1976.

III Realistic Models

    a) Bag models

    b) How to treat fermions

    c) Abelian quark-gluon gauge model

INTRODUCTION

Conventional perturbation theory has provided the foundations and the basic tool in the application of field theory to the weak interactions of leptons. While it is hoped that field theory will also eventually yield a unified and complete description of hadron physics, it is as yet unclear which field theory one should take and what the best point of attack to such a theory should be. The impressive success of the quark model would favor the identification of quarks with the underlying constituents of hadronic matter and the unquestioned success of the principle of local gauge invariance would then lead to the choice of a field theory of quarks and gauge fields. An esthetical difficulty facing this approach is the relationship between quarks and leptons or, better, the lack of it, and a practical problem is the present day unobservability of quarks and gauge mesons as independent entities. One could speculate that these two problems will actually solve each other, e.g., by the identification of quarks with leptons, but such interesting options appear for the moment to be unfeasible. Inevitably, quark confinement (even if temporary, i.e.,to be followed by quark liberation at higher energies) presents itself as an important aspect of hadron dynamics.

The disappearence of a class of states from the particle spectrum of the gauge field theory should be accomp-

annied by another remarkable phenomenon, the appearence
of a very rich structure of bound states, the observed
hadrons. Such bound systems should be "extended", have
"soft" strong interactions, etc... Thus, in contrast to
lepton dynamics, salient features of hadron dynamics
have an intrinsically non-perturbative character.

An approach to quantum field theory that exposes
some of its non-perturbative features at the outset is
the semiclassical method. The first step of this method
consists of finding non-perturbative solutions to the
classical (non-linear) field equations. Finite energy
solutions to these equations which meet certain stability
criteria and possess particle-like properties are called
"solitons". They are the classical counterpart of certain
(bound) particle states in the quantum field theory. Because many soliton properties are reminiscent of properties of hadrons, it is hoped that hadrons may be identified
with the quantum solitons of the field theory of strong
interactions.

In these lectures I will present a brief account
of this program. As implied by the previous discussion,
the program may be roughly devided into two parts: a)
the finding of classical soliton solutions; b) their
quantization. Concerning the first part, I will discuss
several examples including simple model field theories
in one-space-one-time dimension, semi-realistic bag-
type theories of quarks and scalar gluons, and finally
realistic gauge theories of quarks and vector gluons.
The blame for leaving out other interesting examples
is mine, a reflection of my personal biases. The same
comment applies to my choice of a method for a discussion of soliton quantization: the variational method.

The main advantages of this method are its simplicity, intuitive appeal and flexibility. I shall use it for quantizing particle spectra, scattering processes and also to derive "classical" equations in theories with fermions.

I. CLASSICAL SOLITONS

Ia) What is a soliton [1].

Consider the linear, free field equation:

$$-\ddot{\Phi} + \Phi'' - m^2\Phi = 0, \quad \Phi = \Phi(x,t) \ . \tag{I.1}$$

This equation is dispersive: its sinusoidal solutions

$$\Phi(x,t) = e^{ikx + ik_o t}, \quad k_o = \sqrt{m^2 + k^2} \tag{I.2}$$

have different phase and group velocities

$$\frac{k_o}{k} = \frac{\sqrt{m^2 + k^2}}{k}, \quad \frac{dk_o}{dk} = \frac{k}{\sqrt{m^2 + k^2}} \ . \tag{I.3}$$

Therefore, wave packet solutions

$$\Phi(x,t) = \int dk \ \tilde{\Phi}(k) \ e^{ikx + ik_o t} \tag{I.4}$$

do <u>not</u> have the property of localizing permanently the energy density

$$H(x) = \frac{1}{2} \dot{\phi}^2 + \frac{1}{2}\phi'^2 + m^2 \phi^2 . \qquad (I.5)$$

Wave packets do <u>not</u> have a particle interpretation:

$$t = 0 \qquad\qquad\qquad t >> 0$$

Fig. 1

This is to be contrasted with the standard linear wave equation

$$- \ddot{\phi} + \phi'' = 0 \qquad (I.6)$$

for which the general solution

$$\phi(x,t) = \sum_n \phi_n (x \pm t) \qquad (I.7)$$

is a superposition of "solitary waves" travelling to the left and to the right with velocities ± 1 without a change of shape. This provides the simplest but rather untypical example of "soliton" solutions in a linear, dispersion-less system.

Introducing non-linearity without dispersion again does not lead to solitary wave solutions because of harmonic generation: the energy density of the wave is continually injected into higher frequency modes. Thus the equation

$$- \ddot{\phi} + \phi'' + \lambda\phi^n = 0, \qquad n > 1 \qquad (I.8)$$

which, when linearized, is dispersionless, will not exhibit solitons.

But introducing both dispersion and non-linearity, solitons may exist once more. In fact, the equation

$$- \ddot{\phi} + \phi'' = U'(\phi) = - 2 m^2 \phi + \lambda \phi^3 \qquad (I.9)$$

has the soliton solution:

$$\phi_v(x,t) = \pm \frac{m}{\sqrt{\lambda}} \tanh \left[\frac{m(x-vt)}{\sqrt{1-v^2}}\right] . \qquad (I.10)$$

This solution has remarkable particle-like properties:
i) while travelling, it retains its shape at all times

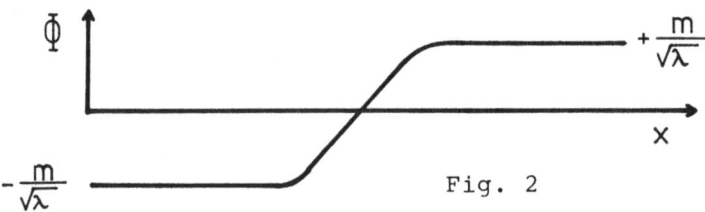

Fig. 2

ii) the energy density is localized

Fig. 3

and leads to a finite energy

$$E = \int dx \, H(x) = \frac{4}{3} \frac{m^3}{\lambda}, \quad (v = 0) \ . \tag{I.11}$$

iii) it carries a charge

$$Q = \int_{-\infty}^{+\infty} dx \, j_0(x) = \int_{-\infty}^{+\infty} \frac{\sqrt{\lambda}}{2m} \, \Phi'(x) = \pm \, 1 \tag{I.12}$$

corresponding to the conserved current:

$$j_\mu(x) = \frac{\sqrt{\lambda}}{2m} \, \varepsilon_{\mu\nu} \, \partial^\nu \, \Phi \ . \tag{I.13}$$

This and analogous charges are sometimes referred to as "topological charges": they are nonvanishing because of the unusual boundary conditions satisfied by certain soliton solutions.

Ib) A mechanical analog

It is illuminating to derive the static soliton solution by use of a mechanical analogy. Rewrite the static equation:

$$\Phi'' = -V'(\Phi), \quad V(\Phi) = -U(\Phi) \tag{I.14}$$

in the form

$$\ddot{y} = -V'(y), \quad y = \Phi, \quad t = x \tag{I.15}$$

528

and interpret it as the equation of motion of a newtonian point particle moving in the conservative field of force generated by the potential V(y). Corresponding to the previous example:

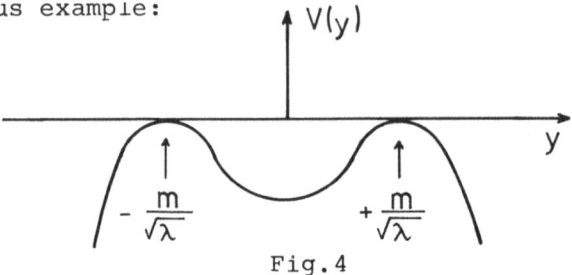

Fig.4

It is immediately recognized that the static soliton solution corresponds to the newtonian point particle starting off on its motion at time $t = - \infty$ from one of the two tops of the potential and ending the motion at $t = + \infty$ on the other top. It should be remarked that:

i) the degeneracy of the potential ground state is crucial to the existence of the solution.

ii) apart from this, the actual, detailed shape of the potential is immaterial to the existence of the solution, only affecting its shape.

iii) the analog problem in more than one-space-one-time dimension has no solution, since

$$\Phi'' \rightarrow \underline{\nabla}^2 \Phi = \partial_r^2 \Phi + \frac{n-1}{r} \partial_r \Phi \tag{I.16}$$

having assumed spherical symmetry, and the mechanical analog equation becomes

$$\ddot{y} = - V'(y) - \frac{n-1}{t} \dot{y}, \qquad 0 \leqslant t < \infty \tag{I.17}$$

which includes a dissipative, time and velocity-dependent (frictional) force.

Ic) A stability argument [2].

The extension of soliton structures to more realistic situations in 3-space-1-time dimension requires the introduction of fields with spin. Consider in fact the following stability argument for static soliton solutions in spinless field theories with non-derivative interactions.

The energy of a solution is given by:

$$E = E_T + E_V = \int d\underline{x} \, [\tfrac{1}{2}(\nabla\Phi)^2 + U(\Phi)] \; . \tag{I.18}$$

If we subject the solution to a scaling transformation,

$$\Phi(x) \to \Phi(\lambda x) \; ,$$

$$E(\lambda) = \lambda^{2-d} E_T + \lambda^{-d} E_V \; , \tag{I.19}$$

where d is the dimensionality of space. Since $\Phi(x)$ is, by assumption, a solution

$$\frac{\partial E}{\partial \lambda}\Big|_{\lambda = 1} = 0 \to E_V = \frac{2-d}{d} E_T \; . \tag{I.20}$$

This virial condition leads to the stability equation

$$0 < \left. \frac{\partial^2 E}{\partial \lambda^2} \right|_{\lambda=1} = 2 \ (2-d) \ E_T \qquad (I.21)$$

which cannot be satisfied for $d \geq 2$ since $E_T > 0$.

Id) Time-dependent soliton solutions

The remarks of sections Ib) and c) suggest that one can easily construct systems with soliton solutions in one-space-one-time dimension. A particularly interesting case, that of the Sine-Gordon equation, corresponds to the choice of potential:

$$U(\Phi) = \frac{m^4}{\lambda} \ [1 - \cos (\frac{\sqrt{\lambda}}{m} \Phi)] \qquad (I.22)$$

Fig. 5

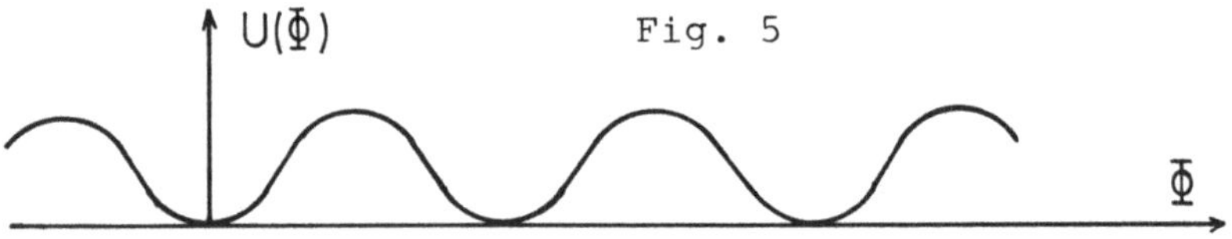

with an infinite number of degenerate ground states. Besides the soliton solution

$$\Phi_v(x,t) = \frac{4m}{\sqrt{\lambda}} \tan^{-1} [\exp (\pm m \frac{x-vt}{\sqrt{1-v^2}})] \qquad (I.23)$$

analytic "multi-soliton" solutions are known in this case. These include bound soliton-anti-soliton pairs

$$\Phi_T(x,t) = \frac{4m}{\sqrt{\lambda}} \tan^{-1} [\zeta \frac{\sin(2\pi\tau/T)}{\cosh(2\pi x/T)}] , \qquad (I.24)$$

$$\zeta = [(\frac{Tm}{2\pi})^2 - 1]^{\frac{1}{2}} \, , \; T > \frac{2\pi}{m} \tag{I.25}$$

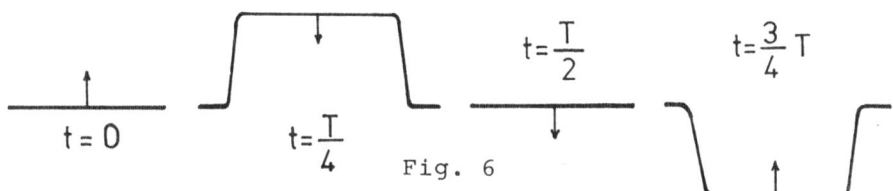

$t = 0 \qquad t = \frac{T}{4}$

$t = \frac{T}{2} \qquad t = \frac{3}{4} T$

Fig. 6

or "breather" modes: Eqs. (I.24-25) define, in the center of mass frame, a one-parameter family of solutions cha- racterized by the period of oscillation, T. Soliton-anti- soliton scattering solutions, which are analytic conti- nuation of the previous solutions above "threshold" are defined by (again, in the center of mass frame):

$$\Phi_u(x,t) = \frac{4m}{\sqrt{\lambda}} \tan^{-1} [\frac{\sinh(u\, mt/\sqrt{1-u^2})}{u\, \cosh(mx/\sqrt{1-u^2})}] \tag{I.26}$$

$t < 0 \qquad\qquad t > 0$

Fig. 7

As a function of the initial velocity u < 1 the scatter- ing process results in a "particle time advance":

$$\Delta = \frac{2}{m} \frac{\sqrt{1-u^2}}{u} \; \ln u \, . \tag{I.27}$$

II. SOLITON QUANTIZATION

Examples of classical model field theories with solitons have been introduced in the previous section. Before proceeding to (more) realistic examples, I would like to discuss, in a technically simpler context, aspects of the quantum theory of solitons from the point of view of a coherent state variational approach in Fock space. In particular I would like to answer, at least partially, the following questions. What is the significance of static classical soliton solutions in the quantum theory? Should they be interpreted as signals for the existence of bound states? Are detailed properties of such bound states, e.g. form factors, related to properties of the classical solutions? What is the spectrum of states corresponding to time-dependent soliton solutions (e.g., "breather modes"). What can one infer about the quantum scattering of solitons from the knowledge of classical scattering solutions? [3]

IIa) Classical equations as quantum variational equations

Let me consider the theory of a real scalar field ϕ in one-space-one-time dimension, such as the $\lambda\phi^4$ or the sine-Gordon model, with classical dynamics characterized by the Hamiltonian density:

$$H(x) = \tfrac{1}{2}\pi^2(x) + \tfrac{1}{2}\phi'^2(x) + U(\phi) . \tag{II.1}$$

Here, $\pi(x) = \dot{\phi}(x)$ is the momentum conjugate to the field $\phi(x)$, and $U(\phi)$ is a selfinteraction potential with absolute minimum at $\phi(x) \equiv 0$. In the quantum theory $\phi(x)$ and

$\pi(x)$ lead to operators acting in the Hilbert space of
states. I shall assume that vectors in this space may
be approximated by vectors in Fock space. The latter
is constructed in the usual way from the free fields
with canonical commutation relations. In the expansion
of these fields in terms of creation and annihilation
operators, at $t = 0$:

$$\Phi(x) = \int \frac{dk}{2\pi} [2\sqrt{m_o^2 + k^2}]^{-\frac{1}{2}} [a(k,m_o)e^{-ikx} + a^\dagger(k,m_o)e^{ikx}] \quad (II.2)$$

$$\Pi(x) = i \int \frac{dk}{2\pi} [\frac{1}{2}\sqrt{m_o^2 + k^2}]^{\frac{1}{2}} [a^\dagger(k,m_o)e^{ikx} - a(k,m_o)e^{-ikx}] \quad (II.3)$$

$$[a(k,m_o), a^\dagger(k',m_o)] = 2\pi\delta(k-k') \quad (II.4)$$

the mass m_o is regarded as a free parameter. I shall de-
note by $|0\rangle$ the Fock space no-particle state:

$$a(k,m_o)|0\rangle = 0 , \quad (II.5)$$

and by colons::normal ordering with respect to $|0\rangle$. As
usual this implies ordering an operator expression with
all the annihilation operators to the right.

A renormalized quantum Hamiltonian density, $:H(x):$
is defined as an operator in Fock space by normal ordering. For any theory of a scalar field in two dimensions
with non-derivative interactions this prescription is
sufficient to remove all divergences which occur in any
order of perturbation theory. A change of basis, i.e.

a change of m, corresponds to a finite renormalization of the Hamiltonian parameters.

For our purposes, it is not convenient to work in the ordinary Fock space basis which diagonalizes the "particle number" operator, but rather in a coherent state basis. Among the many reasons for choosing coherent states, let me mention in particular:

i) the fact that coherent states play a special role in the transition from a quantum theory to its classical limit; this is obviously a useful property when one is, as we are, trying to construct the quantum analog of a classical concept, the soliton

ii) technical reasons, since coherent states are eigenstates of annihilation operators it becomes particularly easy to calculate matrix elements of normal ordered operator expressions.

Let me recall that a coherent state of a real scalar field is defined in terms of two real functions $f(x)$, $g(x)$ by:

$$|f,g\rangle = \exp\left[- i\Pi(f) + i\Phi(g)\right]|0\rangle \tag{II.6}$$

where

$$\Pi(f) = \int dx\, f(x)\Pi(x), \quad \Phi(g) = \int dx\, g(x)\, \Phi(x). \tag{II.7}$$

It follows from the CCR that

$$\Phi^{(-)}(x)|f\rangle = \tfrac{1}{2}f(x)|f\rangle, \quad \Pi^{(-)}(x)|g\rangle = \tfrac{1}{2} g(x)|g\rangle \tag{II.8}$$

where $\Phi^{(-)}$ an $\Pi^{(-)}$ are the annihilation parts in the decomposition of the fields:

$$\phi(x) = \phi^{(-)}(x) + \phi^{(+)}(x) , \quad \Pi(x) = \Pi^{(-)}(x) + \Pi^{(+)}(x) . \quad (II.9)$$

It is then obvious that

$$\langle f | :\phi^n(x): | f \rangle = f^n(x) , \langle g | :\Pi^n(x): | g \rangle = g^n(x) . \quad (II.10)$$

In turn this implies that the expectation value of the renormalized quantum Hamiltonian density between coher-ent states of zero average momentum ($g = 0$) is the classical Hamiltonian density corresponding to the static field configuration f:

$$\langle f | : H : | f \rangle = H(f) . \quad (II.11)$$

Thus the Euler equations of the variational problem de-fined by the minimization of the Hamiltonian expectation value within the family of "trial" coherent states $\{|f\rangle\}$ are the static classical field equations. These equat-ions are therefore seen to acquire a quantum mechanical significance. Their static soliton solutions define co-herent states which may be interpretable as approximate one-particle states in the quantum field theory.

Of course particle states in field theory are characterized as eigenstates of the mass operator

$$M|m\rangle = m|m\rangle , \quad M = \sqrt{H^2 - \underline{P}^2} , \quad (II.12)$$

where $\underline{P}$ is the momentum operator, and, consequently, a variational search for approximate one-particle states

based on the Hamiltonian is only justified when the
trial states are momentum eigenstates or, at least, if
their momentum spread is small as compared to the mass.
Thus one must realize that, from this point of view,
classical soliton solutions loose some of their quantum
mechanical significance away from a non-relativistic
region ($< \underline{p}^2 > \ll m^2$). This region corresponds to the so-
called "weak coupling limit" ($\lambda \to 0$ in the example of
section Id) in which the soliton becomes infinitely
massive as compared to the fundamental meson, the ord-
inary particle exposed in perturbation theory around $|0>$.

IIb) The soliton form factor

In the weak coupling limit the Fourier transform
of the classical soliton "wave function" yields approxi-
mately the static field "form factor" of the quantum
soliton state. To see this, given the solution $|f>$ of
the variational problem discussed in the previous section, let me construct from it momentum eigenstates by
projection:

$$|p> = \frac{1}{\rho(p)} \int da\ e^{i(p-\underline{P})a}|f> = \frac{1}{\rho(p)} \int da\ e^{ipa}|f_a>. \qquad (II.13)$$

The weight $\rho(p)$ is determined up to a phase from the normalization condition

$$<p'|p> = 2\pi\delta(p' - p) \qquad (II.14)$$

to be

$$|\rho\,(p)|^2 = \int da\ e^{ipa} <f|\,f_a> =$$

$$= \int da\ e^{ipa}\ e^{-\frac{1}{2}||\,f\,-\,f_a||} \qquad (II.15)$$

where

$$||f||^2 = <0|\,\pi\,(f)^2|\,0> = \frac{1}{2}\int dp\ \sqrt{m_o^2 + p^2}|\,\tilde f\,(p)|^2\ . \qquad (II.16)$$

The equal time CCR then imply:

$$<p'|\,\phi\,|p> = \frac{\rho^2(p')+\rho^2(p)}{2\rho\,(p')\rho\,(p)}\int da\ e^{i\,(p-p')\,a}\ f(a)\ . \qquad (II.17)$$

In the weak coupling limit, constraining the momentum transfer to a non-relativistic region, Eq. (II.17) reduces to:

$$<p'|\,\phi\,|p> \simeq \int da\ e^{i\,(p-p')\,a}\ f(a)\ . \qquad (II.18)$$

Note that in models with spontaneous symmetry breakdown and soliton wave functions, $f(x)$, interpolating between different vacua and therefore approaching different values as $x \to \infty$ in different directions, this form factors will exhibit a pole at zero momentum transfer. The singularity in question does not correspond to a physical particle pole or a threshold and calls for attention to the peculiar analyticity (locality) properties of certain soliton models.

IIc) Quantum fluctuations in a coherent state basis

The family of trial states $\{|f>\}$ considered so far
is manifestly too restricted to be able to provide a
description of physical systems away from the extreme
weak coupling limit. In particular, a coherent state
allows only for minimal, gaussian free field zero point
fluctuations around a classical field configuration. To
allow for global field fluctuations we must consider
superpositions of coherent states.

It will be illuminating to discuss the problem
first in the context of systems with one rather than
an infinite number of degrees of freedom. Let us denote
by q and p the co-ordinate and conjugate momentum of
such a system, or particle, and by

$$\alpha = \alpha_1 + i\alpha_2 = \sqrt{\omega}\, q + i\, p/\sqrt{\omega} \qquad\qquad (II.19)$$

the complex co-ordinate of a point in its phase-plane.
A one-particle coherent state is defined by

$$|\alpha> = e^{\frac{1}{\sqrt{2\hbar}}(\alpha a^\dagger - \alpha^* a)}|0> \;=$$

$$= e^{-\frac{1}{4\hbar}|\alpha|^2}\, e^{\frac{\alpha}{\sqrt{2\hbar}}a^\dagger}|0> \qquad\qquad (II.20)$$

where

$$[a, a^\dagger] = 1, \qquad a\,|0> = 0 . \qquad\qquad (II.21)$$

The Schrödinger, coordinate representation of such a
state is of course a gaussian centred around α. The

inner product between two such states is

$$\langle \alpha '|\alpha \rangle = \exp \left\{ - \frac{1}{4\hbar}[\, |\alpha '-\alpha |^2 + (\alpha '^*\alpha -\alpha '\alpha^*)]\right\} \qquad (II.22)$$

and the family $\{\,|\alpha \rangle\}$ would manifestly yield an overcomplete basis for the Fock space. Intuitively, this conclusion directly follows from the uncertainty principle. Let us now observe that, apart from normalization factors, coherent states are entire analytic, in the sense that their inner product with a fixed state $|S\rangle$

$$\langle S|\alpha \rangle = e^{- \frac{1}{4\hbar}|\alpha |^2} \; S(\alpha) \qquad (II.23)$$

yields an entire analytic function of α, $S(\alpha)$. It then follows that any open set, or convergent sequence of points in the phase plane will define a complete coherent state basis in the Fock space. Thus, for example, if we take such an open set on the unit circle in the phase plane, the ordinary, particle number basis is generated by:

$$|n\rangle \propto \int_{0}^{2\pi} d\theta \; e^{in\theta} \; |e^{i\theta}\rangle \; . \qquad (II.24)$$

For nonperturbative variational calculations, families of trial states as superpositions of coherent states may therefore be formed from a sequence of sets of points in the phase plane. If the limit of the sequence is an open set, the corresponding limiting variational calculation becomes exact.

This approach generalizes to the case of field theory in an obvious way: the unit circle becomes the unit cylinder,... Notice, in particular, that a complete basis may be formed from coherent states of zero average momentum. In fact it is sufficient to retain only those states which are generated by curves on a strip of arbitrary width:

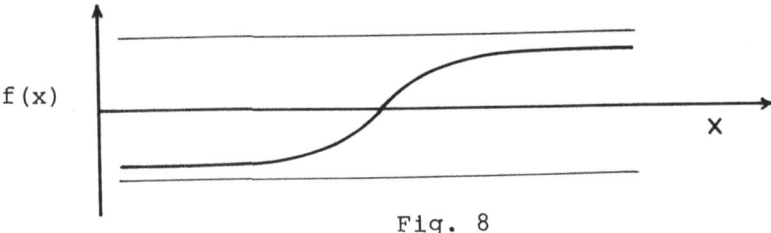

Fig. 8

Families of trial states may be generated by the linear span of states obtained by joining points of a lattice formed on the strip.

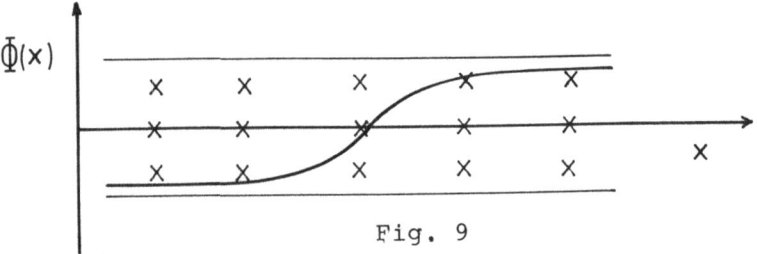

Fig. 9

As the lattice spacing goes to zero, a variational calculation should become exact. Notice that the lattice spacing provides a short distance cut-off; a volume cut-off is also provided by the finite length of the strip. As these cut-offs are removed, a renormalization program must be carried out to assure the convergence of physical observables.

IId) Construction of an effective potential [4]

It is obvious however that not all of the a
priori allowed field configurations will be excited
as quantum fluctuations to a non-negligeable degree
in a given physical situation. This consideration will
now lead me to a drastic short cut in the form of a
variational approximation.

Let

$$\Phi(x,t = \tau) = f(x,\tau), \quad \dot{\Phi}(x,t = \tau) = g(x,\tau) \qquad (II.25)$$

denote a time (τ)-dependent extended solution to the
classical field equations. For reasons which will become clear, I will refer to $f(x;\tau)(g(x;\tau))$ as a "classical degree of freedom" of the field (of the momentum conjugate to the field) and to τ as its "coordinate". The
solution $\Phi(x,t)$ may correspond to a freely moving soliton or to a bound soliton-antisoliton pair ("breeder
mode") or to a scattering solution involving an arbitrary
number of solitons. Thus the coordinate τ specifies the
state of an "effective" degree of freedom which generally
describes a system of solitons.

I will consider classes of effective classical
degrees of freedom and quantize them in the following
way. From the solutions (II.25) via Eq. (II.6) I will
form a class $\{|\tau\rangle\}$ of coherent states. An element of
the class is specified by its co-ordinates. The linear
span of all elements of $\{|\tau\rangle\}$ yields a subspace of the
original Hilbert space, associated with a projection
operator P. We define a (variational) restriction of

the original theory by replacing the Hamiltonian operator $:H:$ by $H' = P:H:P$. Such a restriction may be studied as a quantum theory in its own right. It will be a relativistic quantum theory provided that the class of effective classical degrees of freedom considered is Poincare' invariant. The restricted quantum theory may also be regarded as an extension of classical field theory, for which the only allowed states may be identified with pure coherent states (no superpositions allowed).

The inner products $<\tau'|\tau>$ may be calculated from the following identities:

$$<\tau'|\tau> = \exp \{- \frac{1}{4\hbar} [\|f' - f\| + \|g' - g\| +$$

$$+ 2i ((f',g) - (g',f))] \} \qquad (II.26)$$

$$\|f\| = \int dk\ k_o |\tilde{f}(k)|^2, \quad \|g\| = \int \frac{dk}{k_o} |\tilde{g}(k)|^2, \qquad (II.27)$$

$$(f,g) = \int dx\ f(x)\ g(x) . \qquad (II.28)$$

The Hamiltonian H is normal ordered with respect to $|0>$ and matrix elements of H' are, therefore, easily calculated with the help of the identities:

$$<\tau'| :\phi^n(x): |\tau> = [\frac{1}{2}(f'(x) + f(x)) - \frac{1}{2}(g'(x) - g(x)]^n <\tau'|\tau>$$
$$(II.29)$$

$$<\tau'| :\pi^n(x): |\tau> = [\frac{1}{2}(g'(x) + g(x)) +$$

$$+ \frac{1}{2}(f'(x) - f(x))]^n <\tau'|\tau> \qquad (II.30)$$

where

$$\underline{f}(x) = \frac{1}{\sqrt{2\pi}} \int dk \; k_0 \; e^{-ikx} \; \tilde{f}(k) , \qquad (II.31)$$

$$\underline{g}(x) = \frac{1}{\sqrt{2\pi}} \int \frac{dk}{k_0} \; e^{-ikx} \; \tilde{g}(k) . \qquad (II.32)$$

Knowledge of $<\tau'|\tau>$ and of $<\tau'|H'|\tau>$ suffices to enable one to derive the exact spectrum of H' and scattering properties of the restricted quantum theory under consideration. For instance, in the τ-representation, the evolution equation for an arbitrary state

$$|\Psi(t)> = \int d\tau \; \psi(\tau, t)|\tau> \qquad (II.33)$$

is the integral equation:

$$i\hbar\partial_t <\tau|\Psi(t)> = \int d\tau' <\tau|H'|\tau'> \Psi(\tau',t) . \qquad (II.34)$$

But, for the moment, let us concentrate on an approximation which is suggested by the weak coupling or $\hbar \to 0$ limit. The inner product $<\tau'|\tau>$ of Eq. (II.26) undergoes fast exponential fall-off with increasing separation $\varepsilon = \tau' - \tau$ in that limit. This suggests an ε-expansion of $<\tau'|\tau>$ and retention only of the leading terms in the exponent in Eq. (II.26). I obtain

$$<\tau'|\tau> \underset{\hbar \to 0}{\approx} \exp\{-\frac{1}{4\hbar}[A(\tau)(\tau'-\tau)^2 +$$

$$+ 2i \; B(\tau)(\tau'-\tau)]\} , \qquad (II.35)$$

544

where

$$A(\tau) = \int dk[k_o |\tilde{\dot{f}}(k,\tau)|^2 + \frac{1}{k_o} |\tilde{f}(k,\tau)|^2] \quad , \tag{II.36}$$

$$B(\tau) = \int dk[|\tilde{\dot{f}}(k,\tau)|^2 - \tilde{\ddot{f}}(k,\tau) \tilde{f}(k,\tau)] \quad . \tag{II.37}$$

Let me now remark that the inner product given by Eq. (II.35) is reminiscent of the inner product given by Eq. (II.22) between coherent states of a hypothetical system with one rather than an infinite number of degrees of freedom. In fact, for states (II.20) corresponding to points on a phase plane path defined by $\alpha = \alpha(\tau)$ and for $\bar{h} \to 0$, Eq. (II.22) reduces to:

$$<\alpha(\tau')|\alpha(\tau)> \underset{\bar{h}\to 0}{\approx} \exp \{- \frac{1}{4\bar{h}} [|\dot{\alpha}|^2 (\tau' - \tau)^2$$

$$+ 2i \, \text{Im} \, (\dot{\alpha}^*\alpha) \, (\tau' - \tau)]\}. \tag{II.38}$$

Comparing Eq. (II.38) with Eq. (II.35), we then discover that the isomorphism induced by

$$|\tau> \quad \longleftrightarrow \quad |\alpha(\tau)> \tag{II.39}$$

is an isometry

$$<\tau'|\tau> = <\alpha(\tau')|\alpha(\tau)> \tag{II.40}$$

provided that

$$|\dot{\alpha}(\tau)|^2 = A(\tau) \quad , \tag{II.41}$$

$$\text{Im } \dot{\alpha}^*(\tau) \ \alpha(\tau) \ = \ B(\tau) \quad . \tag{II.42}$$

These equations, supplemented by the specification of
(compatible) initial conditions[+] ($\alpha(0)$), uniquely define the one-particle phase plane path $\alpha = \alpha(\tau)$[++]. As
implied by the previously discussed analyticity properties of coherent states (or, equivalently, by the
completeness of families of one-particle coherent states
parametrized by continuous labels), the condition

$$<\tau'|H|\tau> \ =<\alpha(\tau')|H|\alpha(\tau)> \tag{II.43}$$

also uniquely defines a Hamiltonian operator H in the
one-particle Hilbert space.

We have therefore seen that, in the weak coupling
limit, it is possible to reformulate "restricted quantum
field dynamics" as quantum dynamics of point particles.
Intuitively, this equivalence is a consequence of the
high degree of coherence of the soliton field in the
weak coupling limit: while the soliton is extended and
built from an infinite number of degrees of freedom,
their coherent excitation may be described in terms of
one effective coordinate. In fact, if $f(x;\tau)$ corresponds
to a classical system of solitons, such a coordinate will
describe the entire system. Needless to say, if the restricted field theoretical quantum dynamics under con-

---

[+] These are often determined by symmetry requirements.
[++] In fact, Eq.(II.41) specifies for any τ the absolute
magnitude $|\underline{W}|$ of the phase plane velocity $\underline{W} = \dot{\underline{\alpha}}$, $\underline{\alpha} \equiv$
$\equiv(\alpha_1,\alpha_2)$, while Eq. (II.42) determines its orientation
by specifying the curl $\underline{W} \times \underline{\alpha}$.

sideration was obtained by simultaneously retaining N different classical degrees of freedom, $f^1(x;\tau)$, the above construction will yield, as an equivalent problem, the quantum dynamics of a system of N interacting point particles.

What are the effective interactions of these fictious particles? As implied by Eqs. (II.29-30-43), I know that

$$<\alpha(\tau)|H|\alpha(\tau)> = H(f) = E \qquad (II.44)$$

where E is the energy of the classical solution $f(x;\tau)$. Therefore, $\alpha(\tau)$ defines a phase plane path of constant classical energy. I may then infer an effective classical potential $V(q)$ for our fictious point particle, from the first integral of motion:

$$V(q) - E = - \frac{1}{2M} P^2(q,\omega) = - \frac{\omega}{2M}|\alpha_2(\alpha_1)|^2 . \qquad (II.45)$$

Here $P(q,\omega)$ is the Cartesian representation in terms of coordinate and momenta of the path

$$\alpha = \sqrt{\omega}q + i P/\sqrt{\omega} = \alpha(\tau) .$$

Notice that a non-relativistic energy momentum relation has been adopted in Eq. (II.45). This is in line with my use of a harmonic oscillator Fock space for the mapping (II.39). Such a choice is also motivated by the observation that a relativistic field is constructed from a sequence of non-relativistic oscillators: their coherent (collective) excitations are naturally described in terms of a similar one-degree-of-freedom system.

Also notice that the coherent state "width parameter" ω and the fictious particle mass M, which appear
in the definition of the potential V(q), are still undetermined, as they are not constrained by the isometry
condition (II.40) and by the diagonal condition (II.44).

The potential V(q), obtained up to these ambiguit-
ies from Eq. (II.45), defines a Schrödinger Hamiltonian
operator:

$$H_S = -\frac{\hbar^2}{2M} \nabla^2 + V(q) \; . \tag{II.46}$$

Differences between this operator and the exact Hamiltonian, defined by Eq. (II.43), will manifest themselves
through terms which are of non-leading order in $\hbar$ in
Eq. (II.43). Exploiting knowledge of the off-diagonal
matrix elements in the left-hand side of this equation,
such differences may be partially corrected by properly
tuning the left over parameters ω and M. Thus an optimal
Schrödinger-type Hamiltonian, Eq. (II.46), and a quantum
evolution equation for our duplicate world are established.

Analysis of this Schrödinger equation will generally
yield a spectrum of bound states and scattering solutions.
The bound state spectrum may be directly identified with
the spectrum of the restricted field theory from which
we started, while from the scattering solutions a reconstruction of the S matrix will be possible.

IIe) Quantization of the spectrum

As an interesting exercise, one may apply the method described in the previous section to an analysis of the quantum dynamics in the soliton-anti-soliton channel of the sine-Gordon model. I will outline here the results of this analysis.

The relevant classical solutions were briefly described in section Id). The soliton-antisoliton channel is, in contrast to the soliton-soliton channel, characterized by the presence of attractive forces. This is mirrored in Eq. (I.21) which specifies the time advance suffered by the soliton-antisoliton pair in the scattering process as a function of the initial velocity u in the center of mass frame. The attractive forces also lead to time-dependent bound solutions, Eq.(I.24), describing the stable "breading" oscillations of a soliton-anti-soliton pair. The period of oscillation T can take on a continuum of values ranging from $2\pi/m$ to ∞, depending on the energy of the pair. The period T approaches ∞ as the classical energy approaches $2M = 2.8m^3/\lambda$, which is twice the classical soliton mass and therefore also the scattering "threshold". I will now indicate how the continuum of classical bound solutions quantum mechanically converts into a discrete spectrum of bound states. This result was first obtained, using a WKB method, by Dashen, Hasslacher and Neveu, Ref. [3].

I will consider for simplicity only one classical degree of freedom, $f(x;\tau)$, which I choose to identify with a scattering solution, Eq. (I.26), near threshold (u << 1). Following the effective potential method introduced in the previous section, I then proceed to: construct coherent states and a Hilbert space; calculate, with the help of Eqs. (II.26-32), inner products $<\tau'|\tau>$ and Hamiltonian matrix elements $<\tau'|H|\tau>$; derive from Eqs. (II.41-42) a phase plane path and from Eqs.(II.43-45) an effective potential for the fictious particle. This is approximately given by [4],[5]:

$$V(q) \simeq 2M \tanh^2 (mq) ,\hspace{3cm}(II.47)$$

where $M = 8m^3/\lambda$ (in the notations of section Id)) is the classical soliton "mass". The effective mass of the fictitious particle is $\simeq 2M$ which is also the depth of the effective potential. As expected, this potential is attractive and has a range of the order of the inverse meson mass (m).

I then pick up a book of elementary quantum mechanics and read off the bound state spectrum associated with the potential (II.47):

$$M_n \simeq nm - n^2 \frac{m^2}{\pi M} ,\hspace{3cm}(II.48)$$

where

$$n = 1,2, \ldots \leq \pi M / m .\hspace{3cm}(II.49)$$

As expected, the number of bound states is a function of the coupling constant λ (or, equivalently, m/M approaching ∞ in the weak coupling limit. As λ increases more and more states unbind. The lowest lying bound state is the meson itself.

IIf) Scattering of quantum solitons [4],[5]

I will only briefly discuss soliton-antisoliton scattering. Solitons and antisolitons are distinguishable particles corresponding to opposite values of the topological charge (I.12). In contrast to soliton-soliton scattering, we may therefore distinguish between forward and backward scattering. As evidenced by the solution (I.26), the deterministic classical theory leads to purely forward scattering with a time advance given by Eq. (I.21). The quantum theory leads, also, to reflection and transmission coefficients, R and T.

These quantities are easily calculable with our effective potential method. In fact, the potential (II.47) in the Schrödinger equation yields:

$$|R|^2 = \frac{1}{1+p^2} \ , \quad |T|^2 = \frac{p^2}{1+p^2} \qquad\qquad (II.50)$$

where

$$p \approx \frac{\sinh\ (\pi k/m)}{\sin\ (\pi^2 M/m)} \ . \qquad\qquad (II.51)$$

Here k is the initial momentum of the fictitious particle. The mapping into the duplicate soliton world identifies k with twice the initial soliton (antisoliton) momentum. Let me remark that the reflection coefficient vanishes identically, independently of the initial momentum k for special values of the coupling constant λ or ratio M/m. The occurrence of such "reflectionless" since-Gordon models corresponds to solutions of an eigenvalue equation:

$$\sin\,(\pi^2 M/m) = 0\ ,\quad \pi^2\,M/m \gtrsim 2\,\pi\,. \tag{II.52}$$

It should be noted that the occurrence of the first of these pseudofree sine-Gordon models was already implied by the equivalence [6] with the massive Thirring model. From this equivalence and the identification $\pi^2 M/m=2n+2g$ we may now infer absence of backward scattering in massive Thirring models corresponding to $g = n\frac{\pi}{2}$. Let me also remark that, with the exception of those special cases, the reflection coefficient equals one at threshold, yielding purely backward scattering, and decays exponentially with the energy or the scale set by the meson mass.

III. REALISTIC MODELS

So far I have mainly discussed classical solitons and their quantization in the context of model field theories involving only spinless fields and in one-space-one-time dimension. I will now proceed to generalize the soliton concept to realistic field theories with spin degrees of freedom and in three-space-one-time dimension. This extension will not be obtained by progressive steps but rather in one stroke since, as it was concluded in sections Ib) and c), the existence of static solitons in more than one-space-one-time dimension presupposes the presence of spinning fields.

All of the models (theories) I will consider will include Fermi fields, because their presence is necessary to the quark model. I also like to dismiss the suggestion that half-integral spin may arise as a global quantity not carried by "elementary" constituents.

In the first class of models to be considered, Fermi fields will be accompanied by self-interacting spinless (gluon) fields. I shall generally refer to this class as "bag models", even though the structure of their soliton solutions, which crucially depends upon the non-linearities of the scalar field self-interaction, may totally differ from that of a "bag". These models have many properties which are desirable from a phenomenological standpoint and which motivated their introduction. From a theoretical standpoint, however, they suffer from a high degree of arbitrariness which makes them unsuitable candidates for a fundamental description.

For this reason I will rapidly move on to consider gauge theories of vector and spinor fields. In absence of a fictitious self-interacting scalar glue, the binding forces holding the soliton together do not originate from the vacuum pressure phenomenon but rather the soliton appears as the balance of what may be more easily regarded as fundamental physical forces, essentially familiar to all of us from electromagnetism. I will limit myself to a brief account of the abelian case. The reader may find a more complete discussion including the extension to the relevant case of a non-abelian "color" gauge theory of the strong interaction in Ref.[7].

Note that, mutatis mutandis, many of the following considerations are also applicable to gauge theories of the weak interactions to predict the existence of new particle states in these theories.

IIIa) Bag models

To introduce the idea at the origin of bag models, let me consider the theory of a Fermi field $\Psi(x)$ (the "quark") and of a neutral scalar field $\Phi(x)$ (the "gluon") with Hamiltonian density:

$$H(x) = \Psi^\dagger \frac{\alpha \cdot \nabla}{i} \Psi + \frac{1}{2}(\nabla\Phi)^2 + \frac{1}{2}\dot{\Phi}^2 + U(\Phi) + g_0 \bar{\Psi}\Psi\Phi. \qquad (III.1)$$

The scalar field is subject to a self-interaction defined by the potential $U(\Phi)$

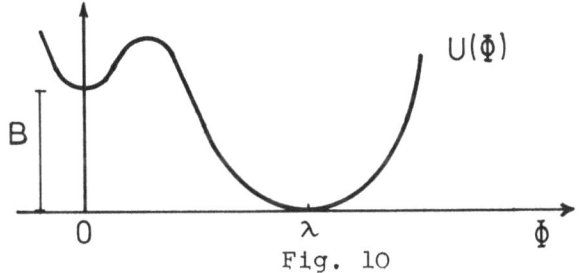

Fig. 10

with absolute minimum at $\Phi = \lambda$ and a local minimum at $\Phi = 0$. Let B denote the potential difference between the two minima. The vacuum state in the tree approximation is defined by the field configurations ($\Phi = \lambda$, $\Psi = 0$) and in perturbation theory the quark acquires a bare mass term $\mu = g\lambda$.

However if the quark bare mass μ is sufficiently large as compared to $B^{1/4}$, $\mu > B^{1/4}$, perturbation theory breaks down. This is evidenced by the appearence of non-perturbative configurations of energy $E < \mu$ describing localized quarks and scalar field amplitudes locally excited to the secondary potential minimum $\Phi = 0$:

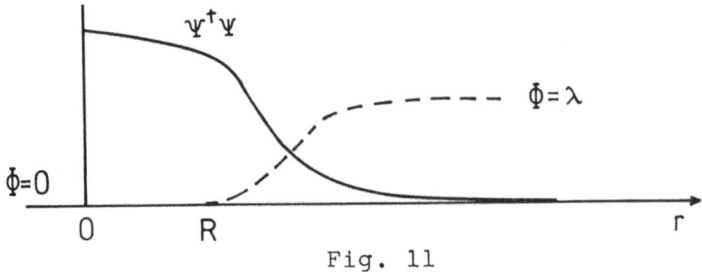

Fig. 11

The energy of such field configurations is in fact easily estimated to be

$$E \sim \frac{1}{R} + B^{\frac{4}{3}} \pi R^3 \sim B^{1/4} \qquad\qquad (III.2)$$

In words, if $B^{1/4} < \mu$, the quark likes to dig itself a hole in the vacuum to turn off its bare mass at the lesser expense of promoting the scalar field to an excited state. The characteristic radius of the system is roughly determined by the equilibrium between two contrasting energy terms, the kinetic energy of the quark (1/R) and the "vacuum pressure" term $(B\frac{4\pi}{3}R^3)$.

The vacuum pressure mechanism [8] sketched above can bind any number of quarks and antiquarks and may therefore be adopted as the first building bloch in the construction of a quark model of hadrons. It must be complemented by other mechanisms, such as a "color" mechanism, selecting the appropriate combinations of quarks and antiquarks to obtain the observed hadronic spectrum.

A similar approach was undertaken by a group at MIT [9]. The MIT bag model was not formulated as a field theory, but may be obtained [10] as a limiting case of field theories with spontaneous symmetry breakdown and

the vacuum pressure mechanism that I have described. The
limiting procedure in question is very singular as it involves the elimination of all parameters in the Hamiltonian density (III.1), including the Yukawa coupling constant $g_0(\to\infty)$ and parameters defining the potential $U(\Phi)$,
with the exception of the potential difference B which is
identified with the MIT "bag constant". The end product,
the MIT bag model in the so called "cavity approximation",
provides a satisfactory description of static properties
of hadrons.

A different implementation of my idea was discovered
by a group at SLAC [11] and goes under the name of SLAC
"bubble model". It turns out that corresponding to the
choice of gluon selfinteraction potential, $U(\Phi)$,

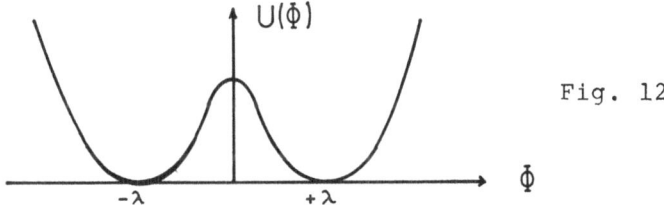

Fig. 12

the vacuum pressure mechanism localizes the quark gas
in a thin shell, rather than inside a bag:

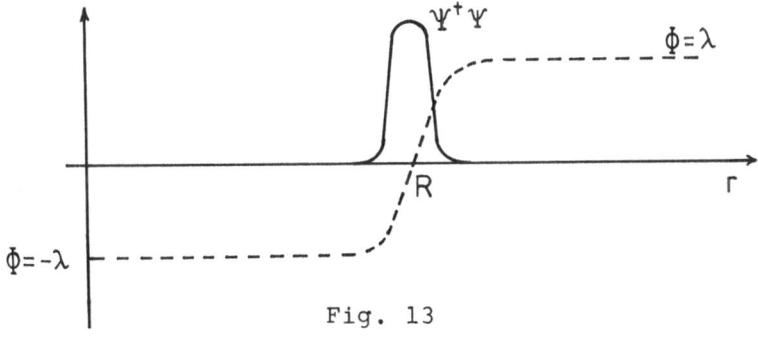

Fig. 13

IIIb) How to treat fermions

In the preceeding discussion I have appealed to your intuition to suggest the existence of extended, bound systems with particle-like properties in a vast class of models characterized by Hamiltonian densities of the type of Eq. (III.1). Do the bound systems in question correspond to solitons?

The answer can be affirmative at the expense of a generalization of the soliton concept. Our working definition of a (classical) soliton has been so far: a non-dissipative particle-like solution to the classical field equations. Unfortunately the classical counterpart of a fermion field operator, in contrast to boson operations, it not a commuting c-number and the field equations which derive from the Hamiltonian, density (III.1)

$$ i \not{\partial} \Psi = g_o \Phi \Psi \ , \tag{III.3} $$

$$ \Box \Phi - U'(\Phi) = g_o \bar{\Psi} \Psi \ , \tag{III.4} $$

are therefore not interpretable as classical field equations. Yet I believe that the soliton approach characterized by the idea of exposing particle-like, non-perturbative properties to a non-linear, dispersive theory, should also be fruitfully applicable to fermion field theories. Since it cannot be applied at a classical level, I will apply it directly at the quantum level. However it is not a priori obvious which steps one should take. I will therefore devote this section to a brief discussion of different methods[+].

In section IIa) I offered a quantum mechanical interpretation of static soliton solutions to classical boson field equations: the soliton wave function was to be identified with the expectation value of the boson field operator within a soliton coherent state, a quantum soliton state to leading order in $\hbar$. The soliton field equations could then be identified with the Hamiltonian variational equations within a class of trial coherent states. This approach is easily generalized to a field theory with fermions, such as bag models.

Let me expand the field operator $\Psi(x)$ in terms of creation and annihilation operators for quarks $(b^{\dagger}, b)$ and anti-quarks $(d^{\dagger}, d)$

$$\Psi(x) = \sum_n (b_n u_n + d_n^{\dagger} v_n) \tag{III.5}$$

in a general orthonormal basis of spinor functions

$$(u_i | u_j) = (v_i | v_j) = \delta_{ij}, \quad (u_i | v_j) = 0 \tag{III.6}$$

$$(u_i | u_j) = \int d^3x \, u_i^{\dagger}(x) u_j(x), \ldots \tag{III.7}$$

The equal time canonical anticommutation relations of the field imply

---

[+] I will not discuss here a method to deal with fermions recently suggested by Dashen, Hasslacher and Neveu, based on the observation that fermion fields always occur bilinearly in the Lagrangian, a property one can exploit to integrate them away in functional integrals.

558

$$\{b_n, b_m^\dagger\} = \{d_n, d_m^\dagger\} = \delta_{nm}, \quad 0 = \{b_n, d_m\} = \ldots \qquad \text{(III.8)}$$

and a Fock space may be constructed in the usual way. It should be noted that, given the arbitrariness of the spinor basis, the no-particle state $|0\rangle$ defined by

$$b_n|0\rangle = d_n|0\rangle = 0 \qquad \text{(III.9)}$$

and the ensuing Fock space construction will in general not be Poincare invariant. I will also consider a Fock space for the boson field $\phi(x)$, as in section IIa), and $|0\rangle$ will henceforth denote the no-particle state of the product of the fermion and boson Fock spaces. Suppose that I wanted to describe in the bag model character-ized by the (unrenormalized) Hamiltonian density (III.1) a one quark state of zero average momentum. I would then consider the family of trial states:

$$|q\rangle = e^{-i\Pi(f)} \, b_0^\dagger \, |0\rangle \quad . \qquad \text{(III.10)}$$

An element of the family is determined by the specificat-ion of the quark spinor wave function $u_0(\underline{x})$ and of the gluon coherent state function $f(\underline{x})$. The expectation value of the (normal ordered) Hamiltonian operator within the state (III.10) is:

$$\int d\underline{x} \langle q | : H(x) : | q \rangle = \int d\underline{x} \, \{ u_0^\dagger \, \frac{\alpha \cdot \nabla}{i} \, u_0 +$$

$$+ \tfrac{1}{2} |\nabla f|^2 + U(f) + g_0 \, \bar{u}_0 \, u_0 f \} \quad . \qquad \text{(III.11)}$$

To obtain the appropriate Euler equations a Lagrange multiplier (or spinor energy eigenvalue E_o) is needed to enforce the normalization condition (III.6) for the spinor wave function:

$$\frac{\delta}{\delta u^\dagger} \{\int d\underline{x} \ [<q|:H(x):|q> - E_o \ u_o^\dagger \ u_o]\} = 0 \qquad (III.12)$$

$$\frac{\delta}{\delta f} \{\int d\underline{x} \ <q|:H(x):|q>\} = 0 \ . \qquad (III.13)$$

The Euler equations which obtain in this case

$$i \ \underline{\gamma} \cdot \underline{\nabla} \ u_o = g_o f \ u_o + E_o \ \gamma_o \ u_o \qquad (III.14)$$

$$\underline{\nabla}^2 f - U'(f) = g_o \ \bar{u}_o \ u_o \qquad (III.15)$$

are formally equivalent to the time-independent restriction of the operator field equations (III.3-4). On the other hand, if I wanted to describe an antiquark state I would consider the family of trial states:

$$|\bar{q}> = e^{-i\Pi(f)} \ d_o^\dagger \ |0> \ . \qquad (III.16)$$

Following an analogous procedure, I would then arrive to the Euler equations:

$$i\underline{\gamma} \cdot \underline{\nabla} \ v_o = - g_o f \ v_o - E_o \gamma_o v_o , \qquad (III.17)$$

$$\underline{\nabla}^2 f - U'(f) = - g_o \ \bar{v}_o \ v_o \ . \qquad (III.18)$$

The approach trivially generalizes to a bound state consisting of any number of quarks and anti-quarks and leads to the Euler equations:

$$i \, \underline{\gamma} \cdot \underline{\nabla} \, u_n = g_o f \, u_n + E_n \, \gamma_o \, u_n \qquad \text{(III.19)}$$

$$i \, \underline{\gamma} \cdot \underline{\nabla} \, v_m = - g_o f \, v_m - E_m \, \gamma_o \, v_m \qquad \text{(III.20)}$$

$$\underline{\nabla}^2 f - U'(f) = \sum_{n,m} g_o f \, (\bar{u}_n u_n - \bar{v}_m v_m) \quad . \qquad \text{(III.21)}$$

Soliton solutions to this set of equations determine, in our variational approximation, the structure of bound states in the theory defined by the unrenormalized Hamiltonian density (II.1). Explicit examples of spherically symmetric solutions may be found in Ref. [12]. String-like solutions with cylindrical symmetry are discussed in Ref. [13].

As already noted, the way sketched above to approach the problem of obtaining soliton equations in fermion field theories may be regarded as an immediate generalization of the Hamiltonian quantum variational approach in Fock space for boson field theories described in section IIa). For boson field theories, there exists however a simple alternative procedure for arriving from the quantum theory to the classical field equations without committing oneself to a Fock space representation. The procedure in question consists of saturating the boson operator field equation

$$\Box \, \phi(x) = U'(\phi) \qquad \text{(III.22)}$$

with only one state $|S\rangle$

$$\square\langle S|\phi(x)|S\rangle = U'(\langle S|\phi(x)|S\rangle) \qquad\qquad (III.23)$$

and interpreting $\langle S|\phi(x)|S\rangle$ as the classical field. Is there an analogous procedure for fermion field theories?

It is obvious that since a Fermi field carries fermion number and its diagonal matrix elements are identically vanishing, a similar approach would require more than one state: the "classical" c-number fermion field would then correspond to an off-diagonal matrix element of the fermion field operator between states with fermion number differing by one. Depending on the particular solution of the c-number field equations obtained, such states would be interpreted as approximations to bound states and auxiliary (constituent) states, which intuitively correspond to what remains of a bound state when one of its fermion constituents is suddenly removed. Because of this interpretation, I will refer to such an approach as the "constituent method". I will illustrate it at work in the following section.

c) Abelian quark-gluon gauge model

Let me consider the U(1) gauge theory of a massive quark field, $\Psi(x)$, and of a massless gluon field, $B_\mu(x)$, with interaction defined by the canonical unrenormalized Hamiltonian density:

$$H(x) = \Psi^\dagger \frac{\alpha \cdot \nabla}{i}\Psi + m_o\bar\Psi\Psi + \frac{1}{2}(G^2_{ij} - G^2_{oi}) + g_o\bar\Psi\gamma^\mu\Psi B_\mu, \qquad (III.24)$$

where

$$G_{\mu\nu} = \partial_\nu B_\mu - \partial_\mu B_\nu \quad . \tag{III.25}$$

The unrenormalized field equations are:

$$(i\not\partial - m_o) \psi(x) = g_o \not B \psi(x) \tag{III.26}$$

$$\partial_\nu G^{\mu\nu} = g_o \bar\psi \gamma^\mu \psi \quad . \tag{III.27}$$

The charge operator is

$$Q = \int d^3x \; \psi^\dagger \psi + \text{c-numb.} = - \int d^3x \; \psi\psi^\dagger + \text{c-numb.} \; . \tag{III.28}$$

To apply the "constituent method", I will consider three states with charges $+$, $-$ and 0^*:

$$Q|q\rangle = |q\rangle \; , \; Q|\bar q\rangle = -|\bar q\rangle, \; Q \; |q\bar q\rangle = 0 \tag{III.29}$$

and the following transformation properties under charge conjugation (C):

$$C|q\rangle = |\bar q\rangle \; , \; C|\bar q\rangle = |q\rangle \; , \; C|q\bar q\rangle = \pm \; |q\bar q\rangle \; . \tag{III.30}$$

I will consider the projection of the operator field equations (III.26-27) onto the Hilbert subspace spanned by any three such vectors. More specifically, I will

---

* To simplify the presentation I will ignore the spin labels of the states. They are introduced and properly taken into account in Ref. [7], where soliton equations for spin 1 and spin 0 $q\bar q$ bound states are derived.

make the variational approximation of evaluating operator products by saturation with states within that
supspace. It follows from the postulated charge and
charge conjugation properties of these states, and of
the quark field:

$$[Q, \Psi(x)] = -\Psi(x), C\Psi_\alpha(x)C^{-1} = C_{\alpha\beta}\bar{\Psi}_\beta(x), C=i\gamma^2\gamma^0 \qquad (III.31)$$

that the only contributing quark field matrix elements
are:

$$\langle\bar{q}|\Psi(x)|q\bar{q}\rangle = u(x) = \langle q\bar{q}|\Psi^\dagger(x)|\bar{q}\rangle^\dagger, \qquad (III.32)$$

$$\langle q|\Psi^\dagger(x)|q\bar{q}\rangle = v^\dagger(x) = \langle q\bar{q}|\Psi(x)|q\rangle^\dagger \qquad (III.33)$$

and are related by

$$u(x) = \pm i\pm \gamma^2 v(x). \qquad (III.34)$$

Needless to say, the only non-vanishing matrix elements
of the (abelian) gluon field within our subspace are
diagonal ones.

Thus we arrive at the following system of c-number
field equations:

$$(i\slashed{\partial} - m_0)u(x) = g_0 \langle\bar{q}|\slashed{B}|\bar{q}\rangle u(x), \qquad (III.35)$$

$$\partial_\nu\langle\bar{q}|G^{\mu\nu}|\bar{q}\rangle = -g_0 \bar{u} \gamma^\mu u, \qquad (III.36)$$

$$\partial_\nu\langle q\bar{q}|G^{\mu\nu}|q\bar{q}\rangle = 0, \qquad (III.37)$$

subject to the normalization condition:

$$\langle \bar{q} | Q | \bar{q} \rangle = - \int d\underline{x} \; u^{\dagger} u = - 1 \; . \tag{III.38}$$

Soliton solutions to this system of equations will describe, in our variational approximation, quark-anti-
quark bound states.

Before proceeding, it will perhaps be useful to
attract your attention to a number of points: a) The
bound states that one is seeking all correspond to $|q\bar{q}\rangle$;
which bound state in the complete Hilbert space of the
theory is supposed to be approximated by $|q\bar{q}\rangle$ depends
on which soliton solution of Eqs. (III.35-38) one is
considering. b) $|q\rangle$ and $|\bar{q}\rangle$ are auxiliary states needed
to construct the quark and anti-quark wave functions $u(x)$
and $v(x)$; physically one could think of them as virtual,
constituent states, of localized quarks and antiquarks.
c) The c-number equations for the quark wave function,
Eqs. (III.35-36), exhibit an important change of sign
with respect to the operator field equations (III.26-27);
physically this mirrors the interaction of the quark
charge distribution with the antiquark charge distribution:

in Eq. (III.35) the quark wave function $u(x)$ in fact
"feels" the gluon amplitude in the antiquark constituent
state; technically, this change of sign is brought about
by the operator ordering of the quark current. d) The
gluon amplitude in the bound state, $\langle q\bar{q} | G^{\mu\nu} | q\bar{q} \rangle$, is,
according to Eq. (III.37), a free field; this occurs

because of the mutual cancellation between the constituent quark and anti-quark current densities in our
self-consistent approximation; I shall henceforth set:

$$\langle q\bar{q}|G^{\mu\nu}|q\bar{q}\rangle \equiv 0 \quad .$$

(III.39)

Soliton solutions to the system of equations (III.35-36)
have been exhibited in Ref.[7]. They are solutions with
maximal spherical symmetry obtained from the ansatz

$$u(\underline{x}) = \begin{bmatrix} u_1(r) \; \chi_k^\mu \\ u_2(r) \; \chi_{-k}^\mu \end{bmatrix} \cdot e^{iEt}$$

(III.40)

where u_1 and u_2 are upper and lower component radial
functions, χ_k^μ are ordinary spin-angular functions, and
$k = \pm (j + 1/2)$ identifies the parity of the two quark
states of angular momentum j. Consistency with the requirement of spherical symmetry forces $j = 1/2$. The
system of Eqs. (III.35-36) is reduced to the system
of radial equations

$$\frac{d}{dr} u_2(r) = \frac{k-1}{r} u_2(r) - (E-m-g_oB^o(r))u_1(r) \; ,$$

(III.41)

$$\frac{d}{dr} u_1(r) = (E+m-g_oB^o(r))u_2(r) - \frac{k+1}{r} u_1(r) \; ,$$

(III.42)

$$\frac{1}{r^2} \frac{d^2}{dr^2}(r^2B^o(r)) = g_o(u_1^2 + u_2^2) \; .$$

(III.43)

These were solved numerically with boundary conditions of
regularity at the origin (r = 0) and sufficiently rapid

fall-off at infinity ($r \to \infty$) to insure integrability of
the soliton energy density. In particular, asymptotically
as $r \to \infty$ u_1, u_2 fall off exponentially, while $B^{\circ}(r)$ exhibits the Coulomb tail characteristic of the charge of the
constituent $|\bar{q}\rangle$ state. Since the effective potential $B^{\circ}(r)$
acting on the wave function $u(x)$ corresponds to a negative
charge distribution, Eq. (III.36), only positive energy
($E > 0$) soliton solutions exist. The ground state solution has the structure indicated in the figure:

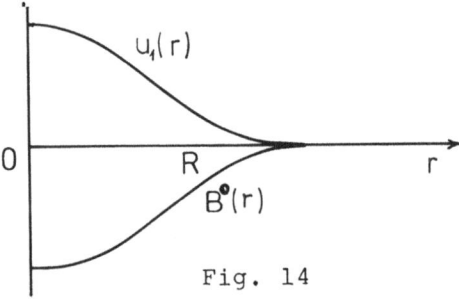

Fig. 14

These solutions correspond to minima of the energy and
are therefore stable. In fact, introducing a parameter
R to characterize the "radius" of the solution, its
energy E as a function of R is easily estimated to be

$$E \sim 2\sqrt{m^2 + p^2} - \frac{g_0^2}{4\pi}\frac{3}{5}\frac{1}{R} \quad, \quad p \sim \frac{\pi}{R} \tag{III.44}$$

where the first and second terms approximate respectively the rest mass and kinetic energy of the constituents and the interaction energy between constituents.
A balance between these contrasting effects obtains for

$$mR \sim \frac{\pi(1-x^2)^{1/2}}{x} \quad, \quad x = \frac{3}{10\pi}\frac{g_0^2}{4\pi} \tag{III.45}$$

which determines the minimum energy to be:

$$E \sim 2m (1 - x^2)^{1/2} .$$
(III.46)

Note that Eqs. (III.45-46), derived in a weak coupling approximation, cease to be meaningful for strong coupling, i.e. as $x \to 1$.

REFERENCES

1. For a review see A. Scott, F. Chu and D. Mc Laughlin, Proc. IEEE 61, 1443 (1974).

2. G.E. Derrick, J.Math. Phys. 5, 1252 (1964).

3. For alternative approaches to soliton quantization see: R. Dashen, B. Hasslacher and A.Neveu, Phys.Rev. D10, 4114; 4130; 4138 (1974).
   J. Goldstone and R. Jackiw, Phys. Rev. D11, 1486 (1974);
   N. Christ and T.D. Lee, Columbia Univ. Preprint;
   J.-L.Gervais and B.Sakita, Phys. Rev. D 11, 2943 (1975);
   C. Callan and D.Gross, Nucl. Phys. B93, 29 (1975);
   V.E. Korepin, P.P. Kulish and L.Fadeev, JETP Lett. 21, 139 (1975);
   E. Tomboulis, MIT preprint;
   A. Klein and A. Krejs, Univ. of Penn. preprint.

4. P. Vinciarelli, "Effective potential approach to the quantum scattering of solitons", Physics Letters in press.

5. For a different derivation, see W. Troost, CERN preprint.

6. S. Coleman, Phys. Rev. D 11, 2o88 (1975) and references therein.

7. P. Vinciarelli, "Solitons in gauge theories of vector and spinor fields", Physics Letters in press; W. Troost and P. Vinciarelli, "Color singlet solitons in SU(3) gauge theory of vector and spinor fields", CERN preprint No. 2162.

8. P. Vinciarelli, Lett. al Nuovo Cimento $\underline{4}$, 9o5 (1972).

9. A. Chodos et al., Phys. Rev. D $\underline{9}$, 3471 (1974).

10. M. Creutz, Phys. Rev. D $\underline{10}$, 1749 (1974).

11. W. A. Bardeen et al., Phys. Rev. D $\underline{11}$, 1o94 (1975).

12. P. Vinciarelli, Nuclear Phys. B $\underline{89}$, 463 (1975).

13. P. Vinciarelli, Nuclear Phys. B $\underline{89}$, 493 (1975).

Acta Physica Austriaca, Suppl. XV, 569–570 (1976)

A CLAIM FOR PRIORITY[+]

by

J.RAYSKI

Institut Fizyki, Krakow

The soliton business is very old, to remind you
the story of a gentleman who, on horse back, observed
a solitary wave travelling along a channel. But the
possibility of applying this idea to the elementary
particle physics is rather new: In his report on the
subject at the Les Houches Summer School 1975 Faddeev
quoted several papers, each of them being at most two
years old. However, it was exactly four years ago
that I presented here, at Schladming, very similar
ideas. My talk has been published together with the
other reports of the XI-th Schladming Winterschool
in Acta Physica Austriaca, Suppl. 9, 87o-873 (1972).
The main difference was that I did not use the ugly
name "solitons" but "clusters". In particular, I
suggested an interaction Lagrangian

$$L' = -g(A^2 - \alpha A^3) \ F(A)$$

[+] A statement of Prof. Rayski after Vinciarelli's talk.

where A is bilinear in the field quantities and $F(A)$ is
a non-polynomial function (similarly as there is a
cosine in the sine-Gordon case). Since $F(A)$ vanishes
for large A, our example exhibits a vacuum degeneracy
as well.

I have mentioned all that not so much to convince
the younger generation that I have contributed to the
development of physics during the last thirty years.
No, certainly not, I don't care about such things.
However, the point is that I became a patriot of the
Schladming Conferences and therefore I would like to
point out that we here, at Schladming, are not back
but often ahead with the new physical ideas. Thank you.

Acta Physica Austriaca, Suppl. XV, 571–589 (1976)
© by Springer-Verlag 1976

SUM RULES FOR ELASTIC $\gamma\gamma$ SCATTERING? MESON COUPLINGS
TO TWO PHOTONS, AND THE f-P RELATIONSHIP[+]

by

P. GRASSBERGER
Laboratoire de Physique Théorique
Université de Nice[++]

ABSTRACT

Very general assumptions are used to derive superconvergence type sum rules for elastic $\gamma\gamma$ scattering.
Among the results is a proof that tensor meson decays
into two photons must involve predominantly photons
with opposite helicities, and estimates for the f $\to \gamma\gamma$
and $\varepsilon \to \gamma\gamma$ widths. Since these sum rules use full (s,
t, u) - crossing, we can also discuss the analytic
continuation of the pomeron from {t $\leq$ 0, s large} to
{t > 0, s $\leq$ 0}. We argue that the pomeron is responsible for the apparent small deviation from ideal mixing

---

[+]Seminar given at XV. Internationale Universitätswochen
für Kernphysik,Schladming,Austria,February 16-27,1976.
[++]Equipe de Recherche Associée au C.N.R.S. Postal Address:
Parc Valrose 06034 Nice Cedex, France.

in the tensor nonet, in pushing the f meson mass down
from its "bare" nonet value, and making its coupling
to other states somewhat larger.

I. INTRODUCTION

The aim of the present talk (which is mainly based
on a paper by R. Kögerler and myself [1], called I in
the following) is threefold. First of all, we shall make
some definite statements about the phenomenology of $\gamma\gamma$
collisions, which might become relevant in e^+e^- collisions or in the Primakoff effect in γ-nucleus collisions.

Secondly, quite apart from the specific process
chosen, it is interesting to see what restrictions on
absorptive parts follow from very general principles.
The principles I shall use are essentially (a) good
high-energy behaviour, as found e.g. in any renormalizable perturbation theory, and (b) crossing symmetry. It
is well known that these two requirements together give
strong constraints in processes involving the exchange
of spinning particles. One solution to these constraints
- at least in 26 dimensions - is the Veneziano model
with its many modifications. Another solution, as has
been shown by Cornwall, Levin and Tiktopolous [2] and
by Llewellyn-Smith [3] is gauge theory.

Both these solutions are "global" approaches in
the sense that one searches a solution for the general
n-point function, which implies among others that it
is not at all trivial to go beyond the tree approximation. The approach described in this talk is much more
modest. Using purely S-matrix methods, we derive sum

rules similar to FESR's, or rather to superconvergence rules. The difference to the latter will be essentially that we use analyticity not only for fixed t, but we use analyticity in all three Mandelstam variables. This is easiest if one deals with completely crossing symmetric process, and this is the reason why - after the extensive work done on pseudoscalar meson scattering [4], [5] - we discuss here the next simple process, $\gamma\gamma \to \gamma\gamma$.

The most important consequence of using analyticity in both s and t is that we can now study the analytic continuation of the pomeron from large s, negative t to small negative s and positive t. The third purpose of this talk is to show that the pomeron should in this sense correspond to part of the f meson, and to elaborate the consequences following from this. Apart from the unesthetic feature that the f becomes then a hybrid object, we find that a number of very attractive features follow. In particular, the "pure" tensor mesons are in this scheme <u>ideally</u> mixed. Notice that in the conventional scheme the mixing is not quite ideal, making the extremely strong suppression [6] of the decay f' $\to$ $\pi\pi$ somewhat a mystery. Together with Harari's [7] recent arguments showing that the pseudoscalar mesons might be nearly ideally mixed in spite of the high η and η' masses, this means that ideal mixing might generally be much better than traditionally believed.

II. AMPLITUDES FOR $\gamma\gamma$ SCATTERING

For simplicity, we shall restrict ourselves here

to the elastic scattering of real photons, assuming P and T invariance. We have then five independent amplitudes. Kinematic factors can be split off by standard arguments, so that we can define the reduced helicity amplitudes

$$\tau_1(stu) = \frac{1}{s^2} T_{++,++}(stu)$$

$$\tau_2(stu) = T_{++,--}(stu)$$

$$\tau_3(stu) = \frac{1}{u^2} T_{+-,+-}(stu) \tag{1}$$

$$\tau_4(stu) = \frac{1}{stu} T_{++,+-}(stu)$$

$$\tau_5(stu) = \frac{1}{t^2} T_{+-,-+}(stu) \quad ,$$

with $s + t + u = 0$. Since all external particles are massless and identical, crossing and Bose symmetry imply that under any permutation of any two Mandelstam variables, the τ_i get simply interchanged:

$$(\tau_1,\tau_2,\tau_3,\tau_4,\tau_5) \longrightarrow \begin{cases} (\tau_5,\tau_2,\tau_3,\tau_4,\tau_1) & (s \leftrightarrow t) \\ (\tau_3,\tau_2,\tau_1,\tau_4,\tau_5) & (s \leftrightarrow u) \quad (2) \\ (\tau_1,\tau_2,\tau_5,\tau_4,\tau_3) & (t \leftrightarrow u) . \end{cases}$$

These amplitudes are also free from kinematical constraints or zeros, except for τ_2 which has a zero at s = t = 0. However, the latter cannot be split off by a polynomial factor unless we keep s = 0 or t = 0 or u = 0 fixed [8].

III. THE BASIC SUM RULES

To estimate the high-energy behaviour of the τ_i at fixed t, we assume here only a Jin-Martin type bound $|T_{\lambda_1\lambda_2;\lambda_3\lambda_4}| < s^{2-\delta}$ for $|t| < \epsilon$. Though such a behaviour cannot be proven rigorously in the present case (with massless particles and/or a linear unitarity relation), it should hold in any renormalizable theory order by order. On a more phenomenological level, it would also follow from vector meson dominance. We see then that τ_1, τ_3 and τ_4 should satisfy unsubtracted fixed-t DR's, while two subtractions might be necessary for τ_2 and τ_5.

The simplest sum rule is obtained if one compares the fixed-t DR's for $\tau_4(stu)$ and for $\tau_4(s,u,t)$ or, equivalently, the fixed-t with the fixed-u DR:

$$\tau_4(s,t,u) = \frac{1}{\pi} \int_{-\infty}^{+\infty} \frac{ds'}{s'-s} \, \text{Im} \, \tau_4(s',t) =$$

$$= \frac{1}{\pi} \int_{-\infty}^{+\infty} \frac{ds'}{s'-s} \, \text{Im} \, \tau_4(s',u) \, . \tag{3}$$

For t = u, setting the two integrals equal gives a trivial identity. But for t ≠ u, this gives a non-trivial sum rule. I shall not write it out in detail (see I), since it is not **very** useful: low spin particles

with known γγ decays (like π^O and η) make non contribution - which might have been anticipated, since I
said that only spinning particle exchange is constrained - and regge contributions at high energy
are hard to estimate.

Similar remarks hold for most other sum rules
which one can write down in a similar way, following
e.g. the work done in ππ scattering [4], [5]. There
are however some sum rules which are useful because
at least the sign of the high-energy contribution can
be controled by unitarity. Unitarity (or, in this weak
form often called positivity) implies that the partial
waves defined by

$$T_{\lambda_1 \lambda_2; \lambda_3 \lambda_4}(s,t) = \sum_J (2J+1) \; d^J_{\lambda,\mu}(\theta) \; f^J_{\lambda_1 \lambda_2; \lambda_3 \lambda_4}(s) \quad (4)$$

satisfy the constraints

$$\text{Im } f^J_{++,++}(s) \geq 0 \quad (5)$$

and

$$\text{Im } f^J_{+-,+-}(s) \geq 0 \; . \quad (6)$$

To use these, let us do the same procedure with the
amplitude τ_1 that we indicated for τ_4. With $u = 0$ and
$t \to 0$, after having taken the derivative with respect
to t, the sum rule

$$\int_{-\infty}^{+\infty} \frac{ds'}{s'-s} \text{Im } \tau_1(s',t) = \int_{-\infty}^{+\infty} \frac{ds'}{s'-s} \text{Im } \tau_1(s',u) \quad (7)$$

becomes

$$\int_{m_\pi^2}^{\infty} \frac{ds}{s^4} \sum_{J\geq 2} (2J + 1) \{J(J + 1) \text{ Im } f_{++,++}^J(s)$$

$$+ (J^2 + J - 7) \text{ Im } f_{+-,+-}(s)\} = 0 . \tag{8}$$

We see that the $J = 2$ contribution to the second term is the only negative contribution to this integral. If we adopt a two-component duality à la Freund-Harari [9], and make a narrow-resonance approximation, this gives

$$\sum_{T=f,A_2,f'} \frac{\Gamma_{T,+-}}{m_T^7} \geq 6 \sum_T \frac{\Gamma_{T,++}}{m_T^7} . \tag{9}$$

That means that the tensor mesons decay predominantly into two photons of opposite helicity, or inversely, that the differential cross section $\gamma\gamma \to T \to \pi\pi$ with unpolarized photons is

$$\frac{d\sigma}{dt} \propto \sin^4\theta + (\cos^2\theta - \tfrac{1}{3})^2 \, 6 \, \left|\frac{g_{T,++}}{g_{T,+-}}\right|^2 \underset{\sim}{\sim} \sin^4\theta . \tag{10}$$

This result is not too surprising if one formulates it differently: it means that in the emission of a virtual tensor meson from a real photon, the photon spin should dominantly not flip: the vertex in Fig.1 should dominate that with $\lambda'=-\lambda$. This is particularly evident for the exchange of a soft virtual tensor meson, i.e. a graviton, which we had neglected until now but which gives a contribution diverging at $t = 0$. I should point out that

daugther contributions are very much suppressed and
should be negligible. Furthermore, using the U(3) quark
model, one finds that the f meson contribution dominates,
since in this model

$$\Gamma_{f,\gamma\gamma} : \Gamma_{A_2,\gamma\gamma} : \Gamma_{f',\gamma\gamma} = 1 : 0.35 : 0.04 . \tag{11}$$

There are various estimates of the $f-\gamma\gamma$ couplings in the
literature, with conflicting results. For instance, there
exists a dual model for $\gamma\gamma\to\pi\pi$ by Levy and Singer [10],
which gives $\Gamma_{f,+-} < \Gamma_{f,++}$ in contradiction to Eq. (9).
The same is found in unbroken $SU(6)_W$ symmetry combined
with vector meson dominance [11]. On the other hand, it
has been pointed out [12] that this symmetry should be
strongly broken so that Eq. (9) might be expected. The
latter agrees also with an explicit FESR calculation
for $\gamma\gamma \to \pi\pi$ by Schrempp, Schrempp and Walsh [13], who
found

$$\Gamma_{f,+-} = 5.6 \text{ keV}, \quad \Gamma_{f,++} = 0.1 \text{ keV} . \tag{12}$$

We have seen that this qualitative behaviour is a
necessary condition in any dual model, but it might
hold even much more generally. The reason is that
there are indications that Freund-Harari duality is
not quite correct, and the pomeron is closely related
to the f-meson. We should stress that we do not doubt
that the pomeron is dual to the background in the sense
that continuing the flat part of T(s,t) from high s,
t ≤ 0 to small s (fixed t ≤ 0) we end up with the background. What concerns us here - and what cannot be

tested in meson-nucleon scattering - is the continuation to $s \leq 0$ and $t > 0$. To see that it is presumably not any low-energy background which cancels the pomeron contribution to Eq. (8), we notice that this background is presumably dominated by two-pion production calculated in spin-0 QED (see fig. 2). The latter is, however, a renormalizable theory, and thus the above sum rule must be fulfilled identically for the graphs of fig. 2. Thus the only candidate left to cancel the pomeron is the f. Taking this f-pomeron "duality" seriously, one can even attempt a crude estimate of the $f \rightarrow \gamma\gamma$ decay width. For the pomeron we make an ansatz

$$\text{Im } \tau_1^P(s,t) = \text{Im } \tau_3^P(s,t) = \frac{1}{32\pi s} \sigma_{tot}^{\gamma\gamma} e^{\frac{1}{2}bt} \theta(s-s_o). \quad (13)$$

Factorization gives the estimates $\sigma_{tot}^{\gamma\gamma} = (\sigma_{tot}^{p\gamma})^2/\sigma_{tot}^{pp} = 0.25 \mu b$ and $b \approx 5 \text{ GeV}^{-2}$. With a cut-off of $s_o = (2 \text{ GeV})^2$, and neglecting the normal regge contributions, this would give

$$\Gamma_{f,+-} - 6 \Gamma_{f,++} \approx 2.1 \text{ keV} . \quad (14)$$

Comparing this with Eq. (12), we see that this might account for about half of the f width. Before discussing the consequences of this result further, let us study one further sum rule following from somewhat weakened assumptions.

IV. FURTHER SUM RULES

The sum rules we have considered so far resulted from the Jin-Martin type estimate $|T_i(s,t)| < s^2$ for fixed $t \gtrless 0$ and $s \to \infty$. A somewhat stronger assumption, corresponding to the Pomeranchuk "theorem", is

$$|T_{ab,ab}(s,t) - T_{\bar{a}b,\bar{a}b}(s,t)| < \frac{s}{(\ln s)^{1+\varepsilon}} \cdot \qquad (15)$$

$$s \to \infty$$

In our case, this leads to the superconvergence relation

$$\int_0^\infty ds \ \mathrm{Im} \ [\tau_1(s,t) - \tau_3(s,t)] = 0 , \qquad (16)$$

which was for $t = 0$ also derived by Roy [14] and Budnev et al. [15]. Checking it for arbitrary t gives of course a strong consistency check. Saturation with pseudoscalar meson, tensor mesons, and an effective scalar meson ε gives

$$\frac{\Gamma_{\eta'}}{m_{\eta'}^3} + \frac{\Gamma_\varepsilon}{m_\varepsilon^3} = - \frac{\Gamma_\pi}{m_\pi^3} - \frac{\Gamma_\eta}{m_\eta^3}$$

$$- \frac{6.78}{m_f^3}[(1 + \frac{6t}{m_f^2} + \frac{6t^2}{m_f^4}) \ \Gamma_{f,++} - \Gamma_{f,+-}] \qquad (17)$$

$$= \begin{cases} 13,4 \ \mathrm{keV/Gev}^3 & \text{(at } t = 0) \\ 13,8 \ \mathrm{keV/Gev}^3 & \text{(at } t = -.5 \ \mathrm{GeV}^2) . \end{cases}$$

Here, we used Eq. (12) for the f-meson. Indeed, we see

that the sum rule is nearly t-independent, which is a consequence of our old inequality $\Gamma_{f,++} << \Gamma_{f,+-}$. This inequality seems also necessary to obtain saturation of the sum rule. Further interesting aspects of Eq.(17) are:

i) If one writes it as a sum rule for Γ_π in the limit $m_\pi \to 0$, one finds that Γ_π has to be suppressed by a factor m_π^3. This is in agreement with the Adler-Bell-Jackiw anomaly, and with the striking smallness of Γ_π as compared to Γ_η.

ii) For the η', the sum rule gives $\Gamma_{\eta' \to \gamma\gamma} < 14$ keV, as compared to the experimental value $\Gamma_{\eta' \to \gamma\gamma} < 19$ keV [16]. This is not very stringent, however, since SU(3) predictions for this width are $\lesssim 1$ keV.

iii) For $\Gamma_{\varepsilon \to \gamma\gamma}$, our result is more interesting, since there the theoretical predictions differ considerably.(Experimental data are not available.) Using a conventional (but recently questioned [17]) ε mass of ~ 700 MeV, we get $\Gamma_{\varepsilon \to \gamma\gamma} \approx 3$ keV. Finite energy sum rules [13], [18] for $\gamma\gamma \to \pi\pi$ give typically $\Gamma_{\varepsilon \to \gamma\gamma} = 20$ keV, while dilation invariance gives $\sim 1 - 5$ keV.

As a final remark let me discuss current-current scattering (at $q^2 = 0$) with currents carrying isospin or SU(3) indices a, b, c and d. For the helicity structure we use the same notation as in Eq. (1). The amplitudes

$$\frac{1}{t-u} (\tau_1^{ab,cd}(s,t,u) - \tau_1^{ab,cd}(s,u,t)) \tag{18}$$

are then free of kinematic singularities, and they allow
for superconvergence relations. The result are sum rules
analogous to Eq. (16), but with an extra factor $1/s$ in
the integrand. This shows that in the limit $m_\pi \to 0$, the
coupling of a pion to two currents must become diagonal
in the SU(3) indices,

$$g^2_{\pi, V^a V^b} = \delta_{ab} \cdot O(m^2) + O(m^4) \quad . \tag{19}$$

V. WHAT ABOUT THE POMERON?

We have already suggested that there might be
some connection between the pomeron and the f meson.
Let us now look for independent evidence in favour
or against this.

First, let us look at theoretical arguments, coming
from crossing sum rules in other processes. Since the f-
meson region has to be physical in all 3 channels, we
cannot draw any conclusion from meson-nucleon scattering. The process most extensively studied in this
respect is elastic $\pi\pi$ scattering. I do not want to go
into detail, but mention just that a detailed analysis
indeed shows [5] - may be not quite unambiguously [19]
- that about half of the f meson contribution to a sum
rule analogous to Eq. (8) has to be cancelled by the
pomeron.

Somewhat simpler to discuss and more interesting
is elastic pseudoscalar-pseudoscalar meson scattering
with unbroken SU(3). Let us consider the amplitude

$T^{(10)}$ (s, t, u) which carries the representation 10 in the direct channel. It is an odd function of $\cos \theta$, thus the function

$$\frac{s \; T^{(10)} (s, t, u) + perm.}{(s - t) \; (t - u) \; (u - s)} \tag{20}$$

is symmetric in t and u, without kinematic singularities, and it obeys an unsubtracted D.R.. Comparing fixed-t and fixed-u D.R.'s, we find in the same way as we found Eq. (8)

$$\sum_{J \geq 2} (2J+1) J (J+1) \int_{4m^2}^{\infty} \frac{ds}{s^2 (s-4m^2)^2} (Im \; f_J^{(1)} (s) - \frac{16}{5} Im \; f_J^{(8_D)}(s))$$

$$= \int_{4m^2}^{\infty} ds \; (background + pomeron)_{J \geq 2}. \tag{21}$$

If we believe Freund and Harari, we should put the right hand side equal to zero, and for degenerate singlet and octet masses we find the ratio of singlet/octet total $J^+ \to 0^- \; 0^-$ widths

$$\Gamma_J^{(1)} / \Gamma_J^{(8_D)} = 16/5 . \tag{22}$$

This is precisely the U(3) quark model prediction, and is also found in Veneziano type models - for good reasons, as we have just seen.

A closer inspection, in particular of the analogous sum rule for $\pi\pi$ scattering, suggests however
that the background is not sufficient to cancel the
pomeron. Thus, we would either predict $m^{(1)} < m^{(8)}$,
i.e. smaller f or f' masses than predicted by the
nonet model, or $\Gamma_2^{(1)} > \frac{16}{5} \Gamma_2^{(8_D)}$, i.e. larger f or f'
widths. Presumably both of these are realised in na-
ture.

Let me first discuss the mass of the f meson.
Experimentally, the f is not exactly degenerate with
the A_2 (in contrast, the ρ and the ω are degenerate),
but $m_{A_2} - m_f = 40 \pm 14$ MeV. This small difference is
responsible for the deviation of the tensor meson
mixing angle $\theta_T = 31 \pm 2°$ (with a quadratic mass
formula) from the ideal angle $35.3°$. It is also responsible for the fact that a linear Regge trajectory
passing through the f and the ω has a slope $\alpha' = 1.00 \pm 0.03$, while all other mass differences and
the differential cross sections for $\pi^- p \rightarrow \pi^0 n, \rightarrow \eta n$
give slopes $0.8 \leq \alpha' \leq 0.9$.

The deviation of the tensor mixing angle from
the ideal one is particularly important if we con-
sider the f' $\rightarrow \pi\pi$ decay. According to folklore, this
deviation reflects an admixture of non-strange quarks
in the f', which could then decay into two pions.
Using the above value of $31°$, we would expect
$\Gamma_{f' \rightarrow \pi\pi} / \Gamma_{f' \rightarrow K\bar{K}} = 1.8\%$, but the experimental value [6]
is $< 0.93\%$ for this ratio!

So we conclude that the difference between the
f and A_2 masses has nothing to do with non-ideal mixing
(the other tensor meson masses are compatible with ideal

585

mixing), but is due to the pomeron. Indeed, a change of 40 MeV in the f-meson mass enhances the contribution to the above sum rules typically by $\sim$ 30% (somewhat less for Eq. (8), somewhat more for Eq. (21)). Thus, there rest some 30% still to be blamed on a deviation of the f width from the ideal nonet prediction. If one believes the large f width found by the CERN-Munich experiment [20], this is indeed seen.

In the last year, a great number of explicit models have been constructed which all give some kind of f-P "duality" [21]. These models have in common that the f trajectory is indeed identical to the pomeron trajectory, i.e. there is only one vacuum trajectory. As an alternative model (which is presumably much too simple), one might assume that the physical f meson is simply a sum of a "bare" f with $m_{f_{bare}} = m_{A_2} = 1310$ MeV, and some narrow background centered around $\sim$ 1230 MeV (to give a sum at 1270 MeV). There are some reactions, as e.g. $K^-p \rightarrow f\Lambda$ and $K^-p \rightarrow A_2^0\Lambda$, where the quark model predicts equal cross sections for f and A_2^0 production. As a direct consequence of our two-component model, we predict the f-meson cross section to be larger. Also, its two components might be produced with different probability in different reactions, so that the properties of the f might change from reaction to reaction. It is amusing that indeed, at 4.6 GeV/c, one has [22]

$$\frac{\sigma_{K^-p \rightarrow f\Lambda}}{\sigma_{K^-p \rightarrow A_2^0\Lambda}} = \frac{(60 \pm 8)\mu b}{(23 \pm 14)\mu b} , \qquad (23)$$

with an f-meson mass found to be $m_f = (1242 \pm 15)$ MeV.

VI. CONCLUSIONS

In the last section, we have seen that there are indeed many hints that the f meson is connected with the pomeron, just as indicated by our sum rule Eq. (8). This is may be the most interesting phenomenological result of this talk, but also the least certain. The other phenomenological predictions of our sum rules are restricted to two-photon couplings to mesons, and they might well be checked in the near future. Quite apart from these specific predictions, it seems amusing that one can get any useful results from the very general assumptions we started from. It might thus be interesting to try a similar approach in other, more complicated but potentially richer, situations.

REFERENCES

1. R. Kögerler and P. Grassberger, CERN reprint TH.2114 (Jan. 1976).

2. J.M. Cornwall, D.N. Levin and G. Tiktopolous, Phys. Rev. Lett. $\underline{30}$, 1268 (1973) and $\underline{31}$, 572 (E) (1973); Phys. Rev. $\underline{10}$, 1145 (1974) and $\underline{11}$, 972 (E) (1975).

3. C.Llewellin-Smith, in Proceedings of the 14th Scottish Universities Summer School in Physics, 1973 (eds. R.L. Crawford and R.Jennings) (Academic Press Inc., New York 1974).

4. G. Wanders et al., Nuovo Cim. $\underline{63A}$, 108 (1969). R. Roskies, Phys. Rev. $\underline{D2}$, 1649 (1970).

5. P. Grassberger, Nucl. Phys. $\underline{B70}$, 141 (1974).

6. W. Beusch et al., Phys. Letters, 60B, 101 (1975).

7. H. Harari, Phys. Letters, 60B, 172 (1976).

8. R.A. Leo, A.Minguzzi and G. Soliniani, Univ. of Lecce preprint UL/IF/26-74/75 (1975).

9. P.G.O. Freund, Phys. Rev. Lett., 20, 235 (1968). H. Harari, Phys. Rev. Lett. 20, 1395 (1968).

10. N. Levy, P. Singer and S. Toaff, Haifa (Technion) preprint (1975).

11. D. Faiman, H.J. Lipkin and H.R. Rubinstein, Phys. Letters 59B, 269 (1975).

12. J.L.Rosner, Phys. Reports, 11C, 189 (1974).

13. B. Schrempp-Otto, F.Schrempp and T. Walsh, Phys. Letters, 36B, 463 (1971).

14. P.Roy, Phys. Rev. D9, 2631 (1974).

15. V.M. Budnev, I.F. Ginzburg and V.G. Serbo, Nuovo Cim. Letters, 7, 13 (1974).

16. A.Duane et al., Phys. Rev. Lett. 32, 425 (1974).

17. D.Morgan, Rutherford preprint RL - 75 - 133 (1975).

18. G. Schierholz and K. Sundermeyer, Nucl. Phys. B40, 125 (1972).

19. B.R. Mac Gregor, Nucl. Phys. B95, 53 (1975).

20. B. Hyams et al., Nuclear Phys. B100, 205 (1975). P. Estabrooks and A.D. Martin, Nucl. Phys. B95, 322 (1975). C.D. Froggatt and J.L. Petersen, Nucl. Phys. B91, 454 (1975).

21. C.F. Chew and C. Rosenzweig, Phys. Lett. 58B, 93 (1975).

588

M. Bishari, Phys. Lett. <u>59B</u>, 461 (1975).

C. Schmid and C.Sörensen, Nucl. Phys. <u>B96</u>, 209 (1975).

22. M.Aguilar-Benitez et al., Phys. Rev. <u>D6</u>, 29 (1972).

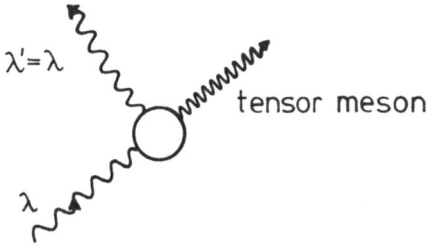

Fig. 1

Scalar QED graphs which should dominate the threshold discontinuities of elastic γγ amplitudes.

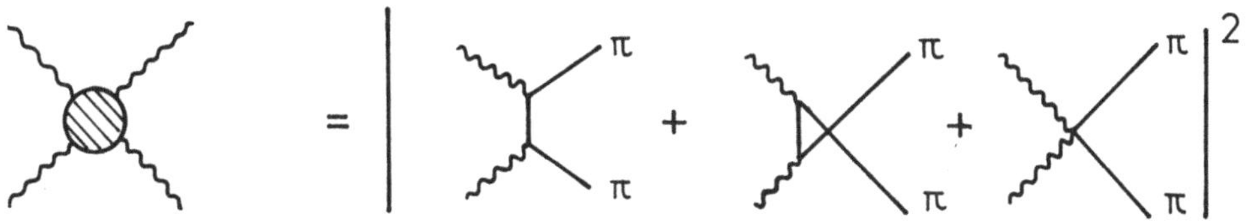

Fig. 2

Acta Physica Austriaca, Suppl. XV, 591–628 (1976)
© by Springer-Verlag 1976

GAUGE THEORY OF DUAL RESONANCE MODELS AND

SPONTANEOUS BREAKING[+]

by

A.D. KARPF

Institut für Theorie der Elementarteilchen

Freie Universität Berlin

Questionable as some of the phenomenological consequences of dual resonance models (DRM) might be, this
theory is on the whole a very successfull one and most
of all a rigorous one. One of the most important properties of the acceptable DRMs is their conformal invariance, a consequence of a ghost-free spectrum. This
invariance, on the other hand, is as stringent as a
gauge invariance and requires likewise massless particles
which physically should have a mass.[*] One therefore has
to devise means to break this symmetry spontaneously
without spoiling the formal symmetry.

It is a long way to this goal and I want to scetch
here only the major steps in easy terms in order to be

[+]Seminar given at XV.Internationale Universitätswochen
für Kernphysik,Schladming,Austria,February 16-27,1976.
[*]A further consequence of the compatibility of this invariance with Lorentz invariance is a strange space-
time dimension of D=26 or 10.This should, however, not
be taken too serious.

able to draw a complete picture of DRMs. We shall therefore talk about the reasons that lead to conformal invariance, we shall mention first and second quantised
versions of DRMs which are necessary to understand the
"string" formulation of DRMs. We shall then discuss
attempts for spontaneous symmetry breaking and shall
mention at the end the vertex (or monopole) formulation as an ansatz to derive the string from an ordinary
field theory.

1. DUAL RESONANCE MODELS

One of the most economic and the most fruitful
formulations of the dual resonant model (DRM) is in
terms of so called Koba-Nielsen variables on the unit
circle $z_i = e^{i\tau_i}$. The amplitude for an N-particle process of the form

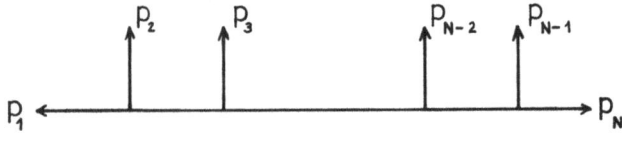

Fig. 1.1

is defined (Schwarz (1973), Karpf (1975)) for the particular case of $\alpha_o = 1$ as

$$A(p_1 \dots p_N) \sim \int_{\tau_2}^{\tau_N} i d\tau_{N-1} \dots \int_{\tau_2}^{\tau_5} i d\tau_4 \int_{\tau_2}^{\tau_4} i d\tau_3 \langle 0| \prod_{i=1}^{N} \mathcal{V}(p_i, e^{i\tau_i}) |0\rangle$$

(1.1)

where z_N, z_1, z_2 or τ_N, τ_1, τ_2 are constant and arbitrary.

A does not depend on their value due to the cyclic invariance of the integrand.

Due to a projective invariance of the integrand one can also set $z_i = e^{t_1}$ making z_i vary along the real axis and one gets the equivalent representation.

$$A(p_1 \ldots p_N) \sim \int_{t_2}^{t_N} dt_{N-1} \ldots \int_{t_2}^{t_5} dt_4 \int_{t_2}^{t_4} dt_3 \; \langle 0| \; \prod_{i=1}^{N} \mathcal{V}(p_i, e^{t_i}) |0\rangle .$$

$$(1.2)$$

This is the Wick rotated representation.

Introducing a "timeordered" product $T\{\ldots\}$ which orders the factors inside the brackets in a decreasing sequence $t_N > t_{N-1} > \ldots > t_2 > t_1$ and includes the necessary step functions

$$\theta(t_N - t_{N-1}) \; \theta(t_{N-1} - t_{N-2}) \ldots \theta(t_4 - t_3)$$

we can also write

$$A(p_1 \ldots p_N) \sim \int_{t_2}^{t_N} \prod_{i=3}^{N-1} dt_i \; \langle 0|T\{ \prod_{i=1}^{N} (p_i, e^{t_i}) \} |0\rangle \quad (1.3)$$

which is very suggestive of a field theoretic pattern in a Dirac picture expansion with the "interaction Hamiltonian"

$$H'_D(t_i) = \mathcal{V}(p_i, e^{t_i}) \tag{1.4}$$

and the final and initial states $<0|$ and $|0>$ being $\xrightarrow{P_N}$ and $\xleftarrow{P_1}$.

The Fock space which one has to introduce is spanned by the vectors $a^{\dagger\mu}(n)$; $n = 1,2,\ldots$ which form a complete denumerably infinite basis

$$[a^{\nu}(n), a^{\dagger\mu}(m)] = -g^{\nu\mu}\delta_{n,m} \tag{1.5}$$

$$|\{N_i\}> = [a^{\dagger\mu_1}(1)]^{N_1}[a^{\dagger\mu_2}(2)]^{N_2}\ldots|0>; \sum_{i=1}^{\infty} N_i = N \tag{1.6}$$

with

$$I = \sum_{\{N_i\}} |\{N_i\}><\{N_i\}| . \tag{1.7}$$

It is furthermore possible to represent the interaction term or vertex[+]

$$\mathcal{U}(p_i, e^{i\tau_i}) \equiv : e^{ip_{i\mu} x^{\mu}(\tau_i,0)} : \tag{1.8}$$

in terms of a field

$$x^{\mu}(\tau,0) = \mathfrak{x}^{\mu} + 2\alpha'p^{\mu}\tau - i\sqrt{2\alpha'} \sum_{n=1}^{\infty} \frac{1}{\sqrt{n}} \{a^{\dagger\mu}(n)e^{in\tau} - a^{\mu}(n)e^{-in\tau}\} \tag{1.9}$$

where $\mathfrak{x}^{\mu}$ and p^{ν} are Hermitian operators

[+] We work alternatively with τ or t whichever is more suggestive of related theories.

$$[\mathfrak{X}^{\mu}, \mathcal{P}^{\nu}] = -ig^{\nu\mu} \quad \text{and} \quad \mathcal{P}^{\mu}|0\rangle = 0 \tag{1.10}$$

commuting with all other operators. α' and α_o from before are slope and intercept of some linear Regge trajectory $\alpha(s) \equiv \alpha_o + \alpha's$. A (free) Hamiltonian can be defined as

$$L_o = -\alpha'\mathcal{P}^2 - \sum_{r=1}^{\infty} r a_{\mu}^{\dagger}(r) a^{\mu}(r) \tag{1.11}$$

which indeed has the property to describe the time evolution of $\mathcal{V}$

$$\mathcal{V}(p, e^{i\tau}) = e^{i\tau L_o} \mathcal{V}(p, \tau = 0) e^{-i\tau L_o} \tag{1.12}$$

which proves that, if at all, this is an interaction term in some Dirac picture.

One can also use this relation and the definition $x_i \equiv \frac{z_{i+1}}{z_i}$ to rewrite the DRM in a Schwinger-Lippman form

$$A(p_1 \cdots p_N) \sim \langle -p_1, 0| \mathcal{V}(p_2, 1) \mathcal{D} \mathcal{V}(p_3, 1) \mathcal{D} \ldots \mathcal{D} \mathcal{V}(p_{N-1}, 1)|p_N, 0\rangle \tag{1.13}$$

where the propagator reads, for

$$\alpha_o \equiv 1 \tag{1.14}$$

$$\mathcal{D} = \int_0^1 dx_i \; x_i^{L_o - 2} = \frac{1}{L_o - 1} . \tag{1.15}$$

The question now arises whether one can develop a field

theory for DRMs, and what the dimension of the fields would be. It will, in fact, turn out that a two-dimensional field theory is necessary in terms of the variables τ and σ, to be introduced later.

Conceptionally such unphysical variables should not pose any problems since the amplitudes will always involve an integration over τ and σ, just like in ordinary field theory there is always an integration over t and x over the whole space so that one is actually independent of these variables.

The necessity of a two dimensional field theory will have to do with the invariance under two dimensional conformal transformations which are necessary to get rid of ghosts. So this dimension has a quite natural origine.

Let us therefore first investigate the pole spectrum of $A(p_1 \ldots p_N)$ and see where the ghosts come into the game: The operator L_o has the eigenvalues

$$L_o \, e^{ipx}|\{N_i\}> \; = \; (-\alpha'p^2 - \sum_{i=1}^{\infty} N_i) \, e^{ipx}|\{N_i\}> \; . \tag{1.16}$$

Therefore upon insertion of (1.7) left and right of any propagator we exhibit the pole structure

$$<\{N_i\}|\mathscr{D}e^{ipx}|\{N_i\}> \; = \; <\{N_i\}|\frac{-1}{(1+\alpha'p^2)+\sum\limits_{i=1}^{\infty}N_i} \, e^{ipx}|\{N_i\}>$$

$$\tag{1.17}$$

of the amplitude

$$P_1 \xleftarrow{\quad\overset{\displaystyle P_2\Big\uparrow\Big\uparrow \;\cdots\; \Big\uparrow P_s \Big\uparrow P_{s+1}\;\cdots}{\underset{p}{}}\quad} P_N \qquad (1.18)$$

and the Regge trajectory $\alpha(p^2) = 1 + \alpha'p^2$. That is to say, there is a pole in the variable p^2 whenever $\alpha(p^2)=J$ with $J = \sum\limits_{i=1}^{\infty} N_i$. Note that many different states, in fact all partitions of J, contribute to one and the same pole. This is called the spectrum of the pole or the spectrum of physical states. It grows exponentially with p.

Since $|\{N_i\}>$ includes $|0>$ there will also be a pole at $\alpha(p^2) = 0$ which arises for $p^2 = - m^2$ i.e. for imaginary masses. This is the tachyon of a DRM.

The choice $\alpha_0 = 1$ which would not have been necessary so far is unavoidable if one requires the absence of ghost states or negative norm states. Obviously, like in QED, a state $a_\mu^\dagger(s)|0>$ has no positive definite norm since (without summation over μ)

$$\| a_\mu^\dagger(s)|0>\| = <0|a_\mu(s)a_\mu^\dagger(s)|0> = <0|[a_\mu(s),a_\mu^\dagger(s)]|0>$$
$$(1.19)$$
$$<0|[a_\mu(s),a_\mu^\dagger(s)]|0> = - g_{\mu\mu} =$$
$$= \{-1 \text{ for } \mu = 0, + 1 \text{ for } \mu = 1,2,3\}. \qquad (1.20)$$

By replacing s by $\underline{k}$ one recovers the corresponding in-definite metric of QED. There, the way out of this di-lemma were subsidiary conditions (one for each $\underline{k}$!)

$$k_\mu \, a^\mu(\underline{k})|\psi> = 0 \qquad (1.21)$$

which guaranteed by this linear combination that there
were always sufficiently many positive norm states in
$|\Psi>$ to cancel exactly the contributions of the negative
norm states in any physical expectation value (such as
the norm itself or e.g. the energy).

It is well known that the subsidiary conditions
can not be realised as operator equations but only in the
weak sense as (Ward identity) equations for the state
vectors.

In DRMs one can define similar subsidiary conditions. That is to say one has to restrict to those DRMs
which allow for these conditions. As we have said, it
is sufficient to define a field theory in a two dimensional configuration space R_2 introducing a variable σ
beside τ or t. This means that the actual field has to
be more general than the field $X_\mu(\tau,1)$ which appears in
the vertex (1.8) and which is but a boundary value. In
view of the conformal invariance to be established we
define the field in terms of the variables

$$z \equiv e^{i(\sigma+\tau)} = e^{t+i\sigma} = e^{i(\sigma-it)} \equiv e^{i\zeta} \qquad (1.22)$$

as

$$X^\mu(\sigma,\tau) \equiv X^\mu(z) = x^\mu - 2\alpha'p^\mu \ln z - i\sqrt{2\alpha'} \sum_{n=1}^{\infty} \frac{1}{\sqrt{n}}\{a^{+\mu}(n)z^n - a^\mu(n)z^{-n}\}$$

$$(1.23)$$

from which the old field follows for $\sigma = 0$.

This field operator is a general analytic function

in the complex variable z since it is nothing but a Laurent expansion[+]. Usually a derivative field is still introduced by[++]

$$P^\mu(z) \equiv \frac{1}{2\alpha'} \, iz \, \frac{dX^\mu(z)}{dz} = \frac{1}{2\alpha'} \, \frac{dX^\mu(e^\zeta)}{d\zeta} = \qquad (1.26)$$

$$= P^\mu + \frac{1}{\sqrt{2\alpha'}} \sum_{n=1}^{\infty} \sqrt{n}\{a^{\dagger\mu}(n) z^n + a^\mu(n) z^{-n}\}. \qquad (1.27)$$

Accepting a general analytic function for $X_\mu(z)$ as given by the Laurent expansion (1.23) in terms of unconstrained expansion parameters $a_\mu(n)$ would not yield a ghost-free DRM. One has to restrict the fields to analytic functions that have definite transformation properties under certain conformal transformations. These restrictions can be expressed in terms of $P_\mu(z)$ as

$$P_\mu(z) \, P^\mu(z) = 0 \; . \qquad (1.28)$$

Introducing the expansion (1.26) for $P_\mu(z)$ one gets in terms of the as yet unquantised Laurent transforms

---

[+]   in contrast to the more familiar Fourier expansion of ordinary field theory

$$A^\mu(x) = (2\pi)^{-3/2} \int \frac{d^3k}{k_0} \{a^{\dagger\mu}(\underline{k}) e^{ipk} + a^\mu(\underline{k}) e^{-ipk}\} \qquad (1.24)$$

[++]   with

$$\frac{d}{dz} X(z) = (\frac{\partial}{\partial \, \text{Re}z} - i \, \frac{\partial}{\partial \, \text{Im}z}) \, \text{Re} \, X(z) \qquad (1.25)$$

$$a^\mu(n) = \oint \frac{dz}{2\pi i z} z^{-n} p^\mu(z) \qquad \text{and}$$

$$\alpha_o^\mu \equiv 2\alpha' p^\mu, \quad \alpha_n \equiv \sqrt{2\alpha'}\, \sqrt{n}\, a(n), \quad \alpha_{-n} \equiv \alpha_n^\dagger \qquad (1.29)$$

$$p^\mu P_\mu = -4\alpha' \sum_{m=-\infty}^{+\infty} z^{-m} L_m = 0$$

and hence
$$L_m = 0 \qquad (1.30)$$

with

$$L_m = -\frac{1}{4\alpha'} \sum_{n=-\infty}^{+\infty} \alpha_{m-n}^\mu \alpha_{n\mu}, \quad m = 0, \pm 1, \ldots \qquad . \qquad (1.31)$$

These are the constraints which can be shown to be sufficient for ghost removal.

As in QED these equations cannot be taken over as operator equations as they stand. One can either use them in the present form to express all states α_n^μ in terms of transverse ones (e.g. α_n^1 and α_n^2) and to quantise only the latter. This would correspond to the noncovariant quantisation. Or one imposes the constraints in the weak sense in eq. (1.21), now for infinitly many values of m instead of k

$$L_m|\Psi> = 0, \text{ for } m = 0,1,2, \ldots \text{ only}, \qquad (1.32)$$

the operator version of L_m being defined as

$$L_m = - \frac{1}{4\alpha'} \sum_{n=-\infty}^{\infty} : \alpha^\mu_{m-n} \alpha_{n\mu} :, \quad m = 0, \pm 1, \ldots \quad . \qquad (1.33)$$

Clearly the physical states $|\psi\rangle$ in addition still have to obey the pole condition (1.16) $L_0 |\psi\rangle = |\psi\rangle$ (where L_0 is (1.11) or (1.33) for $m = 0$). They, however, no longer contain all states $|\{N_i\}\rangle$ of (1.6), but only the ghost-free ones.

2. CONFORMAL TRANSFORMATIONS

It turns out that the operators L_n form an infinite Lie algebra

$$[L_n, L_m] = (n-m) L_{n+m} + \delta_{m,-n} \frac{D}{12} n (n^2-1) \qquad (2.1)$$

and can therefore be regarded as generators of a transformation group. Their operatorial representation can best be given in terms of the "Laurent transformed" field

$$:P^2(z): = \sum_{n=-\infty}^{\infty} L_n z^n \qquad (2.2)$$

with

$$L_n = \oint \frac{dz}{2\pi i z} z^{-n} :P^2(z): = - \frac{1}{4\alpha'} \sum_{m=-\infty}^{\infty} : \alpha_{n-m} \alpha_m : . \qquad (2.3)$$

If we now re-examine the physical states $|\psi\rangle$ it becomes clear that the fields $\vec{V}(p_\mu, z)$ and $X^\mu(z)$ which build up the dual amplitude are bound to have a well defined transformation behaviour under the generators L_n in order to satisfy (1.32). Indeed one finds that[+]

$$[L_n, \vartheta(p,z)] = z^n (\alpha' p^2 n + z \tfrac{d}{dz}) \, \vartheta(p,z) \qquad (2.4)$$

and

$$[L_n, x^\mu(z)] = z^n z \tfrac{d}{dz} x^\mu(z) \qquad (2.5)$$

$$[L_n, p^\mu(z)] = z^n (n + z \tfrac{d}{dz}) \, p^\mu(z) \qquad (2.6)$$

so that a state

$$\mathcal{D}\vartheta \, (p_{s+1}, 1) \; \mathcal{D}\vartheta \, (p_{s+2}, 1) \; \mathcal{D} \ldots \mathcal{D}\vartheta (p_{N-1}, 1) |0, p_N\rangle$$

as it appears in (1.13) would be a physical state provided $\alpha_o = 1$ as assumed.

[+] Proof: For $L_n - L_o$ especially we get the simple form

$$[L_n - L_o, \vartheta(p,z)] = n\alpha' p^2 \, \vartheta(p,z)$$

so that with $(L_n - L_o + 1) \frac{1}{L_o - 1} = \frac{1}{L_o + n - 1}(L_n - L_o + 1 - n)$ we find

$$(L_n - L_o + 1)\frac{1}{L_o - 1} \vartheta(p_{s+1}, 1) \frac{1}{L_o - 1} \ldots \vartheta(p_{N-1}, 1) |0, p_N\rangle =$$

$$= \frac{1}{L_o + n - 1} \vartheta(p_{s+1}, 1)\frac{1}{L_o + n - 1} \ldots \vartheta(p_{N-1}, 1) (L_N - L_o + 1) |0, p_N\rangle = 0$$

since $(L_n - (L_o - 1)) |0, p_N\rangle = 0$. This is the proof that $L_n |\Psi\rangle = 0$, $(L_o - 1) |\Psi\rangle = 0$.

What is more important here is the fact that $X^\mu(z)$ must be a representation operator for the generators L_n.

It should therefore be possible to find a derivative representation for this Lie algebra. For it is always possible to represent the generators of a Lie algebra in terms of differential operators

$$a(x_i) + f(x_i) \frac{\partial}{\partial x_i} \quad . \tag{2.7}$$

The best known example for this is the momentum or the angular momentum generator of the Lorentz group in spinor (spin $J = \frac{1}{2}$) or in vector ($J = 1$) representation . The corresponding covariant field operators are here called $\phi^J(z)$ and are defined to yield with $z \equiv e^{i\zeta}$

$$\delta\phi^J(z) = \alpha[L_n, \phi^J(z)] = \alpha z^n[Jn + z \frac{d}{dz}]\phi^J(z) =$$

$$= \alpha e^{in\zeta} [Jn-i \frac{d}{d\zeta}]\phi^J(e^{i\zeta}) \tag{2.8}$$

with a small parameter α. They are said to have con-formal spin J. Note that all our fields (2.4) - (2.6) have this transformation property[+].

[+] A more general basis has been investigated by Gervais and Sakita (1971) defined by

$$\delta\phi^{J,d}(z) = \alpha[L_g, \phi^{J,d}(z)] = \alpha[dReg'(\zeta) + iJ \, Im \, g'(\zeta) +$$

$$+ \, g(\zeta) \frac{d}{d\zeta}] \, \phi^{J,d}(z) \tag{2.9}$$

with $z \equiv e^{i\zeta}$ which is not necessary here. Our's follows with $d = J$.

We now want to investigate what (conformal) transformations go along with the Virasoro algebra L_n whose
differential representation we have just found. This
correspond to e.g. finding the Lorentz transformation
from the angular momentum operators.

Instead of investigating the transformations induced by each L_n we shall construct the most general
algebra element by the linear composition

$$L_g = - \sum_{n=-\infty}^{\infty} g_n L_n \qquad (2.10)$$

where g are complex numbers. The equs. (2.8) can then
be summed up to give

$$\delta\phi^J \equiv \phi'^J(z) - \phi(z) = \alpha[L_g, \phi^J(z)] = \alpha[Jg'(\zeta) + g(\zeta)\frac{d}{d\zeta}]\phi^J(z) \qquad (2.11)$$

with

$$g(\zeta) = + i \sum_{n=-\infty}^{\infty} g_n e^{in\zeta} = + i \sum_{n=-\infty}^{\infty} g_n z^n \equiv f(z) \qquad (2.12)$$

$$g'(\zeta) = - \sum_{n=-\infty}^{\infty} n g_n e^{in\zeta} . \qquad (2.12')$$

From this we infer immediately the infinitesimal transformation

$$\phi'^J(e^{i\zeta}) \approx \phi^J(e^{i\zeta}) + \alpha[L_g, \phi^J(e^{i\zeta})] =$$

$$= [1+\alpha Jg'(\zeta)][\phi^J(e^{i\zeta}) + \alpha g(\zeta)\frac{d}{d\zeta} \phi^J(e^{i\zeta})] \qquad (2.13)$$

as being $\qquad$ $(\zeta' - \zeta) \approx \alpha g(\zeta)$ $\qquad$ (2.14)

or $\qquad$ $\zeta' \approx \zeta + \alpha g(\zeta)$. $\qquad$ (2.14')

Integration of (2.11) yields

$$\phi'^J(e^{i\zeta}) = e^{\alpha L_g} \phi^J(e^{i\zeta}) e^{-\alpha L_g} = (\frac{g(\zeta')}{g(\zeta)})^J \phi^J(e^{i\zeta'}) \quad (2.15)$$

and

$$\frac{d\zeta'}{d\zeta} = \frac{g(\zeta')}{g(\zeta)} \qquad (2.16)$$

or

$$\eta(\zeta') = \eta(\zeta) + \alpha \qquad (2.17)$$

which is equivalent to a constant integral

$$\int_{\zeta}^{\zeta'} \frac{1}{g(\zeta)} d\zeta = \alpha = \text{const} . \qquad (2.18)$$

In the η-plane this conformal transformation is therefore a mere translation whereas the mapping to the ζ-
and ζ' -plane will yield similar singularity structures.
The Virasoro algebra therefore generates a rather
restricted class of conformal transformations.

We note that these conformal transformations are
described by an infinite Lie algebra in contrast to conformal transformations in four dimensions (Maxwell
theory) which have a 15 dimensional Lie algebra. This
is due to the fact that while the latter algebra corres-

ponds to a __global__ O(2,4) group, the two dimensional one
is an exceptional case which yields a group that can be
realised only __locally__. This means that the group parameters change with the location in space where we perform
the transformation so that we find the group structure
realised only for the differentials

$$
\begin{bmatrix} dx' \\ dy' \end{bmatrix} = \begin{bmatrix} \frac{\partial x'}{\partial x} & \frac{\partial x'}{\partial y} \\ \frac{\partial y'}{\partial x} & \frac{\partial y'}{\partial y} \end{bmatrix} \begin{bmatrix} dx \\ dy \end{bmatrix} = \mu(x,y) \begin{bmatrix} \cos\theta\,(x,y)\,\sin\theta\,(x,y) \\ -\sin\theta\,(x,y)\,\cos\theta\,(x,y) \end{bmatrix} \begin{bmatrix} dx \\ dy \end{bmatrix} ,
$$

$$
(2.19)
$$

with $\zeta \equiv x + i\,y$, $\zeta' \equiv x' + i\,y'$ and where the latter
formulation is due to the Cauchy-Riemann relations.
μ and θ are general harmonic functions in the case of
general conformal transformations. These are hence seen
to correspond locally to a direct product $D \otimes O(1,1)$
of a dilatation times a two dimensional rotation group.

A general representation operator is then defined
as

$$
\phi'^{J,d}(e^{i\zeta}) = (\frac{d\zeta}{d\zeta'})^{\frac{1}{2}(d+J)} (\frac{d\zeta}{d\zeta'})^{*\frac{1}{2}(d-J)} \phi^{J,d}(e^{i\zeta'}) \quad (2.20)
$$

since this has the right group properties. In our case
(2.8) we set $d = J$.

Having found the transformation $\zeta \rightarrow \zeta'$ induced by
the Virasoro algebra and the covariance properties of
field operators $\phi^J(z)$ one naturally asks oneself which
of the many possible operator fields will yield a (ghost-

free) dual model. For the covariance of a field is only a necessary condition.

Our obvious aim is to connect the DRA (1.3) with an S-matrix element in the Dirac picture. Hence one has to develop a Lagrangian formalism with a free and an interaction Lagrangian which will yield the DRA.

The action integral S of such a Lagrangian formalism obviously has to be conformally invariant

$$[L_n, S] = 0 . \tag{2.21}$$

Otherwise there is no hope to get rid of ghosts. Let us therefore first look for possible free Lagrangians $\mathcal{L}_o$ (without interaction terms). One possibility is obviously $\phi^J(z)$ with $J = 0$ and the action integral

$$S = \int dx \, dy \, \frac{d}{d\zeta^*} \, \phi^o \, (e^{i\zeta}) \, \frac{d}{d\zeta} \, \phi^o \, (e^{i\zeta}) \tag{2.22}$$

which we prefer to present in terms of the variables $\zeta = x + iy$ rather than σ, τ. In view of this Lagrangian one would call $\phi^o(z) = X^\mu(z)$ a "massless" field since there are no nonderivative terms.

The conformal invariance of S follows immediately from (2.15) since in[+]

$$\frac{d}{d\zeta^*} \, \phi'^J (e^{i\zeta}) = e^{\alpha L_g} \, \frac{d}{d\zeta^*} \, \phi^J (e^{i\zeta}) \, e^{-\alpha L_g} =$$

[+] with $\zeta \equiv x + iy \equiv \sigma + \tau$.

$$(\frac{d\zeta}{d\zeta'})^{J}\frac{d\zeta'}{d\zeta}\frac{d}{d\zeta'^{*}}\phi^{J}(e^{i\zeta'})+J(\frac{d\zeta}{d\zeta'})^{J-1}\frac{d^{2}\zeta'}{d\zeta^{*}d\zeta}\phi^{J}(e^{i\zeta'}) \quad (2.23)$$

the last term vanishes due to the Laplace equ.

$$\frac{d^{2}\zeta'(\zeta)}{d\zeta^{*}d\zeta} = (\partial_{x}^{2} + \partial_{y}^{2})\, \zeta'(x,y) = 0 \; . \tag{2.24}$$

Hence $\dfrac{d}{d\zeta^{*}}\phi'^{J}(e^{i\zeta})$ behaves like $\phi'^{J-1}(e^{i\zeta})$.

Since the action integral is

$$S = \int\mathcal{L}\; d\sigma\; d\tau = i\int\frac{d}{d\zeta^{*}}\phi^{J}(e^{i\zeta})\frac{d}{d\zeta}\phi^{J}(e^{i\zeta})\; dxdy \tag{2.25}$$

with $\qquad\qquad d\sigma\; d\tau = i\quad dxdy \qquad\qquad\qquad (2.26)$

and since

$$dxdy = \frac{\partial(x,y)}{\partial(x',y')}\; dx'dy' = \frac{d\zeta}{d\zeta'^{*}}\frac{d\zeta}{d\zeta'}\; dxdy \tag{2.27}$$

it is clear that

$$e^{\alpha L_{g}}\, Se^{-\alpha L_{g}} = \int\frac{d}{d\zeta^{*}{}'}\phi^{J}(e^{i\zeta'})\frac{d}{d\zeta'}\phi^{J}(e^{i\zeta'})\, dx'dy' \tag{2.28}$$

only if $J = 0$ as is the case for the field $x^{\mu}(z)$ of the generalised Veneziano model. But also the Shapiro-Virasoro model (for the pomeron) follows from such a Lagrangian. Both fields are scalars ($J = 0$) with respect to conformal spin.

The only other possible ansatz is in terms of 2 complex fields $\psi_1(z)$ and $\psi_2(z)$. Let their transformation properties be

$$\psi_1(z) \rightarrow \psi_1'(z) = \left(\frac{d\zeta}{d\zeta'}\right)^{*J} \psi_1(z') \quad \text{with} \quad z \equiv e^{i\zeta}, \; z' \equiv e^{i\zeta'}$$

$$\psi_2(z) \rightarrow \psi_2'(z) = \left(\frac{d\zeta}{d\zeta'}\right)^{J} \psi_2(z') \; . \tag{2.29}$$

Then the action

$$S = \int [\psi_2^*(z) \frac{d}{d\zeta} \psi_1(z) + \psi_1^*(z) (\frac{d}{d\zeta} \psi_2(z))^*] dxdy =$$

$$\rightarrow \int [\psi_2^*(z') \frac{d}{d\zeta'} \psi_1(z') (\frac{d\zeta}{d\zeta'})^{*J} (\frac{d\zeta}{d\zeta'})^{-1} (\frac{d\zeta}{d\zeta'})^* (\frac{d\zeta}{d\zeta'}) dx'dy' + h.c.]$$

$$\tag{2.30}$$

is invariant provided the conformal spin is $J = -\frac{1}{2}$. This will eventually yield the Ramond or Neveu-Schwarz model depending on the boundary conditions for $\psi_i(z)$.

3. FREE RELATIVISTIC STRING MODELS

The previously found Lagrangian, though having the right conformal invariance, lacks a relativistic formulation: We have stated that a field theoretic version of the dual model leads necessarily to a two dimensional field theory where $i\tau$ or t can be regarded as a time expansion parameter. The other parameter σ must

then be a spatial extension parameter in some configuration space. It is clear that unless one finds such
a field theoretic formulation there is no hope to derive
unitarity corrections in any consistent way.

The new variables are introduced as $z \equiv e^{i\zeta}$ with
$\zeta \equiv \tau + \sigma$ in a Euclidean field theory and/or as $\zeta \equiv \sigma - it \equiv$
$\equiv \zeta^1 - i\zeta^0$ leading to a non-Euclidean theory with
$\zeta_\alpha = g_{\alpha\beta}\zeta^\beta$ and $g_{\alpha\beta} = \begin{pmatrix} 1 & 0 \\ 0 & -1 \end{pmatrix}$.

The natural interpretation of the field $X_\mu(\sigma,t)$
is that of the position of a string of point particles
whose positions along the string are denoted by the
parameter σ and whose world lines are parametrised by t.
They therefore cover a two dimensional surface in the
space-time continuum. The most general relativistic invariant action for such a string is given by

$$S = \int d\sigma \ d\tau \ \sqrt{-\sigma_{\mu\nu}\sigma^{\mu\nu}} \qquad (3.1)$$

with

$$\sigma_{\mu\nu} = \frac{\partial X_\mu}{\partial \tau}\frac{\partial X_\nu}{\partial \sigma} - \frac{\partial X_\mu}{\partial \sigma}\frac{\partial X_\nu}{\partial \tau} . \qquad (3.2)$$

This action is seen to reduce to the previous one (2.22)
or in the alternative form

$$S = \int d\sigma \ d\tau \ [\ (\partial_\sigma X)^2 + (\partial_\tau X)^2] \qquad (3.1')$$

if one imposes the subsidiary conditions (1.28) leading
to conformal invariance.

The previous solutions $X_\mu(\sigma,\tau)$ for the generalized

Veneziano model follow from this Lagrangian or the corresponding equations of motion by requiring special boundary conditions for a finite string. We shall not discuss them in detail.

There is still another solution if we assume a closed string. This is (cf. e.g. Schwarz (1973)(5.6))

$$Y^\mu(\sigma,\tau)=\mathbf{x}^\mu-2\alpha'i\mathbf{p}^\mu\ln z\hat{z}-i\sqrt{2\alpha'}\sum_{n=1}^{\infty}\frac{1}{\sqrt{n}}\{a^{+\mu}(n)z^n +$$

$$+ \hat{a}^{+\mu}(n)\,\hat{z}^n - a^\mu(n)z^{-n} - \hat{a}^\mu(n)\hat{z}^{-n}\} \qquad (3.3)$$

and it yields the Shapiro-Virasoro model. Here $\hat{z}$ is the image of z reflected on the line $z = e^{i\tau}$ or z = $= e^t$ (real axis, then $\hat{z} = z^*$). The interactions (vertex) for external scalar particles is very similar to the one mentioned before and so is the expansion into a time ordered product yielding the DRM. Only the Fock space is somewhat richer due to the new operators $\hat{a}$. But clearly it is again ghost free due to conformal invariance. The spectrum is as in fig. 3.1

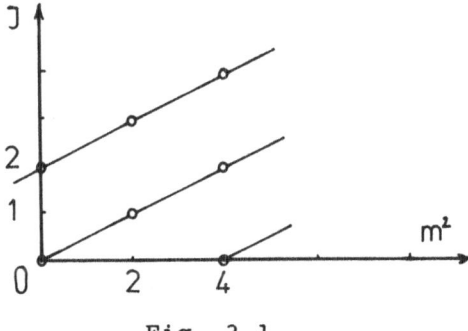

Fig. 3.1

there being now two massless particles given by the
linear combinations of the states $a^{+\mu}(-1)\hat{a}^{+\nu}(-1)|0\rangle$.
They correspond to spin 2 (tensor) and spin 0 (scalar)
particles and can be identified with a pomeron and
dilaton.

It is one of the miracles of DRM that the pomerons
can also be created by nonplanar loops, fig. 3.2, from
the generalised Veneziano model (which defines the slope
in fig. 3.1 to be $\alpha'/2$). It is this intimate connection
that allows lateron to combine both fields in an attempt
to break the symmetry spontaneouslv.

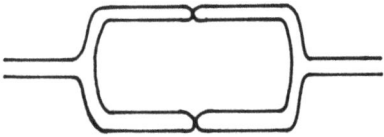

Fig. 3.2

The second sort of Lagrangian (2.30) can be interpreted
as a "spinning" string, that is beside the position
X_μ (σ,τ) of the particles along the string we still
define a function $\psi(\sigma,\tau) = \{ \begin{matrix} \psi_1(\sigma,\tau) \\ \psi_2(\sigma,\tau) \end{matrix}$ (which can have
vector or spinor indices with regard to ordinary con-
figuration space) which can be interpreted as defining
some sort of (conformal) spin forthe point particles
along the string. These new fields read with $z=e^{i(\sigma+\tau)}$
either as

$$\psi^\mu(z) \equiv \tfrac{1}{2}\Gamma^\mu(z) = \tfrac{i}{2}\{\gamma^\mu + i\sqrt{2}\gamma_5 \sum_{n=1}^\infty [d^{+\mu}(n)z^n + d^\mu(n)z^{-n}]\} \qquad (3.4)$$

for the Ramond model or as

$$\psi^\mu(z) \equiv \frac{1}{\sqrt{2}} H^\mu(z) = \sum_{n=1/2}^{\infty} [b^{\dagger\mu}(n) z^n + b^\mu(n) z^{-n}] \qquad (3.5)$$

for the Neveu-Schwarz model. They are both quantised by anticommutation relations

$$[d^\nu(n), d^{\dagger\mu}(m)] = -g^{\mu\nu} \delta_{m,n} \quad \text{or} \quad [b^\nu(n), b^{\dagger\mu}(m)] = -g^{\mu\nu} \delta_{m,n} .$$

$$(3.6)$$

The Fock space of these two models consists of $a^\mu(m)$, $b^\nu(n)$ or $a^\mu(m)$, $d^\nu(n)$, respectively. Again, due to the conformal invariance negative norm states do not appear. The generators L_n are now represented in terms of a_μ and b_μ operators.

The enlarged Fock space has again a richer spectrum of physical particles which now, for the NS model, has the appearance :

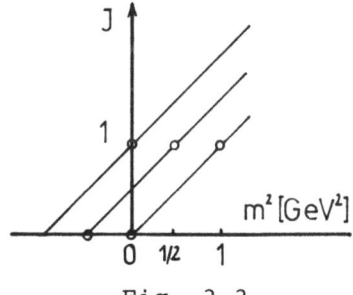

Fig. 3.3

For the Ramond model the state $|\Psi\rangle$ contains spinor states due to the Dirac matrices which correspond to fermions of half integer spin :

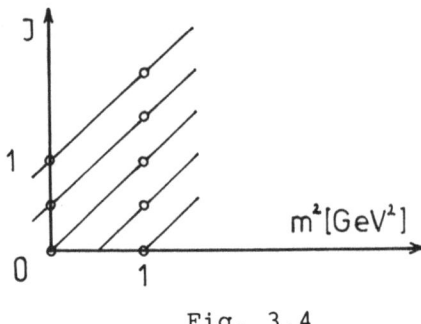

Fig. 3.4

It is an interesting feature of the last models that in addition to the conformal generators one gets another set G_n of generators that introduce supersymmetry transformations (Zumino (1973)), i.e. transformations among meson and fermion fields which leave the Lagrangian invariant. This is, however, beyond the scope of the present review of gauge symmetries.

4. INTERACTING STRING

So far we have been concerned with the freely propagating string only which does not emit particles. Therefore the Hamiltonian $H_o = L_o$ mentioned in the introduction will follow from the free Lagrangian considered so far. As an interaction Hamiltonian one should now choose a term which is proportional to

$$H' \sim e^{i p_i^\mu X_\mu (t_i, 0)} \sim \vartheta (p_i^\mu, e^{t_i}) \tag{4.1}$$

(the exact form is given by Ademollo et al. (1974a) eq. (4.1) but is irrelevant here). This corresponds to the emission of ground state particles with four momentum $p_{i\mu}$ from the ends of the string $\sigma = 0$ or π and at times t_i (with $z_i = e^{t_i}$, fig. 4.1).

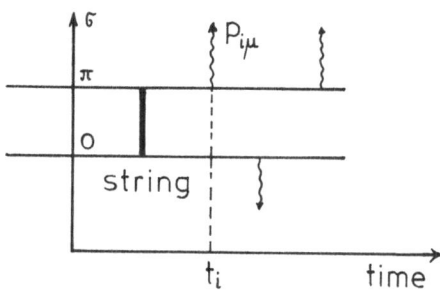

Fig. 4.1

The Dirac picture expansion of this interaction can then be shown to yield the N-point amplitudes of section 1 (Ademollo et al. (1974a,b)).

The interaction introduced upstairs is, however, too simple since it allows to emit only ground state particles. Unfortunately an extension to the emission of arbitrary resonance (hence the interaction with another complete string) is not possible in this formalism.

The reason seems to be that despite appearance the quantised string formalism with its infinitely many operators $a^\mu(n)$ is not a second quantised theory. In fact, in the derivation of the string Lagrangian one starts with a string of classical point particles. Their quantisation (via the Poisson bracket formalism) thus corresponds to the usual first quantisation of position and momentum. The appearance of the set $a^\mu(n)$ is entirely due to the infinitely many classical particles one starts with. The previous "Dirac picture expansion" is therefore the quantum mechanical analogue for <u>extended</u> objects.

As a further evidence of this concept one can take the loop amplitudes (unitary corrections) which do not follow automatically but have to be introduced and calculated one by one separately (cf. e.g. Karpf (1974a,b)). A reformulation of unitary correction in terms of the string picture (string-string interactions, string loops) was given by Mandelstam (1974) in terms of the quantum mechanical Feynman path formalism adapted to DRMs by Hsue, Sakita, Virasoro (1970) and Gervais, Sakita (1971). The connection with the actual DRM expressions is then no longer as transparent as in section 1 (but is in principle on the same footing).

In order to achieve all unitary corrections automatically one has to develop a truely second quantised version which was achieved by Kaku, Kikkawa (1974a,b).

Up to this point we have reviewed DRMs which despite their mathematical rigour and physical relevance suffered of one big defect: The tight conformal invariance, required by the absence of ghosts, necessitated massless particles where physics demands finite masses.

Since one cannot jettison conformal invariance without introducing ghosts one is forced to look for other means, like spontaneous symmetry breaking, to circumvent these unpleasant consequences.

5. SPONTANEOUS SYMMETRY BREAKING

We mentioned already that because of the purely derivative Lagrangian $\mathcal{L}_o$ in the action (2.22) and (3.3) one would call all these fields "massless". This also means that the free Hamiltonian is

$$L_o = -\alpha' \, \mathbf{p}^2 - \sum_{r=1}^{\infty} ra_\mu^\dagger(r) a^\mu(r) \tag{5.1}$$

and not e.g.

$$L_o = -\alpha' \, \mathbf{p}^2 - \sum_{r=1}^{\infty} (r-c_o) a_\mu^\dagger(r) a^\mu(r) \tag{5.2}$$

which we would like much better in order to get poles at

$$L_o |\psi\rangle = \alpha_o |\psi\rangle \tag{5.3}$$

for general α_o.

It is clear that a massterm such as $X_\mu^*(z) X^\mu(z)$ would spoil conformal invariance since

$$\int X_\mu^*(z) X^\mu(z) \; dxdy \tag{5.4}$$

for a $J = 0$ field is not invariant under

$$\delta^{L_g} X_\mu(z) = (Jg'(\zeta) + g(\zeta)\frac{d}{d\zeta}) X_\mu(z) \; . \tag{5.5}$$

In the same way a mass term spoils conformal and gauge invariance with respect to

618

$$\delta A_\mu(x) = \frac{d}{dx_\mu} \Lambda(x) \tag{5.6}$$

of the Maxwell-equations of QED.

In gauge theories there is, however, a device called spontaneous symmetry breaking and Higgs mechanism which allows to introduce massive fields by maintaining formally at least the full gauge symmetry of the Lagrangian.

This mechanism essentially amounts to redefining a free Lagrangian $\mathcal{L}$ of at least two sorts of fields Φ and B^μ, say, in terms of other fields ζ_i that are combinations of the old ones. The crux is that the (same) Lagrangian $\mathcal{L}$ rewritten in terms of the new fields ζ_i now has derivative <u>and</u> mass terms for some of the new fields. Of course it is no longer a free Lagrangian for these new fields, but this is of minor importance since the mere existence of $\mathcal{L}_{o\zeta}$ as a part of $\mathcal{L}$ allows to quantise in terms of new operators a_ζ^+ which belong to massive fields now. The remainder of $\mathcal{L}$ is simply swallowed by an overall interaction Lagrangian.[+] Defining and introducing these new operators a_ζ^+ or generally this <u>new</u> second quantisation it is clear that one changes the Fock space and the vacuum alltogether, the latter not being gauge invariant any longer.

A necessary condition for this mechanism to work is a nontrivial minimum of the potential part of the Lagrangian i.e. a minimum which does not appear for $\Phi = B^\mu = O$ but e.g. for a value $\hat{\Phi}$ of the field $\Phi(x)$. This will usually arise if the Φ-field is a tachyon i.e. possesses a negative mass-square. A naive quanti-

---

[+] for more details see e.g. Abers, Lee (1973).

sation of the Φ-field will then always yield that the
vacuum expectation value does not vanish but is

$$<0|\Phi(x)|0> = \hat{\Phi} \; . \tag{5.7}$$

Since in most of the DR's tachyons are present this seems
to be a rather fortunate fact since it seems to guarantee
that spontaneous breakdown of the high symmetry is potentially possible.

So far a general discussion of spontaneous symmetry
breaking in terms of the string fields $X^\mu(\sigma,\tau)$ and the
others is still not possible since the Lagrangians involved are not well known.

This is, of course, the ultimate aim since we want
to shift the whole set of trajectories, discribed by
$X^\mu(\sigma,\tau)$ and its interactions to arbitrary values of α_o.

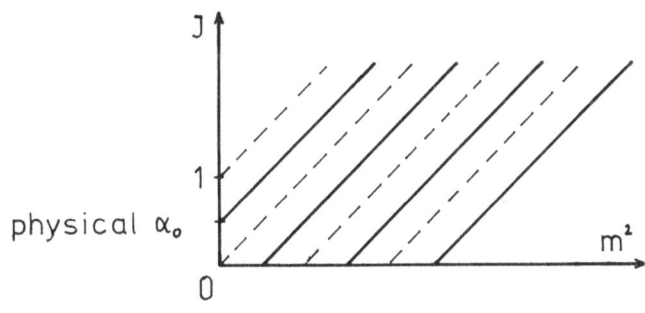

Fig. 5.1

This not yet being possible one settles with the next
best thing by replacing the true string-string interaction by the exchange of single massless particles of
these trajectories. The first attempts have been made

by Kalb and Ramond (1974) who assumed massless vector B_μ and tensor $\Phi_{\mu\nu}$ particles (pomerons) to be exchanged between open and closed strings for which they used the string fields $X^\mu(\sigma,\tau)$ of eq. (1.23) and $Y^\mu(\sigma,\tau)$ of equ. (3.3).

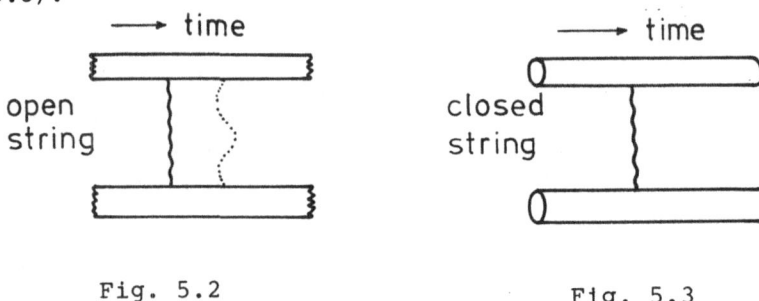

Fig. 5.2   Fig. 5.3

Their full Lagrangian therefore contains the string Lagrangian (3.1) $\mathcal{L}_X$, an interaction term $\mathcal{L}_{XP}$ and a Lagrangian $\mathcal{L}_P$ for the tensor and vector fields

$$\mathcal{L}_P = \tfrac{1}{12}G_{\mu\nu\rho}G^{\mu\nu\rho} - \tfrac{1}{4}F_{\mu\nu}F^{\mu\nu} - \tfrac{1}{4}\frac{g^2}{e^2}\Phi^{\mu\nu}\Phi_{\mu\nu} - \tfrac{1}{2e}g\Phi^{\mu\nu}(\partial_\mu B_\nu - \partial_\nu B_\mu) \quad (5.8)$$

where

$$F_{\mu\nu} = \partial_\mu B_\nu - \partial_\nu B_\mu \qquad\qquad \text{and}$$

$$G^{\mu\nu\rho} = (\partial^\mu\Phi^{\nu\rho} + \partial^\nu\Phi^{\rho\mu} + \partial^\rho\Phi^{\mu\nu}) \ . \qquad\qquad (5.9)$$

The fields $B_\nu(X)$ and $\Phi_{\mu\nu}(X)$ are assumed to depend on the string fields $X_\mu(\sigma,\tau)$ like otherwise fields were functions of space time coordinates x_μ. Clarly, the invariance under the conformal transformations
$z = e^{i(\sigma+\tau)} \rightarrow z' = e^{i(\sigma'+\tau')}$ as well as ordinary

gauge invariance has to hold for these additional fields B_μ and $\phi^{\mu\nu}$.

The Lagrangian $\mathcal{L}_P$ is invariant under the joint gauge transformation

$$\phi_{\mu\nu} \rightarrow \phi_{\mu\nu} + \partial_\mu \Lambda_\nu - \partial_\nu \Lambda_\mu \qquad (5.10)$$

$$B_\mu \rightarrow B_\mu - \frac{g}{e} \Lambda_\mu$$

since only the combination

$$\Psi_{\mu\nu} \equiv \phi_{\mu\nu} + \frac{e}{g} F_{\mu\nu} \qquad \text{with} \qquad \Psi_{\mu\nu} \rightarrow \Psi_{\mu\nu} \qquad (5.11)$$

actually appears in the equations.

One can now choose the Lorentz conditions

$$\partial_\mu \phi^{\mu\nu} = \frac{g}{e} B^\nu \quad , \quad \partial_\mu B^\mu = 0 \qquad (5.12)$$

such that despite the gauge invariance (5.10) B_μ is a massive field with the equations of motion

$$\square B_\mu = - \left(\frac{g}{e}\right)^2 B_\mu \quad . \qquad (5.13)$$

This could thus be achieved by an appropriate _mutual_ **choice** of gauges of vector and tensor fields. The total action integral then reads

$$S = -\mu^2 \int d\sigma\, d\tau\, \sqrt{-\sigma^{\mu\nu}\sigma_{\mu\nu}} + g\int d\tau\, d\sigma\ \sigma_{\mu\nu}\Psi^{\mu\nu} + \int d^4x_\mu\, (\sigma,\tau)\, \mathcal{L}_P \quad . \quad (5.14)$$

622

We note that we have here find an appealing way to introduce massive vectorbosons into gauge invariant theories. But this can only serve as a guide line for DRMs since it certainly does not break spontaneously the conformal invariance of the string fields.

A similar approach was outlined by Cremmer and Scherk (1974). It was argued there that the massless vector particle ρ of the trajectories of fig. 5.1 can acquire a mass through the unitarity sum over non-planar self energy loops

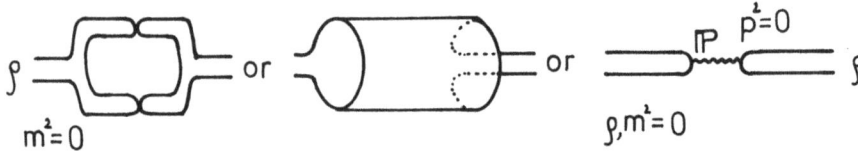

Fig. 5.4

since it can be shown that for d = 25 space dimensions these loop expressions have themselves poles at $p^2 = 0$ (and correspond therefore to massless particles, i.e. pomerons, which are created by these loops). A pole in a self energy term is, however, an exceptional case for a gauge invariant theory which usually guarantees that a massless particle does not acquire mass.

Again this idea cannot be persued in full generality but an example is considered by Cremmer and Scherk how a gauge invariant theory of <u>single</u> massless vector and tensor particles can acquire mass by a clever choice of new fields (a redefinition like usual in spontaneous symmetry breaking).

The model Lagrangian now reads

$$\mathcal{L} = -\frac{1}{4} F_{\mu\nu} F^{\mu\nu} + \frac{1}{6} G_{\mu\nu\rho} G^{\mu\nu\rho} + \frac{1}{3} \lambda \varepsilon_{\nu\alpha\beta\gamma} B^\nu G^{\alpha\beta\gamma} \quad . \tag{5.15}$$

It is gauge invariant under

$$B_\nu \to B_\nu + \partial_\nu f$$

$$\Phi_{\mu\nu} \to \Phi_{\mu\nu} + \partial_\mu \zeta_\nu - \partial_\nu \zeta_\mu \tag{5.16}$$

and can be redefined in terms of the new fields

$$h_\nu \equiv -\frac{1}{6} \varepsilon_{\alpha\beta\gamma\nu} G^{\alpha\beta\gamma} + \lambda B'_\nu \tag{5.17}$$

$$B'_\nu = B_\nu - \frac{1}{\lambda} \partial_\nu \chi \tag{5.18}$$

as

$$\mathcal{L} = -\frac{1}{4} F'_{\mu\nu} F'^{\mu\nu} + \lambda^2 B'_\mu B'^\mu - h_\mu h^\mu \tag{5.19}$$

with the equations of motion

$$-\partial_\mu F'^{\mu\nu} = 2 \lambda^2 B'^\nu \tag{5.20}$$

$$h_\nu = 0 \tag{5.21}$$

which show that B'_μ has become a massive particle.

Even the most ambitious investigations so far by Bardakci and Halpern (1974a,b, 1975) uses only model (effective) Lagrangians (e.g. equ.(4.1) in B.H.(1974a))

624

to study the spontaneous breakdown. These Lagrangians
are, however, built from dual models with external
vector particles. Their results are quite encouraging
since they show explicitely that spontaneous symmetry
breaking is a result of the tachyons of DRMs.

6. VERTEX SOLUTIONS

There is another line of reasoning to tackle
the problem of interaction Lagrangians and symmetry
breaking in dual models which originated from a work
of Nielsen and Olesen (1973). They were able to show
that the gauge invariant Lagrangian of Higgs

$$\mathcal{L} = -\frac{1}{4}F_{\mu\nu}F^{\mu\nu} + \frac{1}{2}|(\partial_\mu + ieA_\mu)\Phi|^2 + c_2|\Phi|^2 - c_4|\Phi|^4 \qquad (6.1)$$

with a nontrivial potential minimum at

$$|\Phi| = \hat{\Phi} = \sqrt{\frac{c_2}{2c_4}}$$

allows for string like (vertex) solutions. These solutions are such that the fields $F_{\mu\nu}(x)$ are concentrated
inside the thin one-dimensional tube (in a three dimensional space) of fig.6.1 with a half width λ. The
ansatz for the potential $\underline{A}$ there is

$$\underline{A}(\underline{r}) = \frac{1}{r}(\underline{r} \times \underline{1}_z)|A(\underline{r})| \qquad (6.2)$$

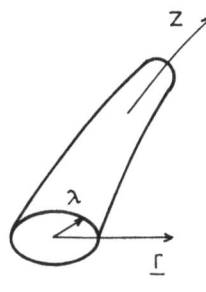

Fig. 6.1

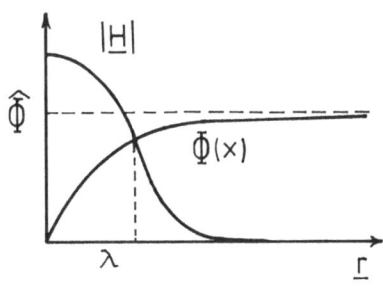

Fig. 6.2

with the flux

$$\phi(\underline{r}) = \oint A_\mu(x)\,dx^\mu \qquad \text{and} \qquad |\underline{H}| = \frac{1}{2\pi r}\frac{d}{dr}\phi(\underline{r}) \qquad (6.3)$$

being concentrated in the tube. Outside the tube the
electromagnetic field decreases exponentially while
the Higgs field $\Phi(x)$ attains practically everywhere
its vacuum value $\Phi = \hat{\Phi}$ except inside the tube of
(fig. 6.2) where it is displaced. Such a solution is a
relativistic generalisation of the Ginzburg-Landau
solution for a superconductor. It seems to be a natural candidate for a relativistic string which indeed
was derived from this theory by Nielsen and Olesen.

In the four dimensional time-space continuum
this string clearly sweeps out the two dimensional
world surface with the Lagrangian (3.37). If the string
solution is not infinite or closed as in fig.6.3

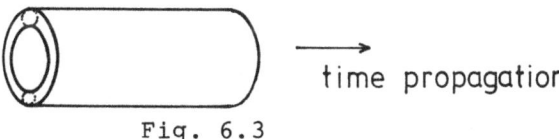

time propagation

Fig. 6.3

then we would expect a magnetic monopole on either end
of the string terminating the magnetic flux. This in
turn matches with the Dirac string solution of electrodynamics with magnetic charges which Dirac (1948) once
had considered. Instead of point particles moving along
worldlines $X_\mu(t)$ Dirac had discussed point particles
strung together like beads along a σ-direction, discribed by $X_\mu(t,\sigma)$. Barut and Förster (1973) have proven
the equivalence of this string with the dual one.

Also Nambu (1975) reconsidered Dirac's extension
of the Maxwell equations and found an effective Lagrangian which resembled very much that of Kalb and
Ramond (5.14).

It was furthermore noted by t'Hooft (1974) that
magnetic monopoles (and hence strings) solutions will
always arise quite naturally if one breaks down spontaneously a higher non-Abelian gauge theory to the
electromagnetic gauge group U(1). This was a firmer
mathematical embedding of what Higgs had found in his
Lagrangian (6.1). One has therefore the hope that
string models can be embedded into such a higher non-
Abelian gauge group which is subsequently broken down
spontaneously. Corrigan et al. (1976) have already
investigated SU(3) as the embedding gauge theory.

REFERENCES

1. E.Abers, B.Lee, Phys. Rep. 9,1, 1973.

2. M. Ademollo, A.d'Adda, R.d'Auria, E.Napolitano,
   S. Sciuto, P.di Vecchia, F.Gliozzi, R.Musto,
   E.Nicodemi, N. Cim. 21A, 77, 1974a.

3. M.Ademollo, A.d'Adda, R.d' Auria, E.Napolitano, P.di Vecchia, F.Gliozzi, S.Sciuto, N.Phys.B77,189, 1974b.

4. K. Bardakci, N.Ph. $\underline{B68}$, 331,1974a.

5. K. Bardakci, M. Halpern, N.Ph. $\underline{B73}$, 295, 1974b.

6. K. Bardakci, M. Halpern, PR.$\underline{D10}$, 4230,1975.

7. A. Barut, D.Förster, IC/73/107, 1973, Relation between the dual and Dirac string. See also

8. A. Barut, Schladming Lectures 1973, Acta Phys. Austriaca Suppl..

9. E.Corrigan, D.Olive, D.Fairlie, J.Nuyts, TH2101, 1976, Magnetic monopoles in SU(3) gauge theories.

10. E. Cremmer, J. Scherk, N. Ph. $\underline{B72}$, 117, 1974.

11. P.A.M. Dirac, PR. $\underline{74}$, 817, 1948.

12. J.-L.Gervais, B.Sakita, PR. $\underline{D4}$, 2291, 1971.

13. C.Hsue, B.Sakita, M.Virasoro, PR. $\underline{D2}$, 2857, 1970.

14. M.Kalb, P.Ramond, PR. $\underline{D9}$, 2273, 1973.

15. A.Karpf, FUB preprint, Vorlesungen über DRM, 1975.

16. A.Karpf, Unitary corrections to dual models, Schladming Lectures 1974, Acta Phys. Austriaca Suppl..

17. A. Karpf, N.Ph. $\underline{B56}$, 565, 1974b.

18. M. Kaku, K.Kikkawa, PR. $\underline{D10}$, 1110,1974a.

19. M. Kaku, K.Kikkawa, PR. $\underline{D10}$, 1823, 1974b.

20. S. Mandelstam, Phys. Rep. $\underline{13}$, 259, 1974.

21. Y. Miyatake, Kyoto Univ. preprint, Hadron model based on the **vertex** in the superfluid, 1976.

22. Y. Nambu, PR.D10, 4262, 1975, see also Miyatake, 1976.

23. H.B. Nielsen, P. Olesen, N. Ph. __B61__, 45, 1973.

24. J. Scherk, Rev. Mod. Phys. __47__, 123, 1975.

25. J. Schwarz, Phys. Rep. __8__, 270, 1973.

26. G.'t Hooft, N.Ph. __B79__, 279, 1974.

27. B. Zumino, TH 1779, Lectures at the 1973 NATO School in Capri, 1973.

Acta Physica Austriaca, Suppl. XV, 629–638 (1976)

META-SUMMARY - SECOND WEEK

by

H. MITTER[+]
Institut für Theoretische Physik
Universität Graz

To give a summary of a week filled with lectures,
which were already summaries (and quite excellent ones)
of research fields, means to give a meta-summary, which
is rather a task for a philosopher. Now I certainly am
a "lover of wisdom", but this is probably a common feature
of all people in the audience as well (we are all physicists, the majority consists even of theorists). So I
dont feel particularly apt for my task. I feel old enough
to have some distance from the subjects discussed here,
but not old enough to be wise. Also with respect to the
entertainment, which a reviewer owes to his audience, I
am certainly inferior to the summary speaker(s) of the
past meetings. So I was not very happy, when I was talked
into this job. You must be content, if my summary will
see matters through the inverse of a considerably magnify-
ing looking-glass with personally tilted optics. My con-

[+] Summary given at XV. Internationale Universitätswochen
für Kernphysik, Schladming, Austria, February 16-27, 1976.

solation is the fact, that summaries are not followed by a discussion, so that I shall not be ruined by the audience.

I shall start with two pessimistic diagrams on our present situation: The first gives the progress P in understanding particle physics in this century as a function of time, as seen from some distance (fig. 1). If you plot the enthusiasm E you get a similar curve, perhaps displaced in time. A portion is magnified (fig. 2): more accurate studies show big steps (in an appropriate scale), which are then called the milestones.

The second diagram (fig.3) gives the development in physics. The abscissa is the orientation: Experimental left, mathematical right. The ordinate is time again. (I was told that in Russia right and left is exchanged: pragmatists are called right wing and orthodoxists left wing. This can be obtained by space reflection). We began with physics as a unique, entire topic long time ago. Today and at this conference, I have found a cluster-structure and could distinguish Experimenters, Phenomenologists, Engineers, Constructors, Mathematicians, whereby each cluster is coupled only to next neighbours. You may decide, where you belong. The bad thing lies in the dynamics of the system. The forces are so, that excitation on one side must be very strong in order that the other side is slightly shaken. Resonances are almost forbidden. In addition, the forces follow an inverse power law with distance and the system expands in time exhibiting a tendency for a sub-cluster-structure. It is the hope, that a meeting like this one here slows down the time expansion. In addition it might act as a box around the system, which may lead to new types of excitations.

But let me now stop joking and begin with the summary. I shall start from left, where the ground is more solid and begin with neutral currents. Theorists of course favour the Salam-Weinberg (SW) model. We have learned from Sakurai, that it is very healthy to be un-prejudiced, and from Faissner, how hard the life of an experimentalist in this field can be. The most interest-ing new experimental results concerned $\nu_\mu e$ scattering, a crucial test for the presence of neutral currents in weak interactions. We have learned that there are not only improved results on $\bar{\nu}_\mu e$ scatt., but that also $\nu_\mu e$-scattering has been observed. These results of the Aachen group were even announced at this conference for the first time. The clear consequence of this (and other) experiments is the fact, that neutral currents are of the same order of magnitude as their charged counter-parts: their ratios amount typically between 10 and 50 %. The SW-model with a parameter of about 1/3 is in good shape:there are so far no contradictions, but other explanations are still possible. Some crucial tests have still to be done. As far as the detailed properties of the interaction are concerned, we are less lucky. The space-time structure is narrowed down, but not uniquely clear. The VA form seems more favourable: if there is no STP part, we cannot have V+A, but either V-A or pure V resp. pure A. The first choice is slightly favoured. About the "heretic" parts we know, that SP alone would not be acceptable, whereas a STP mixture is still possible (but unlikely). Astrophysics seems to rule out T. As far as the internal symmetry structure of

hadronic part is concerned, strangeness changing terms
are absent. An isovector part is needed (see single π-
production). Whether or not an isoscalar part is needed,
remains open at present. In the future the interesting
missing pieces of information will come from inclusive
νN reactions and νp scattering and from effects not involving neutrinos. The time scale set by experts is
$\underset{\sim}{\sim}$ 1980, so that the progress here is made somewhat
faster than in the field covered in the first week.

Concerning charm, we had the rare occasion to hear
a charming lady discussing charm in a charming way. The
main issue here is the Glashow-Illiopoulos-Maiani theory
(GIM-model) (il modello GIM e un prodotto AGI), which is
the most simple way to discuss the new degree of freedom
discovered 1 1/2 years ago. We heard about predictions
and results concerning the excitation of charm in
various Neutrino-Hadron-collisions with production of
2 μ's, probably the best way to "see" (non-hidden)
charm at present. I shall not try to complete with a
lady with respect to charm and shall only remark, that
the GIM is in a similar state as SW: there are no gross
contradictions, (for weak points see the notes), but we
are still in a very exploratory stage both with respect
to experiments and theory. To put it most uncharmingly,
we can be quite sure, that a new degree of freedom has
been touched, but we have to wait for more information
until we can be sure, that the degree of freedom has
really all the properties wanted by the lovers of charmed quark models. If we look from the air down to the
battlefield of elementary particle physics, we see, that
the pince attack of the weak and electromagnetic army on
the enemy has (in contrast to the general belief of,

say, 10 years ago) been much more successful than the bloody fight of strong interaction physicists, in accordance with the old statement of a famous german general ("In a direct attack, you may at most hope for a vulgar victory ").

Let us now turn more towards the right wing of our original drawing. We had lectures on gauge theories and on supersymmetry. The first subject goes more along the party-line of this school, which started with the subject of vector meson theories 15 years ago. Meanwhile we have seen, that renormalizability is by no means an isolated phenomenon. The rules of the "gauge game", which is now very popular, read as follows:

1   <u>"start"</u>: from an internal symmetry

2   <u>"baptize"</u>: turn it into a local symmetry, couple gauge fields.

You $\{^{\text{gain}}_{\text{loose}}\}$ with respect to $\{^{\text{ultraviolet}}_{\text{infrared}}\}$ behaviour.

3   <u>"deform"</u>: introduce spontaneous breakings and Higgs mechanisms.

4   <u>"live happily"</u> for ever.

That at least the first two steps are not only a formal game, is taught by QED and Gravitation. The success of Salam-Weinberg shows, that the procedure may give deeper insight providing for unification of interactions. From Kummer's lecture we have learned that the theory is now also in the non-Abelian case on the level reached long time ago in QED (save for experimental results). We can prove renormalisability, gauge independence and unitarity

of S-matrix elements. Work to be done contains a closer
look on the infrared situation, and, of course, concrete
calculations (The question, whether renormalization constants are finite, is still not settled even in QED!).
Greens functions, Ward identities are the invariable tools.
I personally cannot quite agree with Kummer, that Greens
functions are only tools. We can learn from nonrelativistic
many-body physics, that they may be more. There everything
is stated in terms of Green's functions (expectation values
of physical operators, sum rules etc.). Many body physicists
live to a large extent without an S-Matrix, since they have
to live "in a soup". I am waiting for a larger feedback of
their techniques also with respect to approximation methods.
What makes me feel a little unhappy is, that the most important relations (e.g. Ward-Takahashi-Chang-Mani-Rivers-
Slavnov-Taylor-identities) are very far from experimental
control (as are, by the way, field equations). Perhaps
this should be a challenge for phenomenologists!

Let us move still a little bit more to the right:
we have heard from Wess about supersymmetries (SUSY). The
motivation for combining fields of different spin in a
multiplet can either come from the old SU_6: SUSY is a
good way to generalize, and the difficulties, in which
earlier attempts in this directions ran very soon, are
apparently entirely avoided with SUSYS. Another motivation is, that SUSYS provide for theories with the minimal
amount of divergences: axiomatists should keep that in
mind! We have seen, that SUSYS are very restrictive; this
is attractive for a physicist, who, in contrast to a mathe-
matician, looks for the most restrictive theory just large
enough to fit real life and not for the most general one.
Besides this, there are some concrete models, which are

not even too complicated and entirely nontrivial. In general, the "gauge game" can be played also here without running into contradictions. What is missing so far, is an application to a physical situation. One should not blame the lovers of SUSY for that, however: after all the trouble with the art "how to breed bigger and better groups" like SU_{12} etc. it is healthy to make sure first, that the theory can do what one expects in general.

Let me turn now to the last subject I would like to touch, which is Solitons. It has always been the opinion of some pioneers of modern physics (e.g. W.Pauli and W. Heisenberg), that one should take into account non-linearity as an essential ingredient in quantum field theory, i.e. one should forget about the free fields and build quantum field theory upon the qualitative different behaviour of classical nonlinear theories as compared with linear ones. The quark confinement problem has stimulated new interest in this problem, another stimulans comes from the line of thought renormalization group-phase transitions. Some interesting proposals have been made by Vinciarelli in his lecture, and the subject was also touched by Fröhlich in the first week. Here we all feel, that we are just in the beginning. There is so far little systematics in showing, which types of classical equations may have soliton solutions (I am waiting for R. Thom and his catastrophes to enter the game), quantum theories are in a preliminary stage (which is, however, promising in some cases). We have hardly an idea, how to construct physical operators (except perhaps charges and a "Hamiltonian"), how Lorentz invariance will be established, how to treat scattering or reactions systematically. Anyway it is nice to see a new idea on the horizont, which could turn out to be very useful.

As a summary of the theoretical part I would say, that (in contrast to the belief of most physicists 10 years ago) field theory is more than "in" again: it makes big progress on the whole front! The grain of salt to be added by the pessimist is, that in complicated formalissimi we are always in danger to produce a considerable amount of DEP[+)]. Roughly 20 years ago an old friend and college of mine at Graz made the nasty remark, that the German word for field physics, i.e. Feldphysik is misspelled and should rather read Fehlt Physik. This was certainly exaggerated and the situation has improved, but the statement is unforunately still not entirely disproved.

Before you take up the stones I would like to ask you to thank all those, who have worked hard to organize the conference. They have been extremely efficient: even traditional connections to St.Peter were successfully mobilized, and the local hospital can stand the fact, that this meeting has not added anything to its record. May I end with expressing the wish to see you again next year in Schladming!

[+] DEP = Dehydrated Elephants Powder

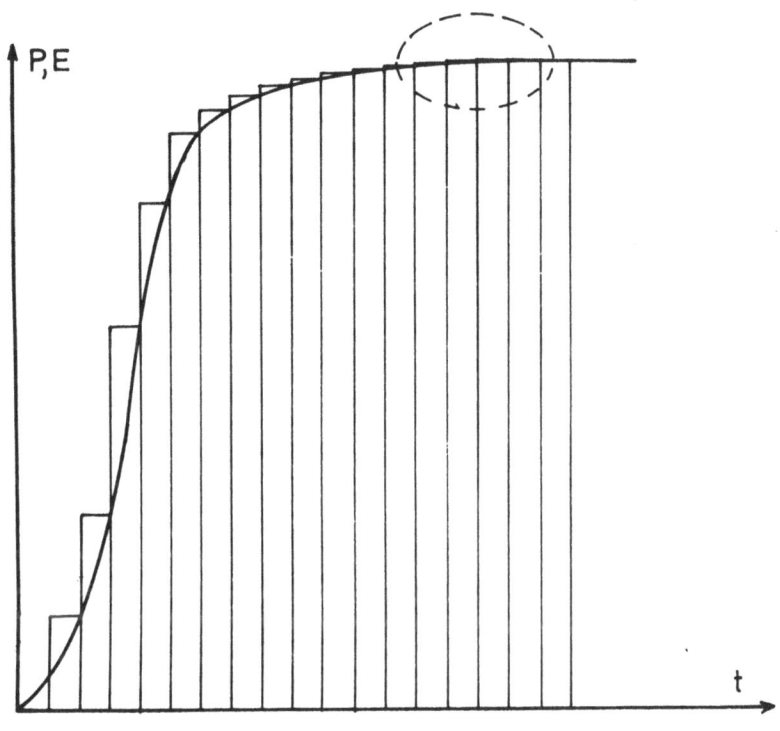

Fig.1

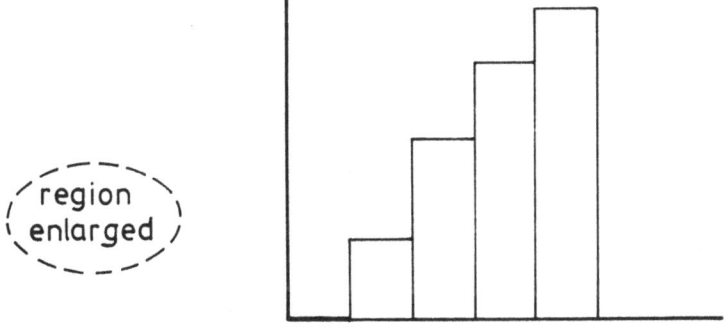

Fig.2

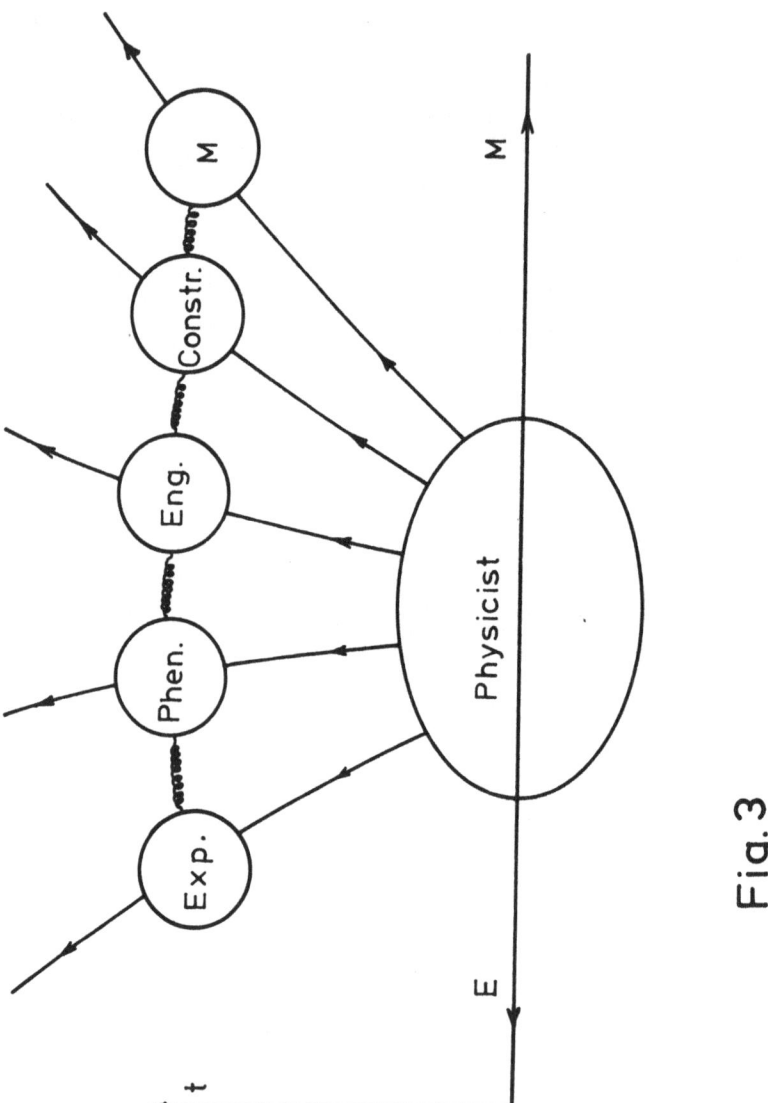

Fig. 3

Druck: Novographic, Ing. Wolfgang Schmid, A-1230 Wien.